TRENDS IN FOOD ENGINEERING

T0172910

FOOD PRESERVATION TECHNOLOGY SERIES

FOOD PRESERVATION TECHNOLOGY SERIES

Trends in Food Engineering

EDITED BY

Jorge E. Lozano
Cristina Añón
Efrén Parada-Arias
Gustavo V. Barbosa-Cánovas

ASSOCIATE EDITOR
M. Marcela Góngora-Nieto

CRC Press
Taylor & Francis Group
Boca Raton London New York

CRC Press is an imprint of the
Taylor & Francis Group, an **informa** business

Trends in Food Engineering

CRC Press
Taylor & Francis Group
6000 Broken Sound Parkway NW, Suite 300
Boca Raton, FL 33487-2742

First issued in paperback 2019

ISBN-13: 978-1-56676-991-4 (hbk)
ISBN-13: 978-0-367-39853-8 (pbk)

Visit the Taylor & Francis Web site at
http://www.taylorandfrancis.com

and the CRC Press Web site at
http://www.crcpress.com

Main entry under title:
    Food Preservation Technology Series: Trends in Food Engineering

Library of Congress Catalog Card No. 00-102585

TABLE OF CONTENTS

SERIES PREFACE

The processing of foods is becoming more sophisticated and diverse, in response to the growing demand for quality foods. Consumers today expect food products to provide, among other things, convenience, variety, adequate shelf life and caloric content, reasonable cost, and environmental soundness. Strategies to meet these demands include modifications to existing food processing techniques and the adoption of novel processing technologies.

This new Technomic Book Series has been conceived to explore the food processing techniques that will facilitate the transformation of the food market to meet consumer expectations. The Series will include titles on fundamental aspects in food processing, such as advances in transport phenomena, as well as titles covering more applied topics, such as the present book, *Trends in Food Engineering*. Other books to follow will discuss innovations in food processing, advances in pulsed electric fields, alternative technologies for processing foods, the interface of food science and biotechnology, and major topics in engineering and food for the 21st century. We expect to publish two to four new titles per year and hope this Series will become a premier reference for readers seeking the latest developments in food processing.

Our challenge, which we face with enthusiasm and optimism, is to bring readers quality publications on what is current, and what may become the food preservation technologies of the future. We hope this effort will be well received by the food industry, food related agencies, universities, and other research institutions.

Gustavo V. Barbosa-Cánovas
Series Editor

PREFACE

The technical papers included in this book are based on presentations made by the invited speakers of the Second Ibero-American Congress on Food Engineering, held at the Universidad Nacional del Sur, Bahía Blanca, Argentina, from March 24–27, 1998. In addition to the invited presentations which gave a wide vision of the state-of-the-art in food engineering, there were more than 200 volunteered contributions, most of them compiled in a proceedings edited by the congress organizers.

The book is divided into three sections. The first part deals with the physical and sensory properties of food. The emphasis of these chapters is on structure-property relationships, food rheology, and the correlations between physicochemical and sensory data. The second part on advances in food processing includes the latest developments in minimal preservation and thermal and nonthermal processing of foods. The book concludes with current topics in food engineering such as applied biotechnology, food additives, and functional properties of proteins—areas which have assumed increasing importance in the food industry over recent years.

It is quite apparent food engineering is gaining worldwide recognition by developing an identity within food and engineering-related programs in academic institutions and the food industry. We are confident this book will help in the consolidation of a much-needed profession and will also provide its readers an incentive for identifying what is to come from the profession in the 21st century.

Jorge Lozano
Cristina Añón
Efrén Parada-Arias
Gustavo V. Barbosa-Cánovas

ACKNOWLEDGMENTS

The Editors want to express their gratitude and appreciation to the following individuals that made possible the publication of this book:

M. Marcela Góngora-Nieto, Associate Editor and Food Engineering Graduate Student, Washington State University, for her decisive contribution in helping the editors and Assistant Editor put together this book.

Dora Rollins, Assistant Editor, Washington State University, for her professionalism and dedication in making possible the completion of this project.

The Editors would also like to thank the following organizations for their support in making this book a reality:

- Organizing Committees of the Second Ibero-American Congress of Food Engineering
- UNS, Universidad Nacional del Sur, Bahía Blanca, Argentina
- UNLP, Universidad Nacional de La Plata, Argentina
- CYTED, Ibero-American Science and Technology Program of Development
- CYTED-RIBIADIR, Ibero-American Network of Food Engineering
- PLAPIQUI (UNS-CONICET), Pilot Plant of Chemical Engineering, Bahía Blanca, Argentina
- CIDCA (UNLP-CONICET), Research and Development Center for Food Cryotechnology, La Plata, Argentina
- CONICET, National Council of Scientific and Technical Research, Buenos Aires, Argentina

CONTRIBUTORS

José Miguel Aguilera, Department of Chemical Engineering and Bioprocessing, Pontificia Universidad Católica de Chile, PO Box 306, Santiago 22, Chile

Stella M. Alzamora, Departamento de Industrias, Facultad de Ciencias Exactas y Naturales, UBA, Ciudad Universitaria, 1428 Buenos Aires, Argentina

María Cristina Añón, CIDCA, (UNLP-CONICET) Calle 47 y 116, PO Box 553, (1900) La Plata, Argentina

José M. Barat, Departamento de Tecnología de Alimentos, Universidad Politécnica de Valenica, Apdo Correos 22012, 46071 Valencia, Spain

Gustavo V. Barbosa-Cánovas, Department of Biological Systems Engineering, Washington State University, Pullman, WA 99164, USA

Silvia Bolado, Department of Chemical Engineering, University of Valladolid, Prado de la Magdalena, s/n 47005, Valladolid, Spain

Floor Boon, TNO-MEP, PO Box 342, 7300 AH Apeldoorn, The Netherlands

María E. Carrín, PLAPIQUI(UNS-CONICET), Camino La Carrindanga Km.7, (8000) Bahía Blanca Argentina

Ramón Catalá, IATA, Apdo Correos 73, 46100 Burjassot, Valencia, Spain

Amparo Chiralt, Department of Food Science, Camino de Vera s/n, Universidad Politécnica de Valencia, Valencia 46022, Spain

Diana T. Constenla, PLAPIQUI (UNS-CONICET), Camino La Carrindanga Km.7, (8000) Bahía Blanca Argentina

Elvira Costell, IATA, Apdo Correos 73, 46100 Burjassot, Valencia, Spain

Guillermo H. Crapiste, PLAPIQUI (UNS-CONICET), Camino la Carrindanga Km. 7 (8000), Bahia Blanca, Argentina

Muthukumar Dhanasekharan, Rutgers University, 1106 Mindy Lane, Piscataway, NJ 08854, USA (LIKE KOKINI)

Luis Durán, IATA, Apdo Correos 73, 46100 Burjasot, Valencia, Spain (LIKE CATALA)

Pedro Fito, Departamento de Tecnología de Alimentos, Universidad Politécnica de Valenica, Apdo Correos 22012, 46071 Valencia, Spain

Rafael Gavara, IATA, Apdo Correos 73, 46100 Burjassot, Valencia, Spain

Lía N. Gerschenson, Departamento de Industrias, Facultad de Ciencias Exactas y Naturales, UBA, Ciudad Universitaria, 1428 Buenos Aires, Argentina

M. Marcela Góngora-Nieto, Department of Biological Systems Engineering, Washington State University, Pullman, WA 99164, USA

Sandra N. Guerrero, Departamento de Industrias, Facultad de Ciencias Exactas y Naturales, Universidad de Buenos Aires, Ciudad Universitaria, 1428 Buenos Aires, Argentina

Nicole Heinsman, Wageningen University, Department of Food Technology and Nutritional Sciences, PO Box 8129, 6700 EV Wageningen, The Netherlands

Guillermo E. Hough, Departamento de Evaluación Sensorial de Alimentos, Instituto Superior Experimental de Tecnología Alimentaria, H. Yrigoyen 931 (6500), 9 de Julio, Argentina

H. Huang, Nabisco Foods Group, 200 DeForest Avenue, PO Box 1944, East Hanover, NJ 07936, USA

Andres Illanes, Escuela de Ingeniería Bioquímica, Universidad Católica de Valparaíso, Av. Brasil 2147, Valparaíso, Chile

Jozef L. Kokini, Department of Food Science, Rutgers University, PO Box 231, New Brunswick, NJ 08903, USA

Asunción Leúnda, Instituto de Ciencia y Tecnología de Alimentos, Universidad Central de Venezuela, Apdo 47097, Caracas 1049, Venezuela

Peter J. Lillford, Unilever, Colworth House, Sharnbrook Bedford MK44 ILQ, UK

Aurelio Lopez-Malo, Departamento de Ingeniería Química y Alimentos, Universidad de las Américas-Puebla, Cholula, Puebla 72820, México

Jorge E. Lozano, PLAPIQUI (UNS-CONICET), Camino la Carrindanga Km.7, (8000) Bahía Blanca, Argentina

Jatal D. Mannapperuma, California Institute of Agriculture and Research, University of California, Davis, CA 95616, USA

Javier Martinez-Monzo, Departamento de Tecnología de Alimentos, Universidad Politécnica de Valenica, Apdo Correos 22012, 46071 Valencia, Spain

Rodolfo H. Mascheroni, CIDCA, Calle 47 y 116, (1900) La Plata, Argentina

Enrique Palou, Departamento de Ingenieria Quimica y Alimentos, Universidad de las Americas-Puebla, Cholula, Puebla 72820, México LIKE WELTI

Efrén Parada Arias, CYTED, Instituto Politécnico Nacional, Apartado Postal 42-186, 06470 México, DF México

María Cecilia Puppo, CIDCA, (UNLP-CONICET) Calle 47 y 116, PO Box 553, (1900) La Plata, Argentina

M. Anandha Rao, Cornell University, Department of Food Science and Technology, NYSAES, Geneva, NY 14456, USA

Ana M. Rojas, Departamento de Industrias, Facultad de Ciencias Exactas y Naturales, UBA, Ciudad Universitaria, 1428 Buenos Aires, Argentina

Amelia C. Rubiolo, INTEC (CONICET, UNL) Güemes 3450, 3000 Santa Fé, Argentina

Alberto M. Sereno, University of Porto, Faculty of Engineering, 4099 Rua dos Bragas, Porto, Portugal

Jos Sewalt, Wageningen University, Department of Food Technology and Nutritional Sciences, PO Box 8129, 6700 EV Wageningen, The Netherlands

R. Paul Singh, Biological and Agricultural Engineering Department, University of California, Davis, CA 95616, USA

Barry G. Swanson, Department of Food Science and Human Nutrition, Washington State University, Pullman, WA 99164, USA

María S. Tapia, Instituto de Ciencia y Tecnología de Alimentos, Facultad de Ciencias, Universidad Central de Venezuela, PO Box 47097, Caracas 1041-A, Venezuela

Albert van der Padt, Wageningen University, Department of Food Technology and Nutritional Sciences, PO Box 8129, 6700 EV Wageningen, The Netherlands

Klaas van 't Riet, TNO-Voeding, PO Box 360, 3700 AJ Zeist, The Netherlands

C.F. Wang, Nabisco Foods Group, 200 DeForest Avenue, PO Box 1944, East Hanover, NJ 07936, USA

Jorge Welti-Chanes, Departamento de Ingeniería Química y Alimentos, Universidad de las Américas-Puebla, Cholula, Puebla 72820, México

Wei Hsiu Yang, Cornell University, Department of Food Science and Technology, NYSAES, Geneva, NY 14456, USA

María Clara Zamora, Laboratorio de Investigaciones Sensoriales (CONICET), Facultad de Medicina, Universidad de Buenos Aires, Argentina

Noemi E. Zaritzky, CIDCA, (UNLP-CONICET) Calle 47 y 116, PO Box 553, (1900) La Plata, Argentina

CHAPTER 1

Structure-Property Relationships in Foods

J.M. Aguilera

INTRODUCTION

Structure-property relations are crucial in material science; for example, in the design of alloys and composites notable for strength, light weight, and resistance to heat for aeronautic applications. These relationships are the basis of three-dimensional structures on earth that are linked to the mechanical functions they serve. By analogy to the previous example, the shell that protects a macadamia nut is stronger than most metals yet is less than half as dense, thus contributing to the light weight the plant must sustain (Niklas, 1992).

The concept "structure-property relationship" involves a connection between a structure and the way a product behaves (Fig. 1-1). In other words, it means that knowledge of the structure would let us understand how and why an object is built and functions the way it does. Conversely, a certain property of a product can be completely understood by the way it is structured.

To facilitate further understanding, assume that an extraterrestrial, having discovered and analyzed the displacement of the human species from outer space, observes people moving around mounted on strange vehicles (bicycles) by turning their legs in a strange rotary fashion. A closer observation (e.g., at a scale of cm) reveals that these bicycles are conformed by two main structural elements: the wheels and a frame. Several properties may be measured: some of them are simple and direct (the speed or the distance), others more elaborated (ratio energy spent/distance). It takes a smaller scaled observation (on the order of mm) to discover that it is the right interaction or linkage (pedals and chain) between the structural elements that propels the bicycles. Once the mechanical aspects are understood this functionality can be modeled mechanically and mathematically (e.g., relating turns of the pedal disc to the back wheel, etc.). The extraterrestrial soon realizes that it has been this understanding of the functional relationship between structure and desired properties which has permitted the human race to dramatically improve the performance of bicycles (use of lighter materials and multiple gears).

WHAT ARE STRUCTURE PROPERTY-RELATIONSHIPS?

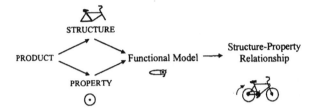

Figure 1-1 A structure-property relationship.

Next examine the very simple case of oil extraction from a soybean flake by a solvent. The property is defined as the "extractability" or amount of oil that is extracted per unit of time. Suppose now that the structure of the flake can be described with two different geometries (Fig. 1-2). In Case 1, the geometry consists of parallel open pores (in black) which connect the upper and lower surfaces of the flake so that the flow of solvent occurs only through the pores (the solid being impermeable). This case is simple to analyze through the effective diffusion coefficient D_{eff} which is related to the true unidirectional diffusivity D by the expression (Cussler, 1997):

$$D_{eff} = D(1 - \phi_F) \tag{1-1}$$

where ϕ_F is the volume fraction occupied by the solid [or $(1 - \phi_F)$ is the void fraction]. Equation 1-1 indicates a drop in extractability because the solvent must transverse a reduced cross-sectional area. Note that it says nothing about the distribution of pores within the solid or their diameter.

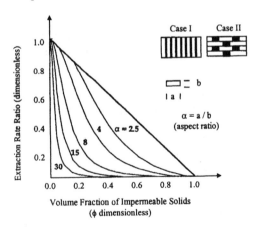

Figure 1-2 Effect of structure on the extractability of a flake for two geometrical systems. The symbol α is the aspect ratio of the impermeable inclusions.

The result is completely different if the solid is arranged as a staggered stack of impermeable slabs (Case 2). In this case diffusion of the solvent is retarded mainly because it is forced to wiggle along a tortuous path between the slabs (Fig. 1-2). For the structure in Case 2 the effective diffusion coefficient is given by:

$$D_{eff} = D \left[\frac{1}{1 + \alpha^2 \phi_F^2 / (1 - \phi_F)} \right] \qquad (1\text{-}2)$$

where α is the aspect ratio of the slabs (in two dimensions, a divided by b), and a large α implies a high tortuosity. To gain an idea of the effect of structure on extractability, take an aspect ratio of 10 and $\phi_F = 0.5$ (e.g., in chemical engineering terms the "porosity" is 0.5). For Case 1 the ratio D_{eff}/D would be one-half, while for the perpendicular arrangement it will be 0.017 or 50 times less. Figure 1-2 shows the decrease in extractability as a function of the void fraction for several geometries.

Implicit in the example was that structure played a role in the phenomenon. Since structural analysis may become time-consuming and costly one ought to make sure first that there is good evidence that structure plays a role in the observed behavior. In this simple example structural effects could explain all changes in extractability, but in reality many problems arise:

1. The microstructure of flakes cannot be seen.
2. The property may not be as simple to define or measure.
3. Even if the microstructure could be seen there probably would not be a simple geometrical model to represent it.
4. Structure may only play a partial role in the observed phenomenon (the rate of dissolution may also play a role).

The objective of this chapter is to discuss how it is possible to "see" food structure (make a sketch and perhaps quantify the main elements in it), probe into it by various means to get complementary data, propose some kind of model, and verify how well it explains the observed property or behavior.

FOOD STRUCTURE

Efforts to see and describe the structure of foods have a long history. Microscopy was originally used to detect adulteration in foods and food microbiologists have amply used it as a standard tool. However, food microstructure is not yet a subject in any major curricula and food materials science is still in its infancy (Stanley, 1994). The first problem from an engineering point of view is to know whether one is dealing with the property of a material or a structure (Vincent, 1990). A material is usually homogeneous and isotropic (properties do not vary with direction), meaning the stiffness is the same in tension, compression, and bending. Materials refer to a pure substance

3

and alloys (as may be the case in some foods) which are homogeneous in composition.

Structures are composed of more than one material or phases. If these materials or phases form elements arranged with clear and discernible regularity it is said to have geometry or "architecture." It is useful also to distinguish between the terms morphology and structure as both are used interchangeably in the literature. The term morphology (from the Greek morphos or shape) is usually associated with a macroscopic property of solids while structure refers to the microscopic view. Heertje (1993) defined structure as the spatial arrangement of the structural elements in a food and their interactions while stressing that visual observation is important in the analysis of food structure. Through different microscopy techniques he identified several structural elements in foods: water droplets, oil droplets, gas cells, fat crystals, strands, granules, micelles, and interfaces. Another definition of structure comes from a discussion paper by Raeuber and Nikolaus (1980) which says: "it is the organization of a number of similar or dissimilar elements, their binding into a unit, and the interrelationship between the individual elements and their groupings." In both definitions the organization of elements and their interactions are emphasized as crucial. Thus, the microstructure of a food can only be understood when its architecture and elements (solid, liquid and gaseous) are considered together. It is in the development of the structural (or physical) model of the food where microscopy becomes irreplaceable because it permits visualization at different scales.

Most foods possess structures and are not materials, although some foods may be regarded as homogeneous alloys upon a first approximation (e.g., soda crackers). This becomes problematic when food elements may belong to one or the other category just by changing the scale of observation. For instance, a starch granule may be regarded either as a filler at a scale of 200 μm or a stratified composite of amylose and amylopectin at 1 μm.

In general, composition gives little information that is relevant to structure. The most notable example is that an apple and milk are largely water (> 85%), but one is a solid and the other a Newtonian liquid. However, the presence of certain food components may reveal their potential contribution to food structure:

1. Sugars can be in amorphous or crystalline states.
2. Fats can be liquid, solid, or in a liquid-crystalline form which shows polymorphism.
3. Polymers can be in an amorphous or semi-crystalline state.
4. The presence of some molecules may imply structural breakdown (e.g., polymers).
5. Starch and gums compete for water and form gels.
6. Aqueous mixtures of food polymers tend to phase-separate.

A problem never addressed by food material scientists is that of micro-heterogeneity in foods. Because foods are generally metastable systems, a material (e.g., a sugar) may be in different phases at different microregions. And since water is not distributed homogeneously, it partitions itself differently between phases, complicating things further. Water content is a critical variable in low and intermediate moisture foods, and is determined as an average value. This problem is typical in the determination of glass transitions in foods and the study of plasticization by water. To recapitulate, a crucial point which makes study of the structure of foods absolutely different from that of engineering materials is that food structure is often in a metastable state and/or belongs to a biologically active material. Structure in foods changes with time perceptively (e.g., during frying) or unnoticed (during storage of a fruit).

FOOD ENGINEERS AND STRUCTURE

Most food engineers have a background in chemical engineering. As such they do not focus on solids or condensed states, but deal much better with liquids and gases. A major unit operation tonnage-wise, milling is explained through formulas derived from mining engineers at the beginning of the 20th century. The emphasis is placed on the energy employed to reduce particle size rather than on the quality or effect of milling. These formulas are no good for predicting how well the adhering coat of a garbanzo bean can be broken and removed or explaining why starch and protein can be fractionated from a bean in almost pure components by fine milling and air classification (Aguilera et al., 1984).

When the unit operation is diffusion-controlled and a solid is involved, the chemical engineer's recourse is to combine all structural effects (as well as other unknown effects) into the previously defined effective diffusivity D_{eff} and introduce the obscure term "tortuosity." This is because more exact descriptions of the structural effects require more exact geometries and this implies more intricate mathematics. For instance, in the case of a simple composite formed by periodically spaced spheres where diffusion takes place both in the interstitial region and through the spheres themselves, the apparent diffusion coefficient starts to get complicated (Cussler, 1997):

$$\frac{D_{app}}{D} = \frac{\dfrac{2}{D_s}+\dfrac{1}{D}-2\phi_s\left(\dfrac{1}{D_s}-\dfrac{1}{D}\right)}{\dfrac{2}{D_s}+\dfrac{1}{D}+\phi_s\left(\dfrac{1}{D_s}-\dfrac{1}{D}\right)} \tag{1-3}$$

where D is the diffusion coefficient in the interstices, D_s the diffusion coefficient through the spheres, and ϕ_s the volume fraction of spheres in the composite. Equation 1-3 is more complicated than Equation 1-2 even though the geometry is still simple; moreover, it requires that D_s be known. This is a surprising result because it implies that diffusion does not depend on the size of the spheres but only on their volume fraction. Chemical engineers have thus simplified their lives by defining an "effective diffusion coefficient" which contains all structural effects as:

5

$$D_{eff} = \frac{D}{\tau}\varepsilon \qquad (1\text{-}4)$$

which includes a correction effect for porosity (ε) as well as tortuosity (τ); the problem now is to determine both parameters. In a porous catalyst they may remain fairly constant during processing, but in fruit drying they change appreciably over time.

An analogous approach is taken when engineers deal with structured liquids. In this case an apparent viscosity η_{app} is used for non-Newtonian liquids which has to be specified at each rate of deformation. Absolute values and variation of η_{app} with shear rate (γ) are often explained qualitatively as changes in the structure of particles and their interactions with other components of the fluid.

Viewing Structures

The desire of engineers to see directly (or indirectly) how structure arises or ruptures dates back to Reynolds' experiment on turbulence and the breakage of a pulse of dye in a stream of flowing water. Microscopy in its many versions is the universal technique for visualizing the unseen in the microcosmos. The application of microscopy techniques to foods has been reviewed by Kalab et al. (1995) and will not be dealt with here in any detail. However, attention is paid to the large number of reviews dealing with advanced microscopy techniques and their application in foods appearing in recent issues of the journal *Trends in Food Science and Technology* [e.g., "Atomic Force Microscopy" by Kirby et al. (1995)].

Important are the trends to use less invasive microscopy techniques like optical sectioning by confocal laser scanning microscopy (CLSM) and environmental scanning electron microscopy (ESEM), perform real-time experiments through videomicroscopy, couple microscopy directly to analytical techniques (X-ray probing, IR and Raman spectroscopy), miniaturize experiments, and interface with computers and software for image processing and data analysis (Aguilera and Lillford, 1996). The primary issue for engineers is that they are moving fast towards quantitative microscopy where elements in images are transformed into numbers and data into graphs, and these into equations.

Probing Structures

Probing structures means examining them at the microstructural level with a probe to recover information other than visual or geometrical. Reference has been made to the coupling of microscopy with sophisticated spectroscopic techniques and at the same time gathering an image and compositional information of the different elements in the sample. Another promising tool is magnetic resonance imaging (MRI) or probing structures through the response of atom nuclei to magnetic fields. The potential of MRI lies in non-invasively following real-time changes in foods and generating spatial data such as temperature or velocity gradients rather than structural information (Hills, 1995). Improvements are expected in the resolution of this technique so that MRI may be used to determine structure as well.

Other ways of probing structures are by mechanical or thermal techniques. Thermal analysis of foods records changes in enthalpy associated with phase (e.g., melting) and order-disorder transitions (e.g., protein denaturation) as a function of temperature or time. It also displays changes in specific heat related to second-order transitions such as the glass-rubber transition. Thus, state diagrams may be constructed showing whether a food material under certain conditions is amorphous, crystalline, rubbery, or flowing (Kokini et al., 1994).

Small-strain dynamic techniques are used to probe structures for their rheology or mechanical responses. The advantage is that structure is not changed under small deformations and fundamental viscoelastic moduli (G' and G'') can be obtained as a function of time, frequency, pressure, or temperature. In small-strain dynamic temperature experiments (curing curves) transitions can also be singled out; in particular, the sol-gel point and Tg. Building of a network structure may be followed by the increase in G' (solid-like behavior). Equilibrium G' values of cooled gels have been used in evaluating network structure, and using theories from polymer physics, the number of crosslinks per molecule has been calculated for protein gels.

A reasonable aspiration of a food technologist is to be able to map the water content in foods at the microscale. This desire may become a reality in the next few years by the application of electron energy loss spectroscopy (EELS) and microscopy.

The Relevant Scale

Earthquake engineers can design complex structures and analyze their behavior under different controlled testing conditions. They can observe (and videotape) cracks and collapse mechanisms as they develop, as well as measure fracture forces and deformations by locating transducers. Unfortunately, to date microstructural food engineers cannot see internal cracks developing during the drying of spaghetti or measure the debonding of cell walls during the cooking of legumes, yet these phenomena are decisive for the properties of finished products. The difference is that the relevant scale at which such phenomena occur in foods is beyond the capacity of the naked eye.

The relevant scale corresponds to the size of the structure which is key to explaining the property or behavior. It is usually found by inference from other analytical techniques or through theoretical analysis. For instance, the toughening of dry legumes under adverse storage conditions can be followed by mechanical analyses and related to changes in the middle lamella of cell walls based on microstructural evidence. In the case of emulsion stabilization, the theory about adsorption of macromolecules to the oil/water interface predicts that the dynamics of adsorption/desorption of macromolecules would be the key phenomenon and thus define the relevant scale. Often a study of key elements in

7

the structure of foods and their hierarchical organization assists in selecting the appropriate microscopy and complementary probing techniques to be used.

DISTINGUISHING PROPERTIES

It is important to determine what kind of properties are of interest. A property is a particular trait of an object. Most physical properties of foods—including mechanical, rheological, electrical, optical, and transport (heat and mass diffusivity)—are quite common to all industrial materials. Additional desirable attributes of foods are more specific; for instance, emulsifying capacity, surface roughness, or sound (Fig. 1-3). Other desirable characteristics are more difficult to define or measure (e.g., flavor, texture, nutritional value, and shelf-life stability). Since all are important and structure may play a role in their manifestation, a structure-property relationship needs to be found.

Clearly most foods are not simple, homogeneous materials: they have complex structures, some of which are biochemically active and change with time (e.g., ripening, dough leavening). Many foods when examined at the microstructural level are highly structured tissues, composites, or complex suspensions. Consequently, physical properties determined in bulk (e.g., considering a food as a "black box") may be appropriate when dealing with a standardized process or food. There are many books and computer data bases listing properties for many foods, though they do not feature uniform values.

Alternatively, individual physical properties may be obtained for key architectural elements and the overall property calculated from a physical model of the structural arrangements and interactions. The latter approach captures the essence of engineering and material science and provides flexibility for changes in formulation and processing. A good physical model is essential for developing relations between structure and mechanical, rheological, or transport properties.

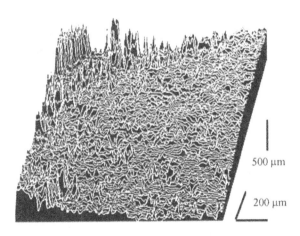

500 μm

200 μm

Figure 1-3 Surface of a fried potato chip (CLSM reflection model at 1,280 x 1,280 μm).

A case in point is texture. The term comprises all physical characteristics sensed by the feeling of touch that are related to deformation under an applied force. In their discussion paper on microstructural bases of plant food texture, Jackman and Stanley (1995) state that the literature is replete with examples of texture having been measured using inappropriate methods. They claim that the relation between objective measurements and sensory properties is poorly characterized and understood. However, that texture is the external manifestation of the arrangement of microstructural elements and their breakdown in the mouth appears to be clear. A simple but illustrative example is the detection of oversized fat or ice crystals by the tongue as defects in poorly tempered chocolate or frozen ice-cream, respectively.

The problem of physical properties of foods does not end in a number as in the biological or materials world regardless of how precise or representative of the property it may be. The ultimate challenge is to find the relationships for the triad structure-property-sensorial effect. This is a difficult task because one must comprehend not only how the microstructure of foods is formed but how it breaks down in the mouth and what perception is derived thereafter. It is first necessary to learn more how the brain functions, another area where foods and biology merge.

DETERMINING RELATIONSHIPS

Assuming that the property and structure are clear, the next question to be asked is what kind of relationship is desired. There are multiple answers for this rather simple question. It is easy to fit a polynomial function to describe whatever property is being measured, but this is not very productive in understanding the relation under investigation. In a few instances—such as in a standardized product or process—this may be all that is necessary.

Kinetic models are frequently used because they provide a relation for variations of a property with time and temperature. The analogy is taken from simple chemical kinetics where concentration is a function of these two variables. Since a change in a property is often due to a chemical reaction (or an overall reaction at least), there is some degree of fundamental principle involved. The justification for using this kind of approach is that foods are too complex to suggest that they can be mechanistically modeled (Lund, 1986). Again, data is fitted to the appropriate kinetic model by a variety of statistical techniques (Levenspiel, 1962). The main problems with this approach are that data for heterogeneous materials show low precision and in many processes the physical property may not change by more than 50% (one half-life), so the order is difficult to determine even for a simple reaction.

The next level of functional dependence is semi-empirical (or semi-theoretical) models. Outstanding are those used in food rheology to describe the behavior of non-Newtonian fluids (e.g., suspensions, pastes, etc.) that translate into equations linking stress, yield stress, and rate of deformation. Constitutive models that aim at relating rheological properties to molecular architecture are more fundamental. Such models are well-developed for dilute solutions of random coils and polymers with rod-like conformations (Kokini, 1994).

Engineers, on the other hand, prefer structural models. They are based on the geometry and structural properties of the intervening elements (matrix, fillers, walls, etc.) and their interactions. Well-studied structural models are the so-called Takayanagi model for solid food structures (Aguilera and Baffico, 1997) and those representing foams and cellular materials (Peleg, 1997). Structural models representing the rheological behavior of suspensions have also been developed (Windhab, 1995).

STRUCTURE-PROPERTY RELATIONSHIPS IN FOODS

Structure-property relationships describe the way in which physicochemical, functional, technological, and even some nutritional properties of foods are related to their structure. There are many examples of structure-property relationships in foods that are not limited only to physical or mechanical properties (Table 1-1). For example, if a desirable property in an emulsion is flavor release, the volatility of the odorous molecule may be higher in a water/oil rather than an oil/water emulsion.

Structure-property relationships can originate at the molecular, ultrastructural, microstructural, or macrostructural level during food processing. The first level corresponds to the size and type of monomers and includes interactions between functional groups. The ultrastructural level includes changes in conformation, association of macromolecules, and aggregation of subunits. The microstructural level corresponds to colloidal dimensions of aggregate associations to yield three-dimensional structures such as networks, stabilizing layers in emulsions, fibers, microgels (swollen starch granules), and crystalline structures. In these three structural levels the food elements are formed by specific and non-specific interactions that take place successively or simultaneously to yield the non-equilibrium structures of real food systems. Thus, knowing the kinetics of these interactions is of great importance, it must be realized that the old concept of "functional property" in reference to isolated proteins and polysaccharides is of little relevance due to the greater role assigned to non-specific interactions and the kinetic component introduced by processing. The concept of structure-building through non-specific interactions appears not to be used by nature since high levels of organization are achieved by highly specific interconnectivity (Baer et al., 1992).

Although advances are being made in conceptualizing how food structures are formed and there are efforts at relating structure to physicochemical properties (mechanical and rheological in particular), this subject is still relatively underdeveloped. As shown in the previous paragraphs, some of the interaction forces participating in structure formation and its stabilization are beginning to be understood at the qualitative level, but quantitative data and models are not yet available.

Table 1-1 Some examples of structure-property relationships in foods.

Class of Property	Measured or Observed Property	Example
Sensory	Texture	Hardening of stored legumes
	Flavor	Encapsulation of unsaturated lipids
	Color	Freeze-dried coffee
Mechanical	Strength	Reinforcement of protein gels
	Structural stability	Caking of amorphous food powders
Rheological	Creaming	Homogenization of milk
	Viscosity	Salad dressing formulation
Transport	Oil uptake	Oil absorption during frying
	Diffusion of solutes	Salt impregnation of cheeses
Functional	Water holding	Water structuring in low-fat spreads
	Heat resistance	Ca deposition in the core of spores
Nutritional	Retention of nutrients	Encapsulation of vitamins

LESSONS FROM NATURE

Nature constructs several structures to perform complex and multiple tasks. As is the case with foods, these structures are made up of a few rather simple macromolecules that consist of even simpler repeating units. Thus, the complexity indigenous to biomaterials lies in the intricacies of their molecular and supramolecular architecture, which constitute the structure. This complexity has been continually unraveled through the use of increasingly powerful analytical techniques such as microscopy, thermal and mechanical analysis, and advanced spectroscopy. This is the direction in which food material science is also moving and where lessons from nature and biology may be applied advantageously.

When nature forms structures it seems to follow the principle of hierarchy in which increasing levels of organization are progressively assembled until the desired properties and functions are achieved (Baer et al., 1992). The majority of soft, pliable tissues in the animal world are made of collagen fibers embedded in a gel matrix of protein-polysaccharide. Tendons acting in uniaxial tension, intestines as a tube under multiaxial tension, and intervertebral discs forming soft pads between rigid bones during compression accomplish these unique functions thanks to the complex architecture of connective tissue.

For example, a tendon is a uniaxial hierarchical structure that serves as the primary linkage between muscle and bone. Collagen microfibrils are possibly arranged in a lattice formed into subfibrils about 30 nm in diameter. These

11

subfibrils are then assembled into the collagen fibril which varies in diameter from 50 to 500 nm. Fibrils are subsequently surrounded by the extrafibrillar matrix and oriented in a hierarchical architecture to form specific soft tissue. The primary macromolecular component of the extrafibrillar matrix is a highly hydrated and swollen proteoglycan consisting of a core protein and numerous pendant mucopolysaccharide units. This proteoglycan aggregate forms a network that connects and maintains the hierarchical architecture of collagen fibrils. Finally, membranes surround the tendon. The resulting structure is one that exhibits properties needed for the function it has to accomplish. Thus, nature compensates for the limited types of molecules by utilizing the same macromolecular design and only varying the hierarchical structure.

In polymers hierarchical structures span from a macromolecular size into a macroscopical or morphological structure. This concept of hierarchy in foods has been adapted in Table 1-2. Nature also uses biomacromolecular assemblies for a wide variety of functions, the best example being cell membranes. Proteins and lipids can self-assemble into complex supramolecular structures capable of performing complex functions (e.g., transport and cell recognition).

CONCLUSIONS

Building a solid food material science demands finding the relations between the structure of a food and its physical, engineering, sensorial, and related properties. Structure-property relationships are more than the numerical description of a physical or mechanical property with an accompanying micrograph. The aim should be to fully understand at the appropriate scale the relationship between the architecture and interactions of structural elements in a food and the desired property or function. To accomplish this one should be able to see how structure is formed and broken down and how this affects the specific property. Structural models representing the complex interplay between structural elements need to be developed so they can represent the behavior of a food over an ample range of conditions. In this respect nature offers a multitude of interesting examples of structure-function relationships.

Table 1-2 Examples of hierarchical food and biopolymer structures.

Size Range	Fruit	Muscle	Gel	Chocolate
Angstroms	Glucose	Amino Acid	Monomer	Fatty acid
nm	Cellulose	Helix	Polymer	Lamellae
μm	Microfibril	Protofibril	Strand	Spherulite
100 μm	Cell wall	Muscle fiber	Network	Fat crystal

ACKNOWLEDGMENTS

Discussion with Dr. J.M. del Valle, partial funding by project FONDECYT 1960389, and a grant from Nestle-Chile are appreciated.

REFERENCES

Aguilera J.M. and Lillford P. (1996). Microstructural and imaging analyses as related to food engineering. In: *Food Engineering 2000*. Fito P., Ortega E., and Barbosa-Cánovas G., Eds. London: Chapman & Hall, pp. 23–38.

Aguilera J.M. and Baffico P. (1997). Structure-mechanical properties of heat-induced whey protein/starch gels. *J. Food Sci.* 62: 1048–1053, 1066.

Aguilera J.M., Crisafulli E.B., Lusas E.W., Uebersax M.A., and Zabik M.E. (1984). Air classification and extrusion of navy bean fractions. *J. Food Sci.* 49: 543–546.

Baer E., Cassidy J.J., and Hiltner A. (1992). Hierarchical structure of collagen composite systems. In: *Viscoelasticity of Biomaterials*. Glasser W. and Hatakeyama H., Eds. ACS Symposium Series 489. Washington, DC: American Chemical Society, pp. 2–23.

Cussler E.L. (1997). *Diffusion: Mass Transfer in Fluid Systems, 2nd ed.* Cambridge: Cambridge University Press.

Heertje I. (1993). Structure and function of food products: A review. *Food Struc.* 12: 343–364.

Hills, B. (1995). Food processing: An MRI perspective. *Trends Food Sci. Technol.* 6: 111–117.

Jackman, R.L. and Stanley D.W. (1995). Perspectives in the textural evaluation of plant foods. *Trends Food Sci. Technol.* 6: 187–194.

Kalab M., Allan-Wojtas P., and Shea-Miller S. (1995). Microscopy and other imaging techniques in food structure analysis. *Trends Food Sci. Technol.* 6: 177–186.

Kirby A.R., Gunning A.P., and Morris V.J. (1995). Atomic force microscopy in food research: A new technique comes of age. *Trends Food Sci. Technol.* 6: 359–365.

Kokini J.L. (1994). Predicting the rheology of food biopolymers using constitutive models. *Carbo. Polym.* 2: 319–329.

Kokini J.L., Cocero A.M., Madeka H., and de Graf E. (1994). The development of state diagrams. *Trends Food Sci. Technol.* 5, 281–288.

Levenspiel, O. (1962). *Chemical Reaction Engineering*. New York: John Wiley & Sons.

Lund D.B. (1986). Kinetics of physical changes in foods. In: *Physical and Chemical Properties of Foods*. Okos M., Ed. St. Joseph, MI: American Society of Agricultural Engineers, pp. 367–381.

Niklas K.J. (1992). *Plant Biomechanics*. Chicago: The University of Chicago Press.

Peleg M. (1997). Review: Mechanical properties of dry cellular solid foods. *Food Sci. Technol. Intl.* 3: 227–240.

Raeuber H.J. and Nikolaus H. (1980). Structure of foods. *J. Tex. Stud.* 11: 187–198.

Stanley D.W. (1994). Understanding the materials used in foods—Food materials science. *Food Res. Intl.* 27: 135–144.

Vincent J. (1990). *Structural Biomaterials*. Princeton, NJ: Princeton University Press.

Windhab E.J. (1995). Rheology in food processing. In: *Physicochemical Aspects of Food Processing*. Beckett S.T., Ed. London: Blackie Academic & Professional, pp. 80–116.

CHAPTER 2

Physical and Microstructural Properties of Frozen Gelatinized Starch Suspensions

N.E. Zaritzky

INTRODUCTION

The importance of starch lies in its dominance as a food component and the significance of its physical properties to food stability and texture. Starch is the most common carbohydrate polymer in foods. It is obtained from cereal and legume seed endosperm, potato tubers, and other plant reserve organs, and exists in the form of granules. The size of starch granules varies depending on their origin over the range of 1 to 100 μm. The granules are insoluble in cold water and composed of two glucose homopolysaccharides forming amylose and amylopectin molecules. Amylose is a linear polymer of 1→4 linked α–D glucopyranosyl units, while amylopectin is a branched polymer of α–D glucopyranosyl units primarily linked by 1→4 bonds with branches resulting from 1→6 linkages. The properties of these two major starch components are summarized in Table 2-1. The amounts of amylose and amylopectin differ significantly in various starches; the amylose content of a number of starches is shown in Table 2-2. In native granular starches, amylose exists in the amorphous noncrystalline state, and crystallinity is attributed to short chained clusters of amylopectin (Whistler et al., 1984; Biliaderis, 1991, 1992).

Most native starches contain 20 to 30% of amylose by weight, but waxy starches contain practically no amylose and certain varieties of hybrid corn such as Amylomaize are made up of a starch with high amylose content. Starch is a dominant component in a food product, significantly contributing to its physical properties, stability and texture. Therefore starch is commonly used to modify food systems texture.

Table 2-1 Properties of amylose and amylopectin components of starch.

Property	Amylose	Amylopectin
General structure	essentially linear	branched
Color with iodine	dark blue	purple
Molecular weight	$2 \times 10^4 - 2 \times 10^5$	$2 \times 10^5 - 2 \times 10^6$

Starch Gelatinization

Gelatinization occurs during the heating of native starch with a sufficient amount of water and includes: a) loss of birefringence of the granules; b) diffusion from ruptured granules and dissolving of linear molecules; c) increasing clarity of the starch water mixture; d) hydration and swelling of granules to several times their original size; and e) a marked increase of consistency and formation of a paste-like mass or gel (Roos, 1995).

Water acts as a plasticizer during starch gelatinization, lowering the melting temperature of starch. Starch gels are composites of swollen gelatinized granules embedded in a continuous amylose network. Swollen granules are mainly fitted with amylopectin due to the preferential leaching of amylose from the granules. Lelievre (1974) related the gelatinization of starch to the melting of homogeneous polymers. Melting of crystallites during gelatinization is supported by X-ray diffraction studies which indicate loss of semicrystalline order and calorimetric measurements that show loss of ordered structure (Evans and Haisman, 1982). Gelatinization temperature ranges for various starches are given in Table 2-2. Factors that govern gelatinization onset temperature and the temperature range over which gelatinization occurs include starch concentration, method of observation, granule type, and heterogeneities within the granule population under observation.

Pasting is the phenomenon that follows gelatinization; it involves granular swelling, exudation of molecular components from the granule, and eventually a total disruption of the granules. The major determination techniques of gelatinization temperatures include polarizing the microscopic hot stage examination of birefringence, detecting changes in viscosity, identifying changes in X-ray diffraction, and measuring thermal properties with differential scanning calorimetry (DSC).

Starch Retrogradation

Starch gels are metastable and non-equilibrium systems (Slade and Levine 1987; Biliaderis and Zawistoski, 1990), and therefore undergo structure

Table 2-2 Amylose content and gelatinization temperature range of various starches (adapted from Roos, 1995).

Starch	Amylose %	Tgel (°C)
Corn (Zea mays)	23–28	62–76
High amylose corn	52	67–86
Hybrid waxy corn	0.8	63–80
Potato	19–23	58–71
Rice	17–21	68–82
Tapioca	17–18	63–80
Wheat	23–26	52–66

transformations during storage and processing. Upon aging, starch retrogradation takes place, a temperature and time-dependent phenomenon that involves partial crystallization of starch components. Starch molecules reassociate depending on the affinity of hydroxyl groups and the attractive forces or hydrogen bonding between hydroxyl groups on adjacent chains (Pomeranz, 1985). The process induces an increase of paste rigidity and phase separation. According to several authors (i.e., Miles et al., 1985; Morris, 1990; Biliaderis, 1992), starch retrogradation kinetics consist of two distinct processes: 1) a rapid gelation of amylose via formation of double helical chain segments that is followed by helix-helix aggregation thermally irreversible below 140–160°C, and 2) a slow recrystallization of short amylopectin chain segments thermally reversible below 100°C. Differential scanning calorimetry only allows the quantification of the reversible process (Roulet et al., 1990).

Glass Transitions

Most food materials exist in an amorphous state which does not reach equilibrium at temperatures below the equilibrium melting temperature of the material. The melting of crystalline polymers results in the formation of an amorphous melt which can be super-cooled to a viscoelastic rubbery state or a solid glassy state. The transition between the rubbery and glassy states is known as the second-order glass transition. According to Slade and Levine (1987), starch is a water compatible polymer which exhibits non-equilibrium melting, annealing, and crystallization behavior characteristic of a kinetic metastable, water-plasticized, partially crystalline polymer with a small extent of crystallinity. Glass transition as a physicochemical phenomenon governs starch processing and stability. Below the glass transition temperature (Tg), polymer material becomes glassy, and the molecular motion is so slow that crystallization does not occur in a finite period of time. For temperatures above Tg and below the crystal melting temperature (Tm), the material is rubbery, allowing sufficient polymer motion for crystallization.

THE ROLE OF STARCH IN FROZEN SYSTEMS

Starch pastes act as protective systems of the solid elements in precooked foods, minimizing dehydration and chemical changes during storage. Precooked frozen foods containing sauces, gravies, pie filling, or desserts undergo freezing damage like rheological changes and syneresis after thawing which may alter desired characteristics and therefore reduce consumer acceptability. The use of hydrocolloids is highly recommended in the food industry in order to restrict syneresis and ice crystal growth, as well as for their traditional role of texturizers and emulsion stabilizers. They also have positive effects on particle suspension, crystallization control, and syneresis inhibition.

The lipid phase improves palatability and modifies the texture of starch pastes. The natural fatty acids of the starch and added monoglycerides modify granule swelling capacity and retard retrogradation. Such effects are attributed mainly to the formation of amylose-lipid complexes and amylopectin-lipid interactions (Evans, 1986; Biliaderis, 1992). Macroscopic modifications (syneresis and textural changes) occurring in frozen gelatinized starch systems can be related to microstructure parameters (starch retrogradation and ice crystal formation).

In the present study, the following aspects related to the stability of a gelatinized starch paste are discussed:

- The effect of adding hydrocolloids to thermal transitions: gelatinization and glass transition temperatures
- The influence of gelatinized paste freezing rates and frozen storage temperatures on exudate production, starch retrogradation, ice crystal formation, and ice recrystallization
- The effect of hydrocolloid and lipid addition on the rheological properties of starch pastes, with emphasis on the viscoelastic behavior of the system

Gelatinization Enthalpy

The effects of different hydrocolloids (xanthan gum, guar gum, locust bean gum, carboxymethylcellulose, sodium alginate) and water concentration on corn starch gelatinization were analyzed by Ferrero et al. (1996) using DSC. Hydrocolloids were added to starch in different proportions (1:10, 1:2, 1:1 g gum/g starch) previous to gelatinization. A tendency for the gelatinization enthalpies to increase with gum concentration was reported. Figure 2-1a shows thermograms of starch water systems with different starch concentrations (0.10 and 0.70 g starch/g mixture), and Figure 2-1b corresponds to starch pastes with a xanthan gum concentration of 0.01 g hydrocolloid/g mixture. In systems with a starch mass fraction of 0.1 (g starch/g mixture), the availability of water was not sufficiently reduced by the presence of gums, and the presence of a second transition peak in the DSC curves was not observed. Differential scanning calorimetry thermograms of starch with low and intermediate water contents often show multiple melting endotherms which reflect the water- and heat-induced disorganization of crystallites (Biliaderis, 1991). Transitions at higher temperatures have been attributed to order-disorder processes of amylose-lipid complexes. The development of a second gelatinization peak as the amount of available water decreases appears initially as a shoulder of the main peak (at a water volume fraction of about 0.60), and at lower water concentration as an independent peak (II). A shift of this peak to higher temperatures as moisture content decreases has also been observed (Ferrero et al., 1996).

Similar shaped endotherms were obtained for starch pastes with and without hydrocolloids. However, at the same starch concentration the shift to higher values in the conclusion temperature of gelatinization was evidenced when the gelatinization occurred in the presence of the gum; thus, the presence of the hydrocolloids was determined to have affected the availability of water for starch during gelatinization.

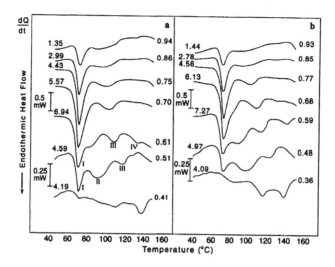

Figure 2-1 Differential scanning calorimetry thermograms of corn starch gelatinization with different volume fractions of water (indicated on the right side): a) starch-water system, and b) starch water with xanthan gum (0.01 g gum/g mixture). The weights of the dry starch (mg) are indicated on the left side (adapted from Ferrero et al., 1996).

Similar shaped endotherms were obtained for starch pastes with and without hydrocolloids. However, at the same starch concentration the shift to higher values in the conclusion temperature of gelatinization was evidenced when the gelatinization occurred in the presence of the gum; thus, the presence of the hydrocolloids was determined to have affected the availability of water for starch during gelatinization.

Biliaderis et al. (1986) and Slade and Levine (1987) have postulated the non-equilibrium nature of this melting process. However, the Flory-Huggins equation can be applied from a phenomenological point of view to describe the effect of a solvent to depress the melting temperature (Tm) of a polymer. This equation relates the Tm of the polymer to the volume-fraction of the plasticizer (water) as follows:

$$\frac{1}{Tm} - \frac{1}{Tm^0} = \frac{R V_u}{\Delta H_u V_1}\left(v_1 - X_1 v_1^2\right) \tag{2-1}$$

where R is the gas constant, V_u/V_1 is the ratio of the molar volume of the repeating unit (glucose) to the molar volume of the solvent, T_m^0 is the polymer equilibrium melting point in the absence of dilution, Tm is the melting point of the polymer solvent mixture, X_1 is the Flory-Huggins polymer solvent interaction parameter, and v_1 is the volume fraction of water. The Flory-Huggins equation reasonably described the DSC melting data for starch, particularly at water volume fractions lower than 0.7 and for starch pastes including hydrocolloids (Ferrero et al., 1996).

STABILITY OF FROZEN STARCH PASTES

Freezing rates can be determined according to the International Institute of Refrigeration (1972) as the minimum distance from the surface to the thermal center, divided by the time elapsed between the moment the surface reaches 0°C and the thermal center reaches a temperature 10°C colder than the temperature of the initial ice formation in the system. In commercial practice, 0.2–0.5 cm/h correspond to slow freezing (bulk freezing in cold chambers), 0.5–3 cm/h to quick freezing (air blast and contact plate freezers), 5–10 cm/h to rapid freezing (individual quick freezing in fluidized beds), and 10–100 cm/h to ultra rapid freezing (cryogenic systems).

The effects of freezing rates and frozen storage on the physical properties of a gelatinized starch paste (Ferrero et al., 1993b) were analyzed on a model system of corn starch (7 to 10% w/w wet basis) with and without the addition of xanthan gum (0.3% w/w wet basis). The rheological behavior of pastes with lipid phases, prepared by adding melted shortening or sunflower oil (5% w/w) to the starch (7% w/w) suspension were also tested (Navarro et al., 1997).

Exudate Production and Texture Modifications

The freezing rate of corn starch pastes has an important effect on exudate production (Ferrero et al., 1993a) in that high freezing rates (> 100 cm/h) lead to lower exudate values. However, the addition of a hydrocolloid like xanthan gum decreases exudate production. During frozen storage at –5°C a spongy matrix is formed, but this structure is not observed when samples are frozen in liquid nitrogen. The spongy structure is attributed to the water release caused by slow freezing, producing local high starch polymer concentrations and interaction between molecular chains. The addition of xanthan gum also avoids the formation of the spongy structure even at high storage temperatures; in this case, exudate levels are higher at –5°C than at –10°C and –20°C.

Ice Crystal Formation and Recrystallization

The effect of freezing rate on ice crystal size distributions in starch pastes was analyzed by Ferrero et al. (1993a, 1994) using the freeze fixation technique at low temperatures to determine the size of the ice crystals. The initial freezing point of the starch paste (10% w/w wet basis) was –0.6°C. Histograms of the relative frequencies of the crystal diameters as a function of equivalent diameter were obtained for different freezing rates and storage conditions, and ice crystal size distributions were analyzed from the micrographs.

As the freezing rate decreased, mean crystal size increased, leading to microstructure deterioration. Crystal size distributions obtained with rapid freezing rates (15 cm/h) had a narrow mean equivalent diameter. The addition of xanthan gum to corn starch pastes did not show a significant effect on ice crystal size at different freezing rates compared to pastes without the gum. During the frozen storage, ice recrystallization took place, visualized as an enlargement of ice crystal sizes at the expense of the smaller ones due to storage time, high temperature levels, and thermal fluctuations (Martino and Zaritzky, 1988, 1989). Histograms of ice crystal relative frequencies showed that at high temperatures and long storage times mean that equivalent diameters increased and that xanthan gum did not avoid ice recrystallization. Curves of equivalent ice crystal diameters vs. storage time (Fig. 2-2) showed a tendency to reach different limit equivalent diameters depending on storage temperatures. A mathematical model was fitted under the assumption that the driving force of this phenomenon is the difference between the instantaneous curvature of the system and the limit curvature (Martino and Zaritzky, 1989):

$$\frac{dD}{dt} = k\left(\frac{1}{D} - \frac{1}{D_l}\right) \tag{2-2}$$

where D is the mean equivalent ice crystal diameter at time t, D_l is the limit equivalent diameter, and k is the kinetic constant. The integration of Equation 2-2 led to the following expression, with D_o = initial equivalent diameter:

$$\ln\left(\frac{D_l - D_o}{D_l - D}\right) + \frac{(D - D_o)}{D_l} = \frac{k\,t}{D_l^{\,2}} \tag{2-3}$$

Calculation of the kinetic constants and activation energy values (Ea = 55.43 KJ/mol (sd = 13.69) for corn starch pastes without xanthan gum and 57.90 KJ/mol (sd = 12.98) for pastes with xanthan gum allowed for the conclusion that xanthan gum does not alter ice recrystallization, nor its activation energy. Hydrocolloids are commonly recommended as ice crystal inhibitors, but some studies (Reid et al., 1987) have shown that the stabilization character of hydrocolloids should be explained on another basis, such as their possible capability to undergo molecular entanglement in the freeze concentrated matrix surrounding ice crystals (Ferrero et al., 1993b).

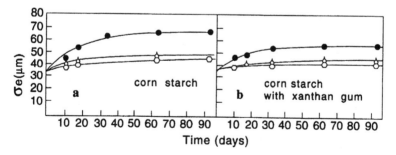

Figure 2-2 Mean equivalent ice crystal diameter vs. storage time, showing the recrystallization of ice in a) corn starch paste, and b) corn starch paste with xanthan gum (0.03% w/w). The freezing rate is 2 cm/h, and storage temperatures ● –5°C, Δ –10°C, and ○, –20°C (adapted from Ferrero et al., 1993b).

Amylopectin Retrogradation

Differential scanning calorimetry measurements were performed to analyze starch retrogradation during freezing and frozen storage (Ferrero et al., 1993a). Transition enthalpies corresponded to the heat involved in melting the retrograded amylopectin. Amylopectin retrogradation was only detected for the corn starch pastes frozen at low rates (<1 cm/h); a marked macroscopic modification (sponge structure) was observed under these conditions. High freezing rates did not produce detectable retrogradation peaks, but lower freezing rates (1 cm/h and 0.3 cm/h) led to measurable retrogradation peaks (enthalpy values $\Delta H = 2.05 \pm 0.26$ J/g and 2.76 ± 0.36 J/g, respectively). The presence of xanthan gum did not modify the effect of freezing on the amylopectin retrogradation.

With reference to storage temperatures, amylopectin retrogradation was detected at -1 and $-5°C$, but not at higher temperatures. Thermograms of frozen samples stored at -1 and $-5°C$ showed an increase of the transition enthalpy and a shift of the peak temperature to lower values with increasing storage times (Fig. 2-3a). The starch retrogradation kinetic was determined by fitting the Avrami equation to ΔH values obtained at different storage times:

$$\theta = \frac{\Delta H_t - \Delta H_o}{\Delta H_\infty - \Delta H_o} = 1 - \exp\left(-k\, t^p\right) \qquad (2\text{-}4)$$

where θ is the retrograded fraction, ΔH_∞ is the maximum enthalpy (limit value), and ΔH_o is the enthalpy of the just frozen sample. The Avrami exponential factor was considered $p = 1$. The amylopectin retrograded fraction as a function of storage time is shown in Figures 2-3b and c. For corn starch pastes stored at $-5°C$, the obtained parameters were $k = 0.114$ days^{-1} (sd = 0.064) and $\Delta H_\infty = 9.288$ J/Kg (sd = 0.836); in systems including xanthan gum the corresponding values were $k = 0.216$ days^{-1} (sd = 0.063) and $\Delta H_\infty = 6.132$ J/Kg (sd = 0.271). These values show that xanthan gum addition did not avoid amylopectin retrogradation during frozen storage.

Textural Changes

The textural characteristics of sponginess observed at low freezing rates and high storage temperatures can be attributed to the type of amylose retrogradation described by Morris (1990) as the coarsening of the fibrillar network. These conditions enhance water release and the formation of high amylose concentration zones with the thickening of the fibrillar structure.

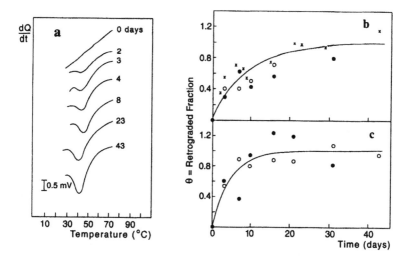

Figure 2-3 Amylopectin retrogradation of corn starch pastes during frozen storage at −5°C: a) DSC thermograms at different storage times of pastes frozen at 12 cm/h, b) without xanthan gum, and c) amylopectin retrograded fraction with xanthan gum. The freezing rates are ● 0.3 cm/h, ○ 1.0cm/h, and X 12 cm/h.

Xanthan gum inhibited the development of this spongy matrix microstructure, which can be explained by considering that amylose-hydrocolloid interaction competes with amylose-amylose aggregation, and thus decreases the probability of amylose retrogradation.

RHEOLOGICAL BEHAVIOR OF UNFROZEN AND FROZEN STARCH PASTES

Among other factors, the rheological characteristics of starch pastes or gels depend on the shape and swelling power of the granules, amount of amylose leached from the granule, network entanglement, and interaction between the paste components (Navarro et al., 1994). Both amylose and amylopectin retrogradation contribute to modifications like rheological changes and syneresis. Pastes thickened by starches are complex viscoelastic materials that fall between the two extremes of elastic solids and viscous fluids. Their flow properties usually depend upon time as well as shear rate. Rheological measurements either by transient rotational viscometry or dynamic oscillatory methods are useful in determining the stability of starch-based products submitted to different process conditions.

Navarro et al. (1995) analyzed the rheological behavior of corn starch pastes by testing the effect of their freezing rate (slow: 0.03 cm/h and rapid: 31 cm/h), the addition of a lipid phase (5% w/w shortening constituted by triglycerides with 76.5% oleic, 14.2% stearic, and 6.2% palmitic acids), and xanthan gum (0.3% w/w wet basis). Transient tests were performed with a rotational viscometer on unfrozen-, rapidly and slowly frozen-, gelatinized corn starch pastes (7 and 10% w/w wet basis) with and without shortening. The curves of shear stress (σ) during shear time at a constant shear rate (D) are shown in Figures 2-4a and b. Two characteristic regions are observed: the first part up to the overshoot represents the elastic response, and the second part corresponds to the structural breakdown. The Bird-Leider model (Kokini and Dickie, 1981) describes these types of curves and includes both the viscous and elastic responses. A modified Bird-Leider model including several exponential terms which represent the structural breakdown was developed by Mason et al. (1982):

$$\sigma = k \cdot D^n \, [1 + (b \, D \, t - 1) \sum w_j \exp (t / c_j)] \qquad (2-5)$$

where k is the consistency and n the flow behavior index of the power law model; b and c_j are adjustable parameters related to the time dependent behavior; and w_j are the weight factors of the exponential terms. At long times this equation converges to the power law model and an equilibrium shear stress is reached ($\sigma_\infty = k \, D^n$). Bird-Leider parameters were obtained for the different experiments.

Rapidly frozen samples maintained the behavior of unfrozen samples, though at long times of shearing the shear stress of the rapidly frozen samples was lower, indicating a decrease in apparent viscosity due to freezing. At short times of shearing the slowly frozen samples showed an overshoot related to the elastic characteristic of the samples (Navarro et al., 1995). Afterwards, a marked stress decay was evidenced with increasing shear time, and the lipid phase was found to increase the viscoelastic characteristics of the corn starch pastes (Fig. 2-4b). The authors also reported that xanthan gum maintained the smooth unfrozen texture regardless of the freezing rate used.

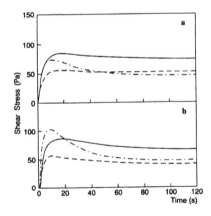

Figure 2-4 Rheological behavior of corn starch pastes, with the effect of freezing rate on shear stress vs. shear time curves in a) corn starch (10% w/w), and b) corn starch (7% w/w) with shortening (5% w/w) pastes. Transient curves at 60°C were obtained at a constant shear rate of D = 16 s^{-1} (——, unfrozen; – – –, rapid freezing (31 cm/h), – · – · –, slow freezing (0.3 cm/h) (from Navarro et al., 1995).

Dynamic oscillatory tests covering a wide range of shear deformation conditions (linear and nonlinear viscoelastic ranges) were also applied on the same starch systems to analyze the effect of freezing rate on structure stability (Navarro et al., 1997). Figure 2-5 shows curves representing the storage modulus (G') and complex modulus (G*) as functions of shear deformation (γ) at a constant frequency of oscillation of 1 Hz. The range of shear deformation where G* is constant corresponds to the linear viscoelastic range, and the deformation encountered by the material is small enough so that the material is negligibly disturbed from its equilibrium state (Kokini et al., 1995).

In the linear viscoelastic range (limited in this case by γ_{max} = 4%), G' and G* showed similar values (G'$_{max}$), indicative of the low contribution of the viscous component (G") to the viscoelastic properties of the system. For $\gamma > \gamma_{max}$ outside the linear region, structural breakdown was produced and the viscous component became important. After slow freezing of the starch samples, dynamic parameters showed an increase in the rigidity of the pastes (high G'$_{max}$ values), and outside the linear viscoelastic range a marked stress decay was observed compared to the unfrozen and rapid frozen samples.

The linear viscoelastic range of the starch pastes (determined by γ_{max}) decreased after freezing. Low values of γ_{max} are usually associated with a less stable system to the applied stress, but in this case unfrozen pastes with lipids showed an increase of the linear viscoelastic range and higher values of G'. In frozen systems the presence of triglycerides produced a decrease in G' over the linear viscoelastic range compared to pastes without triglycerides, showing a higher stability against freezing after structural breakdown. Dynamic tests also demonstrated that the presence of xanthan gum helped to maintain the rheological properties of the pastes after freezing.

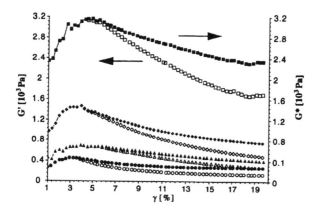

Figure 2-5 Dynamic rheological parameters: G* complex modulus (●, ▲, ■, ◆) and G' storage modulus (○, △, □, ◇) as functions of γ for corn starch pastes CS (7% w/w) and corn starch pastes containing shortening (5% w/w) CS + SH (temperature = 60°C, frequency = 1 Hz). Unfrozen: CS ○ ●, CS + SH △ ▲; slowly frozen: CS □ ■ CS + SH ◇ ◆.

Relationships between transient shear stress data and dynamic measurements for linear and nonlinear viscoelastic ranges were established by Navarro et al. (1997). In particular, high overshoots in the transient test were correlated with large values of G'_{max} and characterized the rigidity of slowly frozen samples. After structural breakdown and outside the linear viscoelastic range, both methodologies quantified the fluid-like character of the system that characterizes the weakness of the resulting structure.

GLASS TRANSITION TEMPERATURES OF STARCH PASTES

Glass formation by the removal of solvent is the most common and natural process of food preservation. The solvent in food is water that can be removed by dehydration or separated by freezing. Ice formation in food materials results in freeze concentration of the solutes, the extent of which depends on temperature according to the melting temperature depression of water caused by the solute. After starch gelatinization, an amylose matrix filled with granules of different degrees of fragmentation is obtained. During the freezing of gelatinized starch suspensions, only part of the total water is converted to ice, and the other remains unfrozen. Nonequilibrium ice formation is a typical phenomenon of rapidly cooled biological materials at low temperatures. Freeze concentration and lowering temperature increases the viscosity of the unfrozen phase until this concentrated solution becomes a glass (Roos, 1995). Rapidly cooled materials can be rewarmed to a devitrification temperature to allow ice formation. The glass transition temperatures of slowly frozen samples are higher than those of rapidly cooled samples and considered to be the glass transition temperature (T'g) of the freeze concentrated solute matrix surrounding the ice crystals in a maximally frozen solution (Levine and Slade, 1986). Roos and Karel (1991) suggested that formation of such maximally frozen solutions with a solute concentration (C'g) in the unfrozen matrix requires annealing slightly below the initial ice melting temperature within the maximally frozen solution. Solutions cooled rapidly to temperatures lower than T'g show nonequilibrium ice formation.

Ferrero et al. (1996) used DSC to analyze the influence of different hydrocolloids on the Tg and T'g values of gelatinized starch suspensions. To obtain T'g of starch pastes, samples were subjected to a slow freezing rate (2°C/min) on the DSC equipment, then annealed for 30 min at a temperature near the estimated T'g and heated at 2°C/min to 20°C. Differential scanning calorimetry thermograms obtained for the starch pastes without hydrocolloids during thawing showed that the characteristic change in Cp due to glass transition was produced before the ice melting peak (Fig. 2-6a). An exothermic peak was observed between the ice melting endotherm and glass transition temperature which was attributed to the additional ice crystallization that could not take place as long as the water-starch matrix was in the glassy state. Maximum ice crystallization was obtained by annealing the samples between T'm (onset of ice melting) and T'g. In this case, the exothermic peak for additional ice formation was not observed. Annealed pastes showed a shift in Tg onset, and the medium and conclusion temperatures led to the corresponding T'g values of the maximally concentrated matrix.

The absence of an exothermic devitrification peak in systems with xanthan gum may be related to the high viscosity of these pastes, which delays the formation of additional ice (Fig. 2-6b). For starch pastes with and without gums, T'g onset values ranged between –4.5 and –5.5°C; other authors (Slade and Levine, 1991) reported similar temperature ranges. This is an important finding and can explain the physical behavior of frozen pastes. Lack of detectable starch retrogradation during frozen storage at –10 and –20°C can be explained by considering that below T'g, amylose and amylopectin chains have a reduced mobility that limits the molecular association responsible for the retrogradation phenomena. Rapid freezing and storage at low temperatures avoided the crystallization of amylose and amylopectin in the concentrated matrix, leading to a homogeneous structure without a spongy network or amylopectin retrogradation peaks.

CONCLUSIONS

The effect of hydrocolloids on starch gelatinization was evidenced by the tendency of gelatinization enthalpies to increase with gum concentration. The shift to higher values in the conclusion temperature of gelatinization was demonstrated when the gelatinization occurred in the presence of the gum at low water contents. Consequently, the presence of hydrocolloids was determined to have affected the availability of water for starch during gelatinization.

The freezing rate was found to be an important factor in paste quality preservation because high freezing rates led to low exudate production and good texture preservation which maintained the characteristics of unfrozen pastes. Low freezing rates led to higher structural changes involving spongy matrix

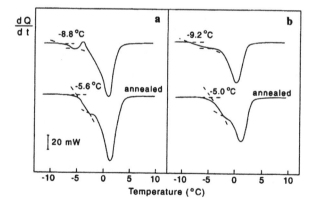

Figure 2-6 DSC thermograms of starch pastes (0.5 g starch/g mixture) during thawing: a) without hydrocolloid, and b) with xanthan gum (0.01 g gum/g mixture). The temperatures shown are the onset values, and the lower curves correspond to the annealed samples (adapted from Ferrero et al., 1996).

formation. However, the addition of a lipid phase moderated the effect of the freezing rate, and adding xanthan gum helped to maintain the rheological characteristics of unfrozen pastes even under low freezing conditions. Xanthan gum did not avoid amylopectin retrogradation, but it did inhibit the development of the spongy structure produced by amylose retrogradation. For starch pastes with and without gums, T'g onset values ranged between –4.5 and –5.5°C. Lack of detectable starch retrogradation during frozen storage at T < –10°C can thus be attributed to the reduced mobility of amylose and amylopectin chains below T'g. Although high freezing rates are recommended for starch-based systems, the addition of low levels of hydrocolloids like xanthan gum would allow the use of low freezing rates to maintain quality characteristics and decrease equipment investment.

REFERENCES

Biliaderis C.G. (1991). Non equilibrium phase transitions of aqueous starch systems. In: *Water Relationship in Foods*. Levine H.and Slade L., Eds. New York: Plenum Press, pp. 251–273.

Biliaderis C.G. (1992). Structures and phase transitions of starch in food systems. *Food Technol.* 46(6): 98–109.

Biliaderis C.G. and Zawistoski J. (1990). Viscoelastic behavior of aging starch gels: Effects of concentration, temperature and starch hydrolysates on network properties. *Cereal Chem.* 67: 240–246.

Biliaderis C.G., Page C.M., Maurice T.J., and Juliano B.O. (1986). Thermal characterization of rice starches: A polymeric approach to phase transition s of granular starches. *J. Agric. Food Chem.* 34: 6–14.

Evans I.D. (1986). An investigation of starch/surfactant interactions using viscometry and differential scanning calorimetry. *Starch* 38: 227–235.

Evans L.D. and Haisman D.R. (1982). The effect of solutes on the gelatinization temperature range of potato starch. *Starch* 34 :224–231.

Ferrero C., Martino M., and Zaritzky N.E. (1993a). Effect of freezing rate and xanthan gum on the properties of corn starch and wheat flour pastes *Int. J. Food Sci. Technol.* 28: 481–498.

Ferrero C., Martino M., and Zaritzky N. (1993b). Stability in frozen starch pastes. Effect of freezing, storage and xanthan gum addition. *J. Food Proc. Preserv.* 17(3): 191–211 .

Ferrero C., Martino M., and Zaritzky N. (1994). Corn starch, xanthan gum interaction and its effect on the stability during storage of frozen gelatinized suspensions *Starch* 46: 300–308.

Ferrero C., Martino M., and Zaritzky N. (1996). Effect of hydrocolloids on starch thermal transitions, as measured by DSC. *J. Therm. Anal.* 47(5): 1247–1266.

International Institute of Refrigeration (1972). Recommendations for the processing and handling of frozen foods. *Paris*, pp. 14–16.

Kokini J.L., Wang C.F., Huang H., and Shrimanker S. (1995). Constitutive models of foods. *J. Tex. Stud.* 26: 421–455.

Kokini J.L. and Dickie A. (1981). An attempt to identify and model transient viscoelastic flow in foods. *J. Tex. Stud.* 12: 539–557.

Lelievre J. (1974). Starch gelatinization. *J. Appl. Polym. Sci.* 18: 293–296.

Levine H. and Slade L. (1986). A polymer physicochemical approach to the study of commercial starch hydrolysis products (SHPs) *Carbohydr. Polym.* 6: 213–244.

Martino M. and Zaritzky N. (1988). Ice crystal size modifications during frozen beef storage. *J. Food Sci.* 53: 1631–1637, 1649.

Martino M. and Zaritzky N. (1989). Ice recrystallization in a model system and in frozen muscle tissue. *Cryobiol.* 26: 138–148.

Mason P.L., Bistany K.L., Puoty M.G., and Kokini J.L (1982). A new empirical model to simulate transient shear stress growth in semi-solid foods. *J. Tex. Stud.* 10: 347–370.

Miles M.J., Morris V.J., Orford P.D., and Ring S.G. (1985) The roles of amylose and amylopectin in the gelation and retrogradation of starch. *Carbohyd. Res.* 135: 271–281.

Morris V.J. (1990). Starch gelation and retrogradation. *Trends Food Sci. Technol.* 7: 1–6.

Navarro A.S., Martino M.N., and Zaritzky N.E. (1994). Swelling and rheological behaviour of starch-lipid systems. *Scanning* 16(4): 76–77.

Navarro A., Martino M., and Zaritzky N. (1995). Effect of freezing rate on the rheological behaviour of systems based on starch and lipid phase. *J. Food Eng.* 26: 481–495.

Navarro A., Martino M., and Zaritzky N. (1997). Correlation between transient rotational viscometry and a dynamic oscillatory test for viscoelastic starch based systems. *J. Tex. Stud.* 28: 365–385.

Pomeranz Y. (1985). Carbohydrates: Starch. In: *Functional Properties of Food Components.* Orlando, FL: Academic Press, pp. 25–90.

Reid D.S., Alviar M.S., and Lim M.H. (1987). The rates of change of ice crystal size in model systems stored at different temperatures relevant to the storage of frozen food. In: *Proceedings of the 16th International Congress on Refrigeration. Vol. C.* Austria, pp. 397–401.

Roos Y. (1995). *Phase Transitions in Foods.* Food Science and Technology International Series. San Diego, CA: Academic Press.

Roos Y. and Karel M. (1991). Applying state diagrams to food processing and development. *Food Technol.* 45: 66–71, 107.

Roulet P., McInnes W.M., Gummy D., and Wursch P. (1990). Retrogradation of eight starches. *Starch* 42: 99–101.

Slade L. and Levine H. (1987). Recent advances in starch retrogradation In: *Industrial Polysaccharides. The Impact of Biotechnology and Advanced Methodologies.* Stivalo S., Cresceni V., and Dea I.C., Eds. New York: Gordon & Breach Science Publishers, pp. 387–430.

Slade L. and Levine H. (1991). Beyond water activity: Recent advances based on an alternative approach to the assessment of food quality and safety. *Crit. Rev. Food Sci. Nutr.* 30: 115–360.

CHAPTER 3

Correlation between Physico-Chemical and Sensory Data

G.E. Hough

INTRODUCTION

The sensory properties (SP) of foods are usually measured by: (a) naive consumer panels who express their acceptability for the food samples, and (b) trained panels who measure sensory properties analytically. In both cases a number of people have to be motivated, assembled, and instructed, and after the evaluations are completed, a large array of data is left behind to be statistically analyzed and interpreted. Physico-chemical measurements (PCM) are carried out by laboratory instruments which are a lot simpler to handle than people because they provide precise data with ease and speed. However, their drawback is that they may not predict sensory properties since these are ultimately perceived by the consumer, and thus this type of evaluation very often becomes useless. Regression tools to analyze the correlation between PCM and SP go from very simple linear equations of the form:

$$\text{Sensory} = a + b \cdot \text{Concentration}$$

to linearized versions of psychophysical laws such as Steven's law (Moskowitz, 1983):

$$\text{Log(Sensory)} = k + n \cdot \text{Log(Concentration)}$$

to multivariate data where a matrix of SP is correlated to a matrix of PCM. Partial Least Square Analysis (**PLS**) is a method especially developed for the correlation of multivariate data.

PRINCIPLES OF PLS

Definitions

The data matrices on which **PLS** operates are **Y** for SP and **X** for PCM. **Y** is a N × Q matrix represented by the rows of N samples (index I = 1...N) and the columns of Q sensory attributes (index k = 1...Q). **X** is a N × P matrix, with N rows and PCM columns (index j = 1...P). Simple examples would be:

$$Y = \begin{bmatrix} \text{sweet}_1 & \text{acid}_1 & \text{thick}_1 \\ \text{sweet}_2 & \text{acid}_2 & \text{thick}_2 \\ \text{sweet}_3 & \text{acid}_3 & \text{thick}_3 \end{bmatrix} \qquad X = \begin{bmatrix} \text{pH}_1 & {}^\circ\text{Brix}_1 & [\text{NaCl}]_1 & \text{poise}_1 \\ \text{pH}_2 & {}^\circ\text{Brix}_2 & [\text{NaCl}]_2 & \text{poise}_2 \\ \text{pH}_3 & {}^\circ\text{Brix}_3 & [\text{NaCl}]_3 & \text{poise}_3 \end{bmatrix}$$

where N = 3 samples, Q = 3 sensory variables, and P = 4 PCM. A caveat to be considered: when *sweet_1* is indicated in the Y matrix above, the numerical value that goes in this slot is the average for the whole sensory panel when scoring Sample 1 on the intensity of sweetness.

Principles

To correlate **Y** vs. **X**, there are a number of techniques. If Q = 1, or only one sensory variable such as overall sensory acceptability, multiple linear regression could be applied (Montgomery, 1991). If there are more than one sensory variable, as in quantitative descriptive analysis (Stone and Sidel, 1985), then canonical correlation analysis could be used (Johnson and Wichern, 1992). However, neither of these methods work unless the number of samples is much higher than the number of variables. In fact, multiple linear regression cannot be used at all if the number of samples is lower than the number of variables. There are additional limitations when the **X** variables are highly correlated between each other, but such problems can be avoided by applying stepwise multiple regression to select a sub-group of the **X** variables. Unfortunately in this case the complete overview of the relationships in the data would be lost.

Partial Least Square Analysis (Martens and Martens, 1986) is a multivariate regression method that handles multicollinearity and small sample sets. **PLS1** is used to predict one dependent variable (**Y** matrix has only one column), and **PLS2** to predict several dependent variables (**Y** matrix has several columns) from a set of independent variables (**X** matrix). The **PLS** approach (Martens and Martens, 1986) combines the PCM from the **X** matrix onto a few factors or latent variables that can be regarded as the main "harmonies" between the **X** variables (similar to principal component analysis). The intensities of these main harmonies in different samples are represented by a reduced factor N x A score matrix **T**, with N rows representing the sample scores on A components. These last are the columns of **T** which are less than the original PCM variables. The result leads to:

$$X = TU + E \qquad (3\text{-}1)$$

and

$$Y = TW + F \qquad (3\text{-}2)$$

where **U** (A x P) and **W** (A x Q) are the loading matrices which transform scores in **T** to scores in **X** and **Y**, respectively. **T** has fewer columns than **X** and tries to explain most of the variation in **X** and **Y**. The variation that cannot be explained is contained in the residual matrices **E** and **F**.

The latent A variables of **T** which are common to both **X** and **Y** are computed such that they are orthogonal and maximally predict the set of dependent variables that represent the sensory properties. Wold et al. (1983), Hoskuldsson (1988), and Martens and Martens (1986) give details on **PLS** calculations. The last authors present a classroom analogy which is helpful in understanding the **PLS** algorithm. The first step in **PLS** calculation is to center the data which is usually scaled to unit variance. Then the parameters of the first factor of **T** are calculated in such a way that they are relevant to **X** and **Y** simultaneously. The effect of this first parameter is subtracted to create the new residuals E_1 and F_1 from which a second factor of **T** is obtained that is orthogonal to the first and subtracted to obtain the residuals E_2 and F_2. This process continues until no more valid factors can be obtained from the data.

A (the number of factors of **T**) cannot exceed N nor P. If A is equal to either N or P then the **X** matrix has been emptied for both information and noise, and consists only of zeros. If N > P so that multiple linear regression can be applied to the same data, then the limiting solution of A = P is identical to the multiple linear regression solution. In many practical cases this means overfitting the data by introducing meaningless noise into the model (Martens and Martens, 1986). In order to guard against overfitting, cross-validation procedures are used. Genstat's **PLS** procedure (Genstat 5 is a general statistics software package developed by Numerical Algorithms Group, Inc. 1400 Opus Place, Suite 200 Downers Grove, IL 60515-5702) uses Osten's (1988) test of significance to determine the number of factors to retain in the T matrix.

EXAMPLE OF A PLS APPLICATION

Martens and Martens (1986) present a number of **PLS** examples. The following was used by the author of the present chapter when searching for correlations between sensory and instrumental measurements of flavor and texture for a grating cheese (Hough et al., 1996; tables and graphs reprinted with permission from Elsevier Science).

Flavor and texture are important sensory attributes with respect to the purchase and use of cheese. In quality control and the classification of cheese there is a need for both sensory and instrumental methods. Instrumental methods tend to be easier to perform, standardize, and reproduce than sensory measurements, yet a clear relationship between them must be established. The objective in this case was to investigate the relationships between the following variables measured during the ripening time of Reggianito grating cheese: a) visual, manual, and oral texture vs. an Instron compression test, and b) aroma and flavor vs. the concentration of organic acids. Oral texture and flavor were considered for this example.

Thirty cheese forms were produced at a local dairy plant (Cabañas y Estancias Santa Rosa Inc., Carlos Casares, Buenos Aires, Argentina) according to standard practices. Approximately every month over a period of 10 months, whole cheeses were analyzed as follows:

(a) *Organic acids:* The following were analyzed using procedures described by Lombardi et al. (1994a): citric, orotic, pyruvic, lactic, formic, uric, acetic, propionic, and butyric.

(b) *Texture:* An Instron 1132 texturometer was used for obtaining compression curves as described by Lombardi et al. (1994b), from which the following parameters were derived: hardness, breaking force, strain at breaking point (from now on: strain), and deformability modulus. Moisture level in the cheese was also added to the texture matrix, as it usually has a high influence on this attribute.

c) *Sensory profile:* A panel of nine trained sensory assessors developed a sensory profile consisting of 52 descriptors divided in appearance, texture, aroma, and flavor attributes. Details of the sensory procedure can be found in Hough et al., 1994, and the descriptors used for texture and flavor in Table 3-2. The profile was used to evaluate the same cheeses for which organic acids and texture were analyzed. For the present study, averages over assessors and replicates were used.

Table 3-1 shows the percent variance explained by the **PLS** correlation. For the individual descriptors, the percent variance of the experimental values accounted for by the first three **PLS** factors are in Table 3-2. The correlation coefficients of sensory descriptors and objective measurements with the first two **PLS** factors are presented graphically in Figures 3-1 and 3-2.

Table 3-1 Percent variance explained by the first three factors of the partial least squares correlation of sensory measurements versus organic acids and instrumental texture.

Y matrix	X matrix	% variance Y			% variance X		
		F-1	F-2	F-3	F-1	F-2	F-3
Oral texture	Compression	48	9	8	44	48	5
Flavor	Organic acids	48	6	(a)	34	26	(a)

(a) This dimension was not a valid predictor according to Osten's F-test for P < 0.05.

Table 3-2 Percent variance of experimental values accounted for by the first three partial least squares factors for individual descriptors. Oral sensory texture was predicted by instrumental compression parameters, and flavor by organic acids.

ORAL TEXTURE		FLAVOR	
Descriptor	% Variance	Descriptor %	Variance
hardness	77	total intensity	81
fracturability	71	cheese	81
cohesiveness	41	salty	81
roughness	67	sweet	52
water absorption	59	bitter	21
cohesiveness of mass	59	acid	49
adhesiveness to teeth	58	lipolysis	1
crystals	69	milky-creamy	36
		tongue-tingling	85
		hot	86
		residual intensity	83

Oral Texture

Sixty-five percent of the variance was explained by the first three factors. Figure 3-1 shows logical relationships between variables: sensory and instrumental hardness are close together as are fracturability, adhesiveness to teeth, water absorption, and roughness, though these are also opposed to moisture and strain. Since oral cohesiveness is defined as the degree of deformation before rupture, it is expected to be correlated with strain. Therefore, a possible explanation for the poor correlation obtained from **PLS** may be an oral saturation effect because above 50% strain, cohesiveness remained practically constant. However, no relation was found between the breaking force and sensory descriptors (Fig. 3-1).

Flavor

Fifty-four percent of the variance was explained by the first two **PLS** factors (see Table 3-1). Osten's test (1988) showed that there were only two significant dimensions as valid predictors; if more dimensions had been included only noise in the data would have been modeled. The following descriptors had over 80% of their variance accounted for by the first three **PLS** factors: total intensity, cheese, salty, tongue-tingling, hot, and residual intensity. All of these descriptors are

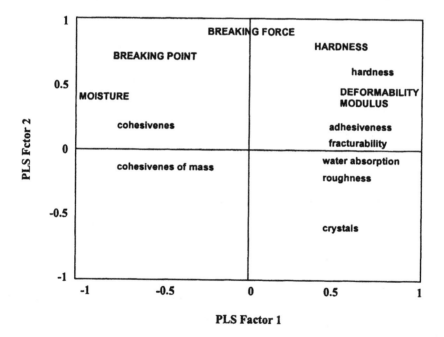

Figure 3-1 Correlation coefficients between the **PLS** X-scores vs. the sensory oral texture values, and **PLS** X-scores vs. moisture and instrumental compression parameters.

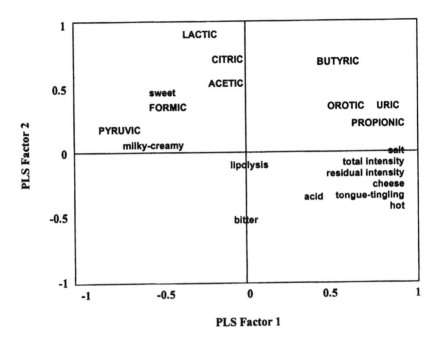

Figure 3-2 Correlation coefficients between the **PLS** X-scores vs. the sensory flavor values, and **PLS** X-scores vs. concentrations of organic acids.

grouped together in Figure 3-2 and relate positively to uric and propionic acids and negatively to pyruvic acid due to its decreasing during cheese ripening (Lombardi et al., 1994a). Organic acids were not responsible for salty taste, yet as this descriptor followed the same trend as total intensity, the **PLS** correlation picked out the indirect relationship. **PLS** was thus a good guide to select appropriate sensory and chemical variables that were highly correlated.

General Conclusions from the Example

Strain was related to fracturability, and considering visual and manual texture (Hough et al., 1996), it was the instrumental parameter which best correlated with sensory texture. Sensory hardness was related to instrumental hardness, although the correlation coefficients were not high. Moisture levels in the cheese were a useful parameters in predicting sensory texture. A number of flavor descriptors were well predicted from organic acids using **PLS**. For this type of grating cheese, propionic acid would be a good indicator of flavor development.

PLS APPLICATIONS IN SENSORY ANALYSIS

Table 3-3 presents a summary of different **PLS** applications to sensory analyses found in the literature by listing the products and type of data analyzed.

Table 3-3 Applications of **PLS** to sensory analysis (A: appearance, T: texture, F: flavor, O: odor).

Product	X matrix	Y matrix	Reference
Peas	Tenderometer, composition	Sensory A-T-F	Martens, 1986
Sausage	Sensory F	% Rancid fat	Sinesio et al., 1990
Dulce de leche	Sensory A-T-F	Acceptability	Hough et al., 1992a
Dulce de leche	Sensory A	L-a-b colorimeter	Hough et al., 1992b
Fermented milk	Sensory T-F	Acceptability	Muir and Hunter, 1992
Whisky	GC volatiles, composition	Sensory O	Piggott et al., 1993
Apple	GC volatiles	Sensory O	Brockhoff et al., 1993
Ham	Composition and rancidity	Sensory F	Careri et al., 1993
Cheese	Instron, composition	Sensory A-T	Jack et al., 1993a
Peppers	Composition	Sensory F	Luning et al., 1994
Cheese	Instron	Sensory T	Jack et al., 1994
Ham	Composition	Sensory F	Parolari, 1994
Peaches	Composition, color L-a-b	Sensory A-T-F	Kluter et al., 1994
Ham	GC volatiles	Sensory F	Hinrichsen and Pedersen, 1995
Oat flour	Fatty acids	Sensory F-T	Molteberg et al., 1996a
Oat grains	Phenolic compounds	Sensory F	Molteberg et al., 1996b
Olive oil	GC volatiles, composition	Sensory F	Monteleone et al., 1996
Pears	Composition, color L-a-b	Sensory A-T-F	Kluter et al., 1996
Cheese	Organic acids and Instron	Sensory A-T-F	Hough et al., 1996
Ice-cream	Composition design variables	Sensory T-F	Koeferli et al., 1996
Hot dogs	Sensory A-T-F	Acceptability	Muñoz et al., 1996
Cheese	Sensory O	Sensory F	Muir et al., 1997
Hot dogs	Sensory A-T-F	Acceptability	Muñoz et al., 1996
Lamb sausages	Sensory T-F	Acceptability	Helgesen et al., 1997

FINAL REMARKS

PLS is a method especially developed for the correlation of multivariate data that handles multicollinearity and small sample sets. In PLS, uncorrelated latent factors of independent variables are computed such that they maximally predict one or a set of dependent variables. An example of PLS application was presented where sensory properties of a hard cheese were correlated with instrumental measures of texture and organic acid concentrations. These correlations facilitated the choice of instrumental parameters that best predicted sensory properties. In addition, a table summarizing PLS applications in sensory analyses found in the literature was provided that showed the wide variety of products and different approaches to the choice of dependent and independent variables.

REFERENCES

Brockhoff P., Skovgaard I., Poll L., and Hansen K. (1993). A comparison of methods for linear prediction of apple flavour from gas chromatographic measurements. *Food Qual. Pref.* 4: 215–222.

Careri M., Mangia A., Barbieri G., Bolzoni L., Virgili R., and Parolari G. (1993). Sensory property relationships to chemical data of Italian-type dry-cured ham. *J. Food Sci.* 58: 968–972.

Helgesen H., Solheim R., and Naes T. (1997). Consumer preference mapping of dry fermented lamb sausages. *Food Qual. Pref.* 2: 97–109.

Hinrichsen LL, Pedersen SB (1995). Relationship among flavor, volatile compounds, chemical changes, and microflora in Italian-type dry-cured ham during processing. *J. Agric. Food Chem.* 43(11): 2932–2943.

Hoskuldsson A. (1988). PLS regression methods. *J. Chemo.* 2: 211–228.

Hough G., Bratchell N., and Wakeling I. (1992a). Consumer preference of dulce de leche among students in the United Kingdom. *J. Sens. Stud.* 7: 119–132.

Hough G., Bratchell N., and MacDougall D.B. (1992b). Sensory profiling of dulce de leche, a dairy based confectionery product. *J. Sens. Stud.* 7: 157–178.

Hough G., Martínez E., Barbieri T., Contarini A., and Vega M.J. (1994). Sensory profiling during ripening of Reggianito grating cheese, using both traditional ripening and in plastic wrapping. *Food Qual. Pref.* 5: 271-280.

Hough G., Califano A.N., Bertola N.C., Bevilacqua A.E., Martinez E., Vega M.J., and Zaritzky N.E. (1996). Partial least squares correlations between sensory and instrumental measurements of flavor and texture for Reggianito grating cheese. *Food Qual. Pref.* 7: 47–53.

Jack F.R., Paterson A., and Piggott J.R. (1993a). Relationships between rheology and composition of Cheddar cheeses and texture as perceived by consumers. *Int. J. Food Sci. Technol.* 28: 293–302.

Jack F.R., Piggott J.R., and Paterson A. (1994). Analysis of textural changes in hard cheese during mastication by progressive profiling. *J. Food Sc.* 59: 539–543.

Johnson R.A. and Wichern D.W. (1992). *Applied Multivariate Statistical Analysis*. Englewood Cliffs, Englewood Cliffs, N.J.: Prentice-Hall, pp. 459–492.

Kluter R.A., Nattress D.T., Dunne C.P., and Popper R.D. (1994). Shelf life evaluation of cling peaches in retort pouches. *J. Food Sci.* 59: 849–854, 865.

Kluter R.A., Nattress D.T., Dunne C.P., and Popper R.D. (1996). Shelf life evaluation of Bartlett pears in retort pouches. *J. Food Sci.* 61: 1297–1302.

Koeferli C.R.S., Piccinali P., and Sigrist S. (1996). The influence of fat, sugar and non-fat milk solids on selected taste, flavor and texture parameters of a vanilla ice-cream. *Food Qual. Pref.* 7: 69–79.

Lombardi A.M., Bevilacqua A.E., and Califano A.N. (1994a). Variation in organic acids content during ripening of Reggianito cheese in air-tight sealed bags. *Food Chem.* 51: 221–226.

Lombardi A.M., Bertola N.C., Giannuzzi L., Califano A.N., Bevilacqua A.E., and Zaritzky N.E. (1994b). Maduración de queso Reggianito en película plástica. *Actas del VI Congreso Argentino de Ciencia y Tecnología de Alimentos*, pp. 206–208.

Luning P.A., van der Vuurst de Vries R., Yuksel D., Ebbenhorst-Seller T., Wichers H.J., and Roozen J.P. (1994). Combined instrumental and sensory evaluation of flavor of fresh bell peppers (Capsicum annuum) harvested at three maturation stages. *J. Agric. Food Chem.* 42: 2855–2861.

Martens M. (1986). Sensory and chemical/physical quality criteria of frozen peas studied by multivariate data analysis. *J. Food Science* 51: 599-617.

Martens M. and Martens H. (1986). Partial least squares regression. In: *Statistical Procedures in Food Research*. Piggot J.R., Ed. Essex, UK: Elsevier Applied Science, pp. 293–359.

Molteberg E.L., Magnus E.M., Bjorge J.M., and Nilsson A. (1996a). Sensory and chemical studies of lipid oxidation in raw and heat-treated oat flours. *Cereal Chem.* 73: 579–587.

Molteberg E.L., Solheim R., Dimberg L.H., and Frolich W. (1996b). Variation in oat groats due to variety, storage and heat treatment. II: Sensory quality. *J. Cereal Sci.* 24: 273–282.

Monteleone E., Caporale G., Carlucci A., Bertuccioli M. (1996). Predizione del profilo sensoriale degli oli vergini di oliva. *Indu. Alimen.* 35: 1066–1072.

Montgomery D.C. (1991). *Design and Analysis of Experiments*. New York: John Wiley & Sons, pp. 498–512.

Moskowitz H.R. (1983). *Product Testing and Sensory Evaluation of Foods, Marketing and R&D Approaches*. Westport, CT: Food & Nutrition Press, pp. 237–260.

Muir D.D. and Hunter E.A. (1992). Sensory evaluation of fermented milks: vocabulary development and the relations between sensory properties and composition and between acceptability and sensory properties. *J. Soc. Dairy Technol.* 45: 73–80.

Muir D.D., Hunter E.A., and Banks J.M. (1997). Aroma of cheese. 2. Contribution of aroma to the flavor of Cheddar cheese. *Milchwissenschaft* 52: 85–88.

Muñoz A.M., Chambers IV E., and Hummer S. (1996). A multifaceted category research study: how to understand a product category and its consumer responses. *J. Sensory Studies* 11: 261–294.

Osten D.W. (1988). Selection of optimal regression models via cross-validation. *J. Chemo.* 2: 39–48.

Parolari G. (1994). Taste quality of Italian raw ham in a free-choice profile study. *Food Qua. Pref.* 5: 129–133.

Piggott J.R., Conner J.M., Paterson A., and Clyne J. (1993). Effects of Scotch whisky composition and flavor of maturation in oak casks with varying histories. *Int. J. Food Sci. Technol.* 28: 303–318.

Sinesio F., Risvik E., and Rodbotten M. (1990). Evaluation of panelist performance in descriptive profiling of rancid sausages: a multivariate study. *J. Sens. Stud.* 5: 33–52.

Stone H. and Sidel J.L. (1985). *Sensory Evaluation Practices.* Orlando, FL: Academic Press, pp. 202–226.

Wold S., Albano C., Dunn III W.J., Esbensen K., Hellberg S., Johansson E., and Sjostrom M. (1983). Pattern recognition: finding and using regularities in multivariate data. In: *Food Research and Data Analysis.* Martens H. and Russwurm Jr. H., Eds. London, UK: Applied Science Publishers, pp. 147–188.

CHAPTER 4

Psychophysical Methods Applied to the Investigation of Sensory Properties in Foods

M.C. Zamora

INTRODUCTION

Psychophysics is a science concerned with the measurement and evaluation of stimulus-response relations by means of human behavior. Stimuli are measured by chemical or physical methods, while the sensations are evaluated by psychophysical techniques. Although psychophysical methods can be applied to all senses, relevant information concerning the sensory properties of foods are mainly related to chemical senses such as smell and taste. Sensory evaluations of food are based on knowledge of the following concepts:

1. Sensation attributes
2. Measurement of absolute sensitivity, or threshold
3. Scaling techniques
4. Psychophysical laws
5. Measures of time, such as reaction and time-intensity registers
6. Descriptive analysis

SENSATION ATTRIBUTES

A stimulus consists of an input of chemical or physical energy which is recognized by a receptor and causes a further response. Nerve impulses then travel from receptors through the nerves and towards the brain where different signals are processed, resulting in the corresponding sensations. All sensations exhibit a multiplicity of aspects, attributes, or dimensions, just as all stimuli are themselves multidimensional. An effective stimulus produces a sensation, the dimensions of which are quality, intensity, duration, and preference. Considering that most stimuli are multidimensional, abstraction is needed for the evaluation of the attributes of a stimulus. Attention must be focused on one aspect or attribute and many other aspects must be discarded. Therefore, what is to be measured must be defined precisely.

Flavor is the result of a complex combination of the sense of smell, taste, pungency, temperature, texture, and consistency. For many foods, smell and taste are the most salient attributes. To avoid the need of abstraction as mentioned above, the stimuli for sensory evaluations should be substances that are mostly odorous, tasteful, or pungent.

MEASUREMENT OF ABSOLUTE SENSITIVITY

The minimum energy required to produce a response corresponds to the absolute threshold or detection threshold. It is intuitively compelling to assume that the amount of stimulation required to activate a sensory system is a direct reflection of the system's sensitivity. The recognition threshold is the level of a stimulus at which the specific stimulus can be recognized and identified. The differentiation threshold or just noticeable difference (JND) is the minimum increase in stimulus intensity required to produce a detectable difference in sensation.

The threshold can be measured by techniques such as constant stimuli, the method of limits, or up-down methods (ASTM, 1968; Doty, 1991). The constant stimuli method may be used for either absolute threshold or difference threshold measurements. For the former, the standard is zero concentration. Each sample is paired with the standard, and the pairs are presented in random order. The subject chooses which sample in each pair is stronger, and the point in the series of concentration where 75% of the judgments are correct is designated as the threshold. For the difference threshold, a series of standards (about four) is used, each one bracketed by its appropriate range of concentrations.

The method of limits is used for absolute threshold determination. Classically, an alternating ascending and descending series of stimulus concentrations are presented. The average concentration of the transition points serves as the threshold estimate. Thresholds for bitter and sour tastes are much lower than those observed for salty and sweet substances. This observation may agree with theories about the harmful potency of bitter substances and the need for their detection at minimal levels. Olfaction is among the chemosensory modalities with the highest absolute sensitivity (lowest detection threshold) as opposed to the common chemical sense. Therefore, only those molecules that vary according to different odorants can be detected by this system.

Signal Detection Theory

Born from engineering concepts developed for radar operation during World War II, signal detection theory allows for the measurement of both sensitivity and the response criterion. The latter is the internal rule that a subject uses to decide whether or not to report the detection of a stimulus. However, it varies among individuals. For example, Subject A may experience the same degree of sensation from a very weak stimulus as Subject B, and yet because of uncertainty, will report that no sensation is perceived. Subject B, on the other hand, may be less cautious and report the presence of a stimulus. Traditional non-forced-choice threshold techniques would lead to the erroneous conclusion that the first of these subjects had less sensitivity than the second (Doty, 1991).

This can be overcome by providing two stimuli: water and a particular concentration of the taste substance. The forced choice in this procedure consists of making the subject choose one of the two samples even though he is not sure (Miller and Bartoshuk, 1991). To detect small differences between samples and distinguish between single and mixed stimuli, another method based on the signal detection theory (R-index) has been developed. The R-index shows the ratio of right answers to total answers and indicates the degree of similitude between each mixture and the corresponding single concentration. This probability calculation reduces the dispersion of perceived intensities among the panelists. Another advantage of this measure is the fact that the panelist has to decide only if he is sure about his answer without the need for any numeric estimation. Thus, the error in scaling that may hide small perceptual differences is avoided (O'Mahony, 1992).

SCALING TECHNIQUES

Scaling techniques allow quantitative evaluation of sensations over the whole dynamic range of intensities from threshold to saturation. These provide information about the magnitude of a stimulus, theoretically making scales of perceived intensity a source of much more information than thresholds. The most important scales are category and ratio.

Category Scaling

Category scaling or partition is a method of measurement in which the subject is asked to rate the intensity of a particular stimulus by assigning it a value (category) on a limited, usually numerical scale. The numbers may serve to reflect the differences or distances between stimuli. According to Pedrero and Pangborn (1989), this scale may be structured or unstructured.

Structured Scales

A structured category scale is a line graph marked into segments with discrete response alternatives or categories associated in one way or another with a number or appropriate adjectives. The optimum number of categories in such a scale is a function of both the subject's ability to make fine distinctions along the stimulus continuum and the amount of discernment inherent in the stimulus continuum itself.

Unstructured Scales

An unstructured scale is just a line on which a subject places a mark to indicate the intensity of an attribute that is being rated. The extremes of such a scale are usually "anchored" to descriptors such as very weak or very strong. An advantage of unstructured scales is that the response can be assigned empirically to nearly any number of categories desired by the experimenter, although the distance from one end of the scale is usually indicated.

Common applications of the rating scale method include a) evaluation of hedonic value (preference) according to reactions of "like" and "dislike", and b) evaluation of the intensity of specific food attributes such as sweetness.

A distinction between unipolar and bipolar dimensions should be kept in mind when using or designing a rating scale. For example, the attribute of intensity is a unipolar dimension, whereas that of pleasantness/unpleasantness is bipolar.

Ratio Scaling

This method is also known as magnitude estimation. In this procedure, the subject assigns numbers relative to the magnitude of the sensations. The numbers may also serve to reflect ratios among stimuli. After the first sample is assigned a number, panelists are asked to assign all subsequent ratings of samples in proportion to the first sample rating. If the second sample appears three times as strong as the first, the assigned rating should be three times that assigned to the first stimulus (Stevens, 1957; 1975).

Ratio and Partition Scaling

Also known as converging limits, this method combines features of both ratio and partition scaling. Subjects are informed in advance of the physical range by sampling standards that correspond to the most and least intense samples, but they define their own numerical range. During the experiment the subjects are free to expand or contract their scale by using numbers larger or smaller than their original endpoint estimates (Guirao, 1991).

PSYCHOPHYSICAL LAWS

Researchers in psychophysics have long grappled with how to measure the basic input-output function of a sensory modality. Psychophysical laws help link this abstract relation of stimulus to sensation.

Weber's Law

In 1834 Weber proposed that the difference threshold increase in proportion to the absolute stimulus intensity perceived could be determined by

$$\Delta\Phi/\Phi = k \qquad (4\text{-}1)$$

where Φ is the absolute intensity of the stimulus, $\Delta\Phi$ is the change in intensity of the stimulus that is necessary for one JND, and k is a constant between 0 and 1. Thus, Weber's law says that the JND grows larger in direct proportion to the size of the stimulus (Stevens, 1975).

This fraction has been used as an index of both the discriminative ability of a subject and the discriminative power of a given sensory system. The smaller the fraction, the greater the discrimination. However, the magnitude of the Weber fraction depends upon a number of factors, and it is unlikely that such a constant holds across a wide range of concentrations.

Fechner's Law

Fechner accepted Weber's law and added a new feature. He proposed that each time a JND is added to the stimulus, the sensation increases by a jump of constant size. He thought he had found the equal units needed to measure the sensation. Fechner's assumption led directly to a logarithmic law for the growth of sensation:

$$\psi = k \log \Phi \qquad (4\text{-}2)$$

where Φ is the stimulus intensity, ψ is the number assigned to the JND, and k is a constant.

Support for Fechner's law is provided by common category scaling. One tangible outcome of his theories was a logarithmic scale of sound intensity known as the decibel scale (Meilgaard et al., 1987).

Stevens' Law

Stevens' Law demonstrates that the sensation magnitude ψ grows as a power function of the stimulus intensity Φ:

$$\psi = k \, \Phi^{\beta} \qquad (4\text{-}3)$$

The constant k depends on the units of measurement, and the value of the exponent β determines curvature in that if it is 1, the function is linear, and the sensation varies directly with the intensity of the stimulus.

Intensity magnitude estimation data are most commonly analyzed by Stevens' law. The exponent for olfaction is nearly always less than 1 and on a linear-linear plot, reflects a slow ascending function of physical concentration. For salty taste the exponent is about 1, while for sour the value frequently observed is about 0.67 (Guirao, 1980). Exponents obtained for different odorants are generally lower than those for tastes, while pungency exhibits the highest exponents of them all (García-Medina, 1981; Cometto-Muñiz, 1981; Cometto-Muñiz and Noriega, 1985). Correlation is usually observed between exponents and threshold values.

MEASURES OF TIME

Reaction Time

Gustatory reaction time (GRT) is defined as the minimum time required by a subject to report any taste changes after the onset of taste stimulation. Taste is a field in which many studies have been developed in order to relate GRTs to the concentration of some substances delivered on the tongue (Piéron, 1952; Yamamoto and Kawamura, 1981; Yamamoto et al., 1982; Kuznicki and Turner, 1988; Bujas et al., 1989). Gustatory reaction time experiments should discriminate the tactile response of any liquid falling on the tongue from the true gustatory response. For that reason water stimuli are generally intermixed within an experimental series (Buratti et al., 1996).

The stimulated surface and stimulus duration and concentration are some of the many variables that may modify GRTs and can be standardized. However, variables such as fluctuations in the subject's attention and differences in preparation time are more difficult to control.

Equipment to measure GRT has been developed at the Laboratorio de Investigaciones Sensoriales (LIS) (Guirao and Zamora, 1997) that consists of a pumping system, an interface between a computer and the pumping system, software to control the interface and measure the time, and a push button to detect the subject's reaction. The subject holds a tube and places it on the middle part of his tongue with his left hand while his right hand lies on the button. When the command for pumping the solution is given from a computer, the subject has to push the button as soon as he feels a stimulus is different from water. The computer measures the time interval between the onset of pumping and button-pushing by the subject.

Reaction time (RT) is useful in the study of sensory systems (Bonnet, 1994; Pins and Bonnet, 1996) because it allows the quantification of subject estimations by motor response without the need of numbers or scales. Taste qualities respective to GRTs allow their different detection, identification, and discrimination in mixture solutions.

Several equations have been proposed to adjust reaction time data, but the Piéron function provides the best fit to RT data as a function of concentration for substances measured for taste (Bonnet et al., 1999). The Piéron function takes the following form (Eq. 4-4):

$$(RT - t_0) = k \, \Phi^{-\beta} \qquad (4-4)$$

where t_0 is an asymptotic reaction time reached at high intensities, Φ is the stimulus intensity, and k and β are parameters. In this function, parameters β and t_0 are metrical characteristics of each modality. The exponent β and the time constant t_0 of such a function appear to characterize each family of tastes (Fig. 4-1).

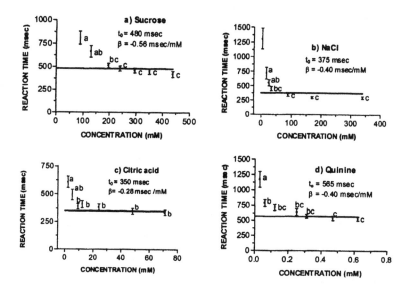

Figure 4-1 Reaction time as a function of concentration. The different small letters indicate significant differences (p < 0.05).

Time-Intensity Registers

Temporal perceptions of sensory attributes influence responses to many psychophysical tests. For example, the perception of maximum intensity for a specific compound is a function of both time and stimulus concentration (Birch et al., 1980). The development of computer technology has increased the ability for sensory scientists to quickly collect and then analyze large time-intensity (TI) data sets (Lundahl, 1992).

Zamora et al. (1998) studied the temporal characteristics of sucrose and fructose with a computerized data collection system. A mouse was used by panelists to move a cursor along a 500 pixel line that represented a 20 cm unstructured line scale on the monitor (500 pixels = 100% intensity). The response to sweetness intensity of each panelist to each sample was automatically recorded by the computer every 0.28 seconds.

Seven parameters were extracted from the TI curves (Fig. 4-2): 1) the maximum intensity (Imax, %), 2) time to maximum intensity (Timax, sec), 3) total duration of the sensation (Tdur, sec), 4) time for the sweetness intensity to decline to half its maximal value (T1/2, sec), 5) area under the curve (AUC, %/sec), 6) rate of increase (Rinc, %/sec), and 7) rate of decrease (Rdec, %/sec). An analysis of variance (ANOVA) of these parameters allowed the determination of which parameters significantly differed among substances and concentrations.

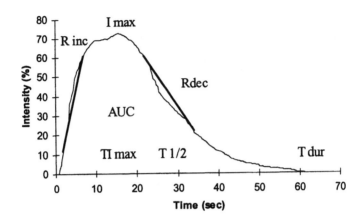

Figure 4-2 Intensity (%) as a function of time, with kinetic parameters indicated.

DESCRIPTIVE ANALYSIS

The global perception of a food product arising from the several simultaneous sensations provoked by eating may be studied by using descriptive analytical methods in association with scaling. A series of sensorial product attributes that may reach different intensity are scored successively for each product. Terms are then listed in order of perception and each attribute is scored using one of the scales (generally a category scale) (Daget, 1977).

Taster Variability

There is always a high variability between tasters due to distinctive thresholds to the different descriptors, ways of using the scales, and understanding of terminology. These last two points can be improved by training. After a subject has finished the first ANOVA, the experimenter should examine the terms for the highest significant difference between tasters and significant no difference between products. The panel must then discuss the reasons for the lack of consensus with the experimenter, and be trained through the tasting of some well-chosen samples. Next they proceed to perform a second ANOVA to determine if these terms are better understood (Calviño et al., 1996; Zamora and Calviño, 1996).

Another way to clarify inconsistencies in taster responses is to segment a panel into three groups. The tactile group should be sensitive to mouthfeel sensations, the visual group sensitive to colors and aspects, and the olfactive-gustative group sensitive to chemical features. Subjects may be trained as experts from this last group since they are sensitive to more subtle features.

It is most convenient to separately evaluate the visual characters that are easily perceived and scored from the aroma and taste which necessitate more concentration. However, it is important they are all on the same score sheet because this compels tasters to use two registers of their sensitivity, return to their discriminating ability in the middle of the assessment, and then adapt to the more subtle characters.

Comparative or Absolute Description

There are two approaches to describing a sample. In the comparative description a sample is scored for each attribute relative to a reference sample which must have intermediate values in all characteristic attributes to prevent distortions; each sample is then compared with the reference for each attribute. The absolute description is used when there is no possibility of having such a reference.

The treatment of data is similar in both cases. Mean scores are calculated for each attribute and represented graphically to produce a profile for each product. Scores are also submitted to several mathematical treatments (Daget, 1977; ASTM, 1981).

Multidimensional Sensory Evaluation

Food chemical profiles have been most successful in the discrimination of product categories among samples of the same commodity. As large sets of descriptive analysis (DA) data arise, the techniques of multidimensional sensory evaluation aid in the extraction of pertinent information such as in the description of a set of food products with several attributes (Vuataz, 1977).

Principal component analysis (PCA) and cluster analysis (CA) are forms of multivariate statistical examinations useful for studying the correlations among a set of measurements of a given number of variables for a determined number of assessors (Resurrección, 1988). Multivariate methodology can be applied to reduce a large number of variables to a smaller subset by discarding the redundant. Further, the overall data structure obtained by DA may be preserved as much as possible by applying the technique of generalized cluster analysis (GCA) (Gains et al., 1988). All of these techniques retain variables that significantly contribute to important components and discard those variables that contribute mainly to unimportant components. This criterion provides a smaller number of variables to compare the performance among samples and makes it much easier to determine in which descriptors a set of samples vary so as to ensure a more precise grouping of samples.

Sensory food evaluation also requires a precise definition of what attributes are to be measured and what range of stimuli concentration and methodology should be used. Proper psychophysical methods must be chosen so as to reduce variability among subjects and obtain only relevant information about the sensory properties of a food. The development of statistical software greatly facilitates the application of new and more accurate mathematical analysis methods. The present use of software highlights the importance of sensory data processing.

REFERENCES

ASTM (1968). *Manual on Sensory Testing Methods*. Special Technical Publication 434. Philadelphia: American Society for Testing and Materials.

ASTM (1981). *Guidelines for the Selection and Training of Sensory Panel Members*. Special Technical Publication 758. Philadelphia: American Society for Testing and Materials.

Birch G.G., Latymer Z., and Hollaway M. (1980). Intensity/time relationships in sweetness: Evidence for a queue hypothesis in taste chemoreception. *Chem. Senses* 5: 63–78.

Bonnet C. (1994). Psicofísica de los tiempos de reacción: Teorías y métodos. *Rev. Lat. Amer. Psicol.* 26: 431–444.

Bonnet C., Zamora M.C., Buratti F., and Guirao M. (1999). Group and individual gustatory reaction times and Piéron's law. *Physiol. Behav.* 66(4): 549–558.

Bujas Z., Szabo S., Ajdukovic D., and Mayer D. (1989). Individual gustatory reaction times to various groups of chemicals that provoke basic taste qualities. *Percep. Psychophys.* 45: 385–390.

Buratti, F.M., M.C. Zamora, M. Otero-Losada, and A.M. Calviño (1996). Tiempo de reacción al gusto (TRG): Su aplicación para la selección de panelistas en análisis sensorial. *Libro de resúmenes del I Simposio Iberoamericano de Análisis Sensorial*. Campinas, Brasil, p. 24.

Calviño A.M., Zamora M.C., and Sarchi M.I. (1996). Principal components and cluster analysis for descriptive sensory assessment of instant coffee. *J. Sens. Stud.* 11(3): 191–210.

Cometto-Muñiz J.E. (1981). Odor, taste and flavor perception of some flavoring agents. *Chem. Senses.* 6: 215–233.

Cometto-Muñiz J.E. and Noriega G. (1985). Gender differences in the perception of pungency. *Physiol. Behav.* 34: 385–389.

Daget N. (1977). Sensory evaluation or sensory measurement? In: *Nestlé Research News 1976/77*. Boella C., Ed. Lausanne, Switzerland: Technical Documentation Center, pp. 43–56.

Doty R.L. (1991). Olfactory system. In: *Smell and Taste in Health and Disease.* Getchell T., Bartoshuk L., Doty R., and Snow J., Eds. New York: Raven Press, pp. 175–203.

Gains N., Krzanowski W.J., and Thomson D.M.H. (1988). A comparison of variable reduction techniques in an attitudinal investigation of meat products. *J. Sens. Stud.* 3: 37–48.

García-Medina M.R. (1981). Escalas psicofísicas de sustancias con los cuatro gustos básicos. *La Aliment. Latinoamer.* 131: 6–18.

Guirao M. (1980). *Los Sentidos, Bases de la Percepción*. Madrid: Alhambra.

Guirao M. (1991). A single scale based on ratio and partition estimates. In: *Ratio Scaling of Psychological Magnitudes in Honor of the Memory of S.S. Stevens*. Gescheider G.A. and Bolanowski S.J., Eds. New Jersey: L. Erlbaum Associates. pp. 59–78.

Guirao M. and Zamora M.C. (1997). *Equipo Para Medir los Tiempos de Reacción al Aabor (Gusto, Pungencia y Otros)*. Instituto Nacional de la Propiedad Industrial. Patent Number P 97 01 02703.

Kuznicki J.T. and Turner L. (1988). Temporal dissociation of taste mixture components. *Chem. Senses* 13: 45–62.

Lundahl D.S. (1992). Comparing time-intensity to category scales in sensory evaluation. *Food Technol.* 46: 98–103.

Meilgaard M., Civille G.V., and Carr B.T. (1987). *Sensory Evaluation Techniques*. Boca Raton, Florida: CRC Press.

Miller I.J. and Bartoshuk L.M. (1991). Taste perception, taste bud distribution, and spatial relationships. In: *Smell and Taste in Health and Disease*. Getchell T., Bartoshuk L., Doty R., and Snow J., Eds. New York: Raven Press, pp. 205–233.

O'Mahony M. (1992). Understanding discrimination tests: A user-friendly treatment of response bias, rating and ranking R-index tests and their relationship to signal detection. *J. Sens. Stud.* 7: 1–47.

Pedrero D.L. and Pangborn R.M. (1989). *Evaluación Sensorial de los Alimentos. Métodos Analíticos*. México D.F.: Alhambra, Mexicana.

Piéron H. (1952). *The Sensations*. New Haven: Yale University Press.

Pins D. and Bonnet C. (1996). On the relation between stimulus intensity and processing time: Piéron's law and choice reaction time. *Percep. Psychophys.* 58: 390–400.

Resurrección A.V.A. (1988). Application of multivariate methods in food quality evaluation. *Food Technol.* 42: 128–136.

Stevens S.S. (1957). On the psychophysical law. *Psychol. Rev.* 64: 153–181.

Stevens S.S. (1975). *Psychophysics*. New York: John Wiley and Sons.

Vuataz S. (1977). Some points of methodology in multidimensional data analysis as applied to sensory evaluation. In: *Nestlé Research News 1976/77*. Boella C., Ed. Lausanne, Switzerland: Technical Documentation Center, pp. 57–71.

Yamamoto T., Kato T., Matsuo R., Araie N., Azuma S., and Kawamura Y. (1982). Gustatory reaction time under variable stimulus parameters in human adults. *Physiol. Behav.* 29: 79–84.

Yamamoto T. and Kawamura Y. (1981). Gustatory reaction time in human adults. *Physiol. Behav.* 26: 715–719.

Zamora M.C. and Calviño A.M. (1996). A comparison of methodology applied to the selection of a panel for sensory evaluation of instant coffee. *J. Sens. Stud.* 11: 211–226.

Zamora M.C., Buratti F.M., and Otero-Losada M.E. (1998). Temporal study of sucrose and fructose relative sweetness. *J. Sens. Stud.* 13(2): 213–221.

CHAPTER 5

Sensory and Instrumental Measures of Viscosity

E. Costell and L. Durán

INTRODUCTION

The relations existing between the results of food sensory analyses and the data obtained from physical and chemical measurements constitute the basis on which the development of instrumental methods for quality control must be sustained. The general process for the selection and development of instrumental methods is comprised of a) identifying the food characteristic or property responsible for the stimulus of human perception; b) selecting the best method to measure this characteristic or property; c) using the most appropriate sensory test to quantify the human sensation resulting from the perceived stimulus; and d) studying the relationship between variations in stimulus and sensation magnitudes (Costell and Durán, 1981).

Ideally the stimulus responsible for the sensation should be identified and the psychophysical relation between them should be established. This simplified working scheme may be useful to determine the traditional psychophysical relationships between a certain sensation and the intensity of the corresponding physicochemical stimulus (Stevens, 1975). However, if applied when the human perception is the result of the integration of several individual sensations (as in the case of most food sensory attributes), then the established relationships are not totally satisfactory. Alternatively, the information integration theory may allow for the modeling of the three linked functions (valuation, integration, and response), and thus open new ways to investigate the nature of the relations between food and humans (McBride and Anderson, 1990).

Because of the limited research in this area, the search for a chemical or instrumental method to evaluate the intensity of a sensory attribute must be done without basic knowledge of the process that governs the human perception of the different aspects of food sensory qualities. Therefore, the characteristic or property to be measured is selected based only on what is known about the food composition and structure or possible response of the food when submitted to certain external conditions like illumination, applied forces, or temperature. The type of relation existing between the measured food characteristic or property and stimulus responsible for the human sensation determines the validity and usefulness of the instrumental method selected to follow variations in the sensory property. If such a relation is not consistent, use of the method on other materials will be erroneous and/or misleading.

SELECTION OF INSTRUMENTAL VISCOSITY INDICES

The viscosity of fluid foods is an important property from a scientific as well as technological point of view. The development and implementation of instrumental methods to measure viscosity that aim to predict consumer response have merited a great deal of research. Most studies have either 1) identified the physical stimulus responsible for the human sensation, or 2) searched for empirical correlations between the values of different rheological parameters and the magnitude of viscosity as perceived by humans.

Identification of the physical stimulus responsible for a certain sensation must be based on knowledge of the process involved in the human perception of a food's physical characteristics from both physiological and psychological points of view. Ascertaining the mechanism of the physiological perception of a certain stimulus is not an easy task. When dealing with viscosity, several techniques have been reported to afford meaningful information on the different aspects of its physiological perception: time-intensity measurements (Pangborn and Koyasako, 1981), monitoring of fluid movements in the mouth (Lee III and Camps, 1991), direct measurement of the physiological response as with acoustic methods that register the sounds produced upon swallowing (Lee III et al., 1992), the electromyographic method in which the electrical potentials in the facial muscles involved in the oral manipulation of liquid and semisolid foods are measured (Dea et al., 1988) as well as the development of a physical theory and solution of the corresponding constitutive equations for a situation that considers the physical structure of the mouth as two flat parallel plates where some sensory attributes of the food can be related to specific forces (Kokini et al., 1977).

As for the psychological aspect, several authors have promoted the use of psychophysical techniques to establish the relations between different rheological parameters and sensory viscosity. According to their results, the perception of viscosity follows the psychophysical power law, but the particular physical stimulus is different depending on the characteristics of a specific food's flow.

Newtonian Fluids

In the case of Newtonian fluids in which viscosity is independent of shear rate, the relationship between this parameter's variation and that of the sensory viscosity fits the Stevens equation ($S = k \, I^n$). In this function, the constant of proportionality (k) depends on the units of measurement and the value of the exponent (n) is a valid index of the relation of instrumental and perceived solution viscosity. When representing instrumental non-oral sensory measures of silicon oil viscosity in log-log plots, Stevens and Guirao (1964) observed that the slope of the function was 0.4 regardless of the sensory evaluation method employed. Relating instrumental and sensory measures of sodium alginate solutions, Christensen (1987) observed slopes slightly lower: 0.35 for non-oral tactile viscosity evaluations, 0.39 for visual evaluations, and 0.35 for oral evaluations. Cutler et al. (1983) found an unusually low power law exponent of 0.22 for such Newtonian fluids as sugar syrups, fresh milk, and some dilutions of honey.

Non-Newtonian Fluids

For non-Newtonian fluids, viscosity is not a constant parameter. Its value varies with the shear rate or shear stress applied, often by several orders of magnitude. In this context, several attempts have been made to identify the range of shear rates and shear stresses that develop in the mouth when sensory viscosity is evaluated. Christensen (1987) explains that the rationale for this type of study is that the shear stresses and rates used by subjects when judging viscosity can be identified by finding that portion of the instrumental shear stress/shear rate curve where both perceptual and instrumental measures of viscosity are matched. Table 5-1 provides information about different rheological indices proposed as physical stimuli of sensory viscosity in non-Newtonian fluids.

Szczesniak and Farkas (1962) selected 30 rpm as the value of apparent viscosity based on their observations about the frequency of tongue movements during eating. Wood (1968) founded his proposal on the information obtained from asking panelists which one of two samples whose flow curves crossed was more viscous. Using the same approach, Shama and Sherman (1973) concluded that the physical stimulus of sensory viscosity varies with fluid characteristics. For less viscous food, the stimulus appears to be the shear rate developed at an approximately constant shear stress, whereas for more viscous foods it appears to be the shear stress developed at an approximately constant shear rate (Table 5-1). Christensen (1979) questioned that viscosity at a constant shear rate is a good measurement of perceived viscosity based on the results obtained when untrained judges assessed the viscosity of different CMC solutions, which suggested that multiple viscosity measurements at a range of shear rates result in some form of average viscosity reading.

Table 5-1 Rheological indices proposed as stimuli of viscosity (non-Newtonian foods).

Rheological Index	Reference
η (30 rpm)	Szczesniak and Farkas, 1962
η (50 s^{-1})	Wood, 1968
Less viscous foods: γ ($\sigma \approx 10^2$ dyn/cm^2) More viscous foods: σ ($\gamma = 10$ s^{-1})	Shama and Sherman, 1973
η (range of γ)	Christensen, 1979
σ (generated in the mouth)	Kokini, 1985

As in the case of Newtonian fluids, Kokini and co-workers (Kokini, 1985; Dikie and Kokini, 1983) applied the physical theory developed to explain the viscosity perception in the mouth and solved for the corresponding constitutive equations for non-Newtonian fluid foods that follow power-law and time-dependent rheological behavior.

An analysis of the results reported in the above-mentioned papers provides valuable information on the different aspects of sensory perceptions of viscosity. Perhaps the first conclusion to be drawn from these studies is that perceived viscosity may be the response to different physical stimuli as a function of the flow characteristics of the particular liquid food under study. More research is needed to unveil the process by which man perceives the viscosity of different foods; only then the real nature of the stimulus wholly or primarily responsible for the sensation will be ascertained in each particular case. This type of study is not at all easy to perform, mainly because geometry in the mouth changes constantly during swallowing.

EMPIRICAL CORRELATIONS

Due to the difficulties encountered when trying to identify the stimulus for viscosity, most researchers have chosen to study the relationships between some rheological parameters and viscosity as perceived in a determined product, and then look for a solution to solve the problem of finding and setting up an instrumental method capable of measuring or controlling viscosity. Table 5-2 provides information about different rheological parameters proposed as instrumental indices of sensory viscosity in non-Newtonian fluids.

Morris and Taylor (1982) and Morris et al. (1984) found that concentrated solutions of random coil polysaccharides show the same type of shear thinning. Thus, to characterize differences in their flow behavior, only two parameters are required. They chose the zero shear viscosity and shear rate at which the measured viscosity was reduced one tenth of this maximum value. These parameters correlated well with perceived thickness. Cutler et al. (1983) investigated the perceived thickness of materials showing different types of

Table 5-2 Some rheological parameters proposed as instrumental indices of sensory viscosity in non-Newtonian foods.

Rheological Index	Reference
η_0 and $\gamma_{0.1}$	Morris and Taylor, 1982
Practical value: η (10 s^{-1})	Cutler et al., 1983
η^*	Baines and Morris, 1988
Proposal: Extensional viscosity	Clark, 1992
Peach nectars: η (500 s^{-1})	Costell et al., 1994
Corn starch pastes: η (36.7 s^{-1})	Zamora, 1995
Lemon pie filling: η^* (50 rad.s^{-1})	Hill et al., 1995

shear thinning and concluded that for solutions of ordered rod-like conformation polysaccharides and a range of fluid foods, viscosity measurements at 10 s^{-1} correlated well with their thickness. Baines and Morris (1988) studied the perceived viscosity of fluids considered as "weak gels" and found that conventional viscosity measurements at a fixed shear rate of 50 s^{-1} as proposed by Wood in 1968 provide a reliable measure of thickness in normal fluids (where $\eta = \eta^*$), but underestimate the thickness of weak gel fluids (where $\eta <$ η^*). They also determined that equivalent small-deformation oscillatory measurements of complex viscosity (η^*) at 50 rad s^{-1} correlated well with the perceived thickness of both types of fluids.

Besides the above-mentioned proposals, several authors have selected some rheological indices of practical validity for control of food sensory viscosity. Costell et al. (1994) studied the relationships between some rheological parameters and the oral viscosity of peach nectars and found that the apparent viscosity measured at 500 s^{-1} can be considered a practical instrumental index for this product and used for routine quality control. With a similar approach, Zamora (1995) selected an apparent viscosity of 36.7 s^{-1} as an instrumental measure of the sensory viscosity of cornstarch pastes. These results were to some extent in agreement with the proposal of Shama and Sherman (1973) in that the value of the shear rate associated with the oral evaluation of viscosity was dependent on the fluid's viscosity: in general, the less viscous the fluid, the higher the shear rate at which to measure the apparent viscosity. The usefulness of complex viscosity as an instrumental index of perceived thickness as proposed by Baines and Morris (1988) in model solutions has also been confirmed by Hill et al. (1995) for lemon pie filling.

STUDY VARIABILITY

The results of different viscosity studies and the valuable information collated in the related review papers (Sherman, 1977; Kokini, 1985; Christensen, 1987; Lee III et al., 1992) clearly show a great variability. It is thus very difficult to draw any basic information about the known relation between physical and sensory measures of viscosity in foods. In this paper, the authors present and discuss the possible sources for these inconsistencies:

- Differences in the concepts of sensory and instrumental viscosity
- Differences in the characteristics of the measuring tools used for sensory evaluations
- Differences among the various instruments and methods used in the rheological characterization of product flow
- Differences in the sensory panels, experimental conditions, and methodology used in the sensory analysis of viscosity
- Differences in the type and characteristics of the products studied

Sensory and Instrumental Viscosity

One of the first questions to consider when analyzing the relations between sensory and instrumental viscosity is their conceptual difference. To avoid use of the same term for both parameters, some researchers have proposed "thickness" to represent sensory viscosity (Kokini et al., 1977; Cutler et al., 1983), while others think "viscosity" clearly stands for the perceived sensation (Sherman, 1977; Christensen, 1987).

An additional difficulty arises when trying to precisely define "thickness" or "sensory viscosity." There are evidently clear differences between what consumers understand as the viscosity of liquid or semiliquid foods and the concept learned by a trained assessor. Many efforts have been directed towards identifying words capable of describing the textural properties that can be perceived in fluid foods (Szczesniak and Skinner, 1973), but the popular vocabulary used to describe liquid texture in the mouth contains some redundancy. Kokini et al. (1977) suggested that a ten attribute vocabulary can be reduced to three words (thick, smooth, and slippery) without great loss of predictive power since consumers are not analytical in their perceptions and tend to act as integrators of stimulus information. For consumers, the sensory viscosity of a liquid food not only depends on the shear and friction forces in the mouth and the food's flow characteristics, but cognitive factors such as knowledge and expectations. For trained assessors the situation can be simpler, so that sensory viscosity can be defined as a mechanical textural attribute that relates resistance to flow. It corresponds to the force required to draw a liquid from a spoon over the tongue, or to spread a liquid over a substrate.

For fluid foods showing a Newtonian type of flow (water, milk, clarified fruit juices, beer, sugar solutions), shear stress (σ) and shear rate (γ) are related linearly, and viscosity (defined as σ/γ) is independent of the rate at which the food is sheared. But more commonly, liquid and semiliquid foods exhibit non-Newtonian flow or viscoelastic behavior and the concept of viscosity cannot be properly applied. Non-Newtonian flow can be time independent exhibiting a Bingham plastic flow (margarine, vegetable fats, chocolate paste, fruit nectars) or a shear thinning flow (concentrated fruit juices, fruit and vegetable purées, protein concentrates, hydrocolloids solutions); or time-dependent, as in thyxotropic flow (mayonnaise, condensed milk, apple sauce). For these types of flow, viscosity is dependent on the shear rate or time with a constantly applied shear rate. The apparent viscosity that changes with shear rate or time can then be used as an index of the product sensory viscosity at a fixed shear rate.

Measuring Tool Differences

One of the most common approaches in research on selecting instrumental methods for evaluating food texture is based on the fact that the closer the instrumental test conditions are to those exhibited by the sensory assessment, the better is the correlation between the two measures (Szczesniak, 1987). Some authors have tried to identify the range of shear rates and shear stresses used in the mouth in the evaluation of viscosity of non-Newtonian fluids (Szczesniak and Farkas, 1962; Wood, 1968; Shama and Sherman, 1973), and others undertook calculation of the shear stress at the surface involved in swallowing (Kokini et al., 1977).

Differences between the experimental conditions of mechanical and sensory assessments can explain some of the problems involved with correlating these types of measurements. Some of the discrepancies are due to the effects of saliva in the mouth, differences in sensor devices (Christensen, 1987; Parkinson and Sherman, 1971), the geometry of measuring systems, and the characteristics of the type of force application or sensitivity (Lee III et al., 1992). Therefore, each of these factors must be considered when sensory and physical measures of viscosity are compared.

Instrumental and Methodological Differences

When studying the validity of rheological indices for predicting the sensory viscosity of fluid foods, the type of equipment and measurement conditions should also be very well defined, especially when dealing with non-Newtonian liquids. Table 5-3 shows some information about viscometers and relevant experimental conditions used by different authors that have worked on correlating sensory and instrumental measures of viscosity.

Table 5-3 Some examples of instrumental measures of viscosity characteristics.

Mode	Viscometer	Geometry	Shear Rate	Reference
Steady Shear	Brookfield	Helipath	Rate of shear: 0.5–100 rpm	Szczesniak and Farkas, 1962
	Weissenberg	Cone-plate	146.8–1000 s⁻¹	Parkinson and Serman, 1971
	Haake Rotovisco	Coaxial Cylinders	0.133–2620 s⁻¹	Shama and Sherman, 1973
	Cannon-Ubbelhode	4 bulb-Capillary	10, 40, 100 s⁻¹	Christensen, 1979
	Ferranti-Shirley	Cone-plate	0.1–100 s⁻¹	Kokini and Cussler, 1983
	Rheometrics	Cone-plate	0.316–1000 s⁻¹	Cutler et al., 1983
	Rheometrics	Cone-plate	0.1–1000 s⁻¹	Morris et al., 1984
	Rheometrics	Cone-plate	0.1–100 s⁻¹	Elejalde and Kokini, 1992
Oscillation	Bohlin	Coaxial Cylinders	Frequency: 0.02–10 Hz	Bohlin et al., 1990
	Carri-Med	Cone-plate	Frequency: 1–10 Hz	Hill et al., 1995

Variations in Sensory Analyses of Viscosity

Upon reviewing the available information on the relationships between instrumental and sensory measures of food viscosity, one of the more debated issues is that of the experimental conditions under which the instrumental measurements must be conducted. In these discussions, sensory viscosity is normally considered unique. However, the sensory method is not usually questioned and the experimental conditions of the test and/or the composition and nature of the panel are not analyzed. The lack of agreement observed among authors when relating instrumental and sensory measures of viscosity could thus be lessened if the sensory methodology and testing conditions were duly analyzed.

In Table 5-4, basic information on the sensory methods used, characteristics of the panels (number of assessors and training), and degree of definition for the experimental conditions (type of service, sample size, instructions to evaluate viscosity, etc.) reported in most of the papers cited in Table 5-3 are given for comparison. These factors may cause differences in the sensory viscosity values obtained and thus affect conclusions about the practical usefulness of the proposed instrumental indices used for their evaluation.

Table 5-4 Some examples of viscosity sensory evaluation characteristics.

Sensory Method	Number of Assessors	Assessor Training	Experimental Conditions	Reference
Seven-point scale	Not specified	Trained	Defined	Szczesniak and Farkas, 1962
Paired comparison	26	Untrained	Well defined	Shama and Sherman, 1973
Magnitude estimation	Not specified	Semi-trained	Very well defined	Christensen, 1979
Magnitude Estimation	18	Untrained	Defined	Kokini and Cussler, 1983
Magnitude estimation	10	Trained	Defined	Cutler et al., 1983
Magnitude estimation	12	Trained	Defined	Morris et al., 1984
Magnitude estimation	Three panels (12 assessors each)	Not specified	Well defined	Elejalde and Kokini, 1992
Scales (not described)	12	Trained	Not specified	Bohlin et al., 1990
Magnitude estimation	11	Selected and trained	Defined	Hill et al., 1995

Consequently, it should be expected that the conclusions obtained when determining differences in perceptible viscosity between two samples from the results of paired comparisons by a 26 untrained student panel (Shama and Sherman, 1973) will not coincide with those obtained when a trained 10 student panel evaluated viscosity using magnitude estimation under well-defined and controlled conditions (Christensen, 1979). The main difference usually observed is that while information on the type of instrument and geometry used is always given when instrumental data are reported, information on the panel's composition and degree of training is frequently lacking when the results of sensory tests are reported. Similarly, when researchers report experimental conditions, the range of shear rates or frequencies is given, but the specific serving conditions and the way sample viscosity was evaluated may be poorly defined. When this happens, reproduction of the results obtained by different authors and interpretation of the conclusions drawn from them are very difficult.

Product Differences

Research on the relation between sensory viscosity and the rheological parameters that characterize product flow has been mostly carried out on model systems such as hydrocolloid solutions (Szczesniak and Farkas, 1962; Christensen, 1979; Morris et al., 1984; Baines and Morris, 1988; Bohlin et al., 1990). However, study of silicones (Stevens and Guirao, 1964) and starch pastes (Zamora, 1995) that possess a wide range of viscosities have also been studied. In addition, joint analyses of data on real foods and model systems (Kokini et al., 1977; Morris and Taylor, 1982; Cutler et al., 1983), as well as data from a variety of foods with very different rheological behaviors can be found (Shama and Sherman, 1973; Kokini et al., 1984). Information obtained when working with all these types of materials can still be compared, but in the case of real foods the possible influence of other sensations (color, aroma, etc.) on the perception of viscosity must also be taken into account.

The influence of viscosity on the perception of sweetness has been widely studied (Launay and Pasquet, 1982; Pastor et al., 1994). Some data can also be found on the effect of viscosity on the perception of other basic flavors (Pangborn et al., 1973; Paulus and Haas, 1980). However, very little information is available on the effect of the type or intensity of a certain flavor on the perception of viscosity. Most papers reporting on this matter deal with the effect of sucrose or other sweeteners on the perceived viscosity in hydrocolloid solutions (Pangborn et al., 1973) or model food systems (Pastor et al., 1996). For some hydrocolloids, viscosity is perceived higher upon an increase in the sweetener concentration. Burgard and Kuznicki (1990) found that the relation between sensory viscosity and sucrose concentration is quadratic, and its increase higher than expected from the addition of sucrose (2.3 to 3.2 mPas). This was also verified when Burns and Noble (1985) increased the sucrose concentration on vermouth. Such studies reveal a higher than expected sensorial viscosity in comparison with the corresponding physical viscosity. In peach nectars, Costell et al. (1994) found that good correlation between sensory viscosity and apparent viscosity (measured at 500 s^{-1}, r = 0.89) was improved when the sugar concentration (°Brix) was considered along with the rheological parameter (r = 0.93), confirming that the sweeter samples were judged as more viscous.

The above-mentioned results show how the interactions among different food attributes can modify the expected relationships between the physical stimulus and corresponding sensory response deduced from studies on model systems. This can be decisive when selecting instrumental indices of viscosity to be used in food quality control.

ACKNOWLEDGMENTS

To CICyT, Spain for financial support (Project ALI 97-0359).

REFERENCES

Baines Z.V. and Morris E.R. (1988). Effect of polysaccharide thickeners on organoleptic attributes. In: *Gums and Stabilisers for the Food Industry. Volume 4.* Phillips G.O., Wedlock D.J., and Williams P.A., Eds. Oxford: IRL Press, pp. 193–201.

Bohlin L., Egelansdal B., and Martens M. (1990). Relationships between fundamental rheological data and mouthfeel for a model hydrocolloid system. In: *Gums and Stabilisers for the Food Industry. Volume 5.* Phillips G.O., Wedlock D.J., and Williams P.A., Eds. Oxford: IRL Press, pp. 111–121.

Burgard D.R. and Kuznicki J.T. (1990). *Chemometrics: Chemical and Sensory Data.* Boca Raton, FL: Academic Press, pp. 75–91.

Burns D.J.W. and Noble A.C. (1985). Evaluation of separate contributions of viscosity and sweetness of sucrose to perceived viscosity, sweetness and biterness of vermouth. *J. Tex. Stud.* 16: 365–381.

Christensen C.M. (1979). Oral perception of solution viscosity. *J. Tex. Stud.* 10: 153–164.

Christensen C.M. (1987). Perception of solution viscosity. In: *Food Texture: Instrumental and Sensory Measurement.* Moskowitz H.R., Ed. New York: Marcel Dekker, pp. 129–143.

Clark R.C. (1992). Extensional viscosity of some food hydrocolloids. In: *Gums and Stabilisers for the Food Industry. Volume 6.* Phillips G.O., Wedlock D.J., and Williams P.A., Eds. Oxford: IRL Press, pp. 73–85.

Costell E. and Durán L. (1981). El análisis sensorial en el control de calidad de los alimentos: I Introducción. *Rev. Agroquim. Tecnol. Aliment.* 21: 1–10.

Costell E., Pastor M.V., and Durán L. (1994). Rheological parameters as stimuli of perceived sensory viscosity in peach nectars. In: *Progress and Trends in Rheology. Volume IV.* Gallegos C., Ed. GmbH & Co. KG, Darmstadt, pp. 218–220.

Cutler A.N., Morris E.R., and Taylor L.J. (1983). Oral perception of viscosity in fluids foods and model systems. *J. Tex. Stud.* 14: 377–395.

Dea I.C.M., Eves A., Kilcast D., and Morris E.R. (1988). Relationship of electromyographic evaluation of semi-fluid model food systems with dynamic shear viscosity. In: *Gums and Stabilisers for the Food Industry. Volume 4.* Phillips G.O., Wedlock D.J., and Williams P.A., Eds. Oxford: IRL Press, pp. 241–246.

Dikie A.M. and Kokini J.L. (1983). An improved model of food thickness from non-Newtonian fluid mechanics in the mouth. *J. Food Sci.* 48: 57–61, 65.

Elejalde C.C. and Kokini J.L. (1992). The psychophysics of pouring, spreading and in-mouth viscosity. *J. Tex. Stud.* 23: 315–336.

Hill M.A., Mitchell J.R., and Sherman P. (1995). The relationship between the rheological and sensory properties of a lemon pie filling. *J. Tex. Stud.* 26: 457–470.

Kokini J.L. (1985). Fluid and semi-solid food texture and texture-taste interactions. *Food Technol.* 39: 86–94.

Kokini J.L. and Cussler E.L. (1983). Predicting the texture of liquid and melting semi-solid foods. *J. Food Sci.* 48: 1221–1225.

Kokini J.L., Kadane J.B., and Cussler E.L. (1977). Liquid texture perceived in the mouth. *J. Tex. Stud.* 8: 195–218.

Kokini J.L., Poole M., Mason P., Miller S., and Stier E.F. (1984). Identification of key textural attributes of fluid and semisolid foods using regression analysis. *J. Food Sci.* 49: 47–51.

Launay B. and Pasquet E. (1982). Sucrose solutions with and without guar gum: Rheological properties and relative sweetness intensity. *Prog. Food Nutr. Sci.* 6: 247–258

Lee III W.E. and Camps M.A. (1991). Tracking foodstuff location within the mouth in real time. *J Tex. Stud.* 22: 277–287.

Lee III W.E., Takahashi T., and Pruitt J.S. (1992). Temporal aspects of the oral processing of viscous solutions. *Food Technol.* 46: 106–112.

McBride R.L. and Anderson N.H. (1990). Integration psychophysics. In: *Psychological Basis of Sensory Evaluation.* McBride R.L. and MacFie H.J.H., Eds. Essex, UK: Elsevier Science, pp. 93–115.

Morris E.R. and Taylor L.J. (1982). Oral perception of fluid viscosity. *Prog. Food Nutr. Sci.* 6: 285–296.

Morris E.R., Richardson R.K., and Taylor L.J. (1984). Correlation of the perceived texture of random-coil polysaccharide solutions with objective parameters. *Carbo. Poly.* 4: 175–191.

Pangborn R.M. and Koyasako A. (1981). Time course of viscosity, sweetness and flavor in chocolate desserts. *J. Tex. Stud.* 12: 141–150.

Pangborn R.M., Traube J.M., and Szczesniak A.S. (1973). Effect of hydrocolloids on oral viscosity and basic taste intensities. *J. Tex. Stud.* 4: 220–227.

Parkinson C. and Sherman P. (1971). The influence of turbulent flow on the sensory assessment of viscosity in the mouth. *J. Tex. Stud.* 2: 451–459.

Pastor M.V., Costell E., and Durán L. (1994). Influencia de la viscosidad en los umbrales de detección, de reconocimiento y diferenciales de la sacarosa y del aspartamo. *Rev. Esp. Ciencia Tecnol. Aliment.* 34: 91–101.

Pastor M.V., Costell E., and Durán L. (1996). Effects of hydrocolloids and aspartame on sensory viscosity and sweetness of low calorie peach nectars. *J Tex. Stud.* 27: 61–79.

Paulus K. and Haas E.M. (1980) The influence of solvent viscosity on the threshold values of primary tastes. *Chem. Senses* 5: 23–32.

Shama F. and Sherman P. (1973). Identification of stimuli controlling the sensory evaluation of viscosity: Oral methods. *J. Tex. Stud.* 4: 111–118.

Sherman P. (1977). Sensory properties of foods which flow. In: *Sensory Properties of Food*. Birch G.G., Brennan J.G., and Parker K.J., Eds. Essex, UK: Elsevier Science, pp. 303–316.

Stevens S.S. (1975). *Psychophysics*. New York: John Wiley & Sons.

Stevens S.S. and Guirao M. (1964). Scaling of apparent viscosity. *Sci.* 44: 1157–1164.

Szczesniak A.S. (1987). Correlating sensory with instrumental texture measurements. An overview of recent developments. *J. Tex. Stud.* 18: 1–15.

Szczesniak A.S. and Farkas E. (1962). Objective characterization of the mouthfeel of gum solutions. *J. Food Sci.* 27: 381–385.

Szczesniak A.S. and Skinner E.Z. (1973). Meaning of texture words to the consumer. *J. Tex. Stud.* 4: 378–387.

Wood F.W. (1968). Psychophysical studies on the consistence of liquid foods. In: *Rheology and Texture of Foodstuffs*. Society of Chemical Industry (Great Britain), London SCI, Monograph No. 27, pp. 40–49.

Zamora M.C. (1995). Relationships between sensory viscosity and apparent viscosity of corn starch pastes. *J. Tex. Stud.* 26: 217–230.

CHAPTER 6

Food Composites

P.J. Lillford

INTRODUCTION

All of the foods people eat, whether natural or fabricated, contain a bewildering arrangement of molecules. And while it would be ideal to describe what each contributes to the final qualities of foods, this simply is not possible. Certainly industrial research is impatient for novelty and will not wait for elegant description. It is therefore important to determine what matters in the food eating process, and how nature or food processors can influence human response (Hutchings and Lillford, 1988; Prinz and Lucas, 1997). This scheme shown in Figure 6-1, suggests immediately that the first questions should focus on food structure, making the microscope a vital tool in all of food science. Using this tool, it is immediately evident that heterogeneity should be the focus. Fibers, crystals, particles, gels, and fluids can be seen, and the greater the magnification, the greater the complexity. All foods are therefore composites, with several levels of hierarchy in their architecture.

Still unanswered, however, is what of understanding is needed. The same problem faces the engineer when designing a bridge, airplane or building, as their structures must not break or fail. Food scientists have a different target since their structures must collapse, but only in the unique environment of the

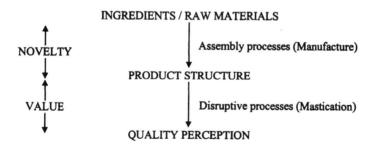

Figure 6-1 Important factors to determine what matters in the food eating process.

mouth. It follows that food science should borrow from the science of structural mechanics and concentrate on design for fracture and failure. The relevant external forces are those of the mouth, which include load, deformation, temperature, and solvents. Since all these conditions are applied to all foods, the arguments should be testable.

FAILURE UNDER MECHANICAL LOADING

Elasto-viscous Fiber Composites

Typical of this class are meat and fish. Elementary observation of perceived quality suggests the following requirements:

(a) Tissue failure must occur, or the food is tough.
(b) Liquids and flavor must be released, or the food is dry and unpalatable.

Mechanical testing shows that muscle origin and toughness correlate with both stress and strain. This is similar to observations with other materials where toughness relates to the work of fracture (area under the stress/strain curve). Clearly for the tougher meats, the crack stopping capability and extensibility of the connective tissue is the limiting mechanism (Purslow, 1987), but this does not explain what actually happens in the mouth.

By examining damage at various stages of chewing, it can be seen that little of the composite fails before swallowing, and detailed analysis shows (Fig. 6-2) the fibrous element perceived in the mouth is mostly that of the high order fiber bundle. This means that the juices providing succulence and flavor must largely be positioned outside the elementary contractile fiber (Lillford, 1991).

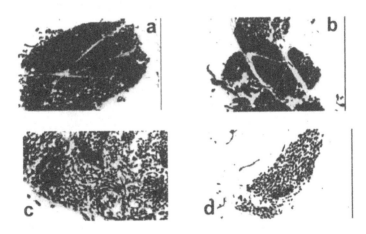

Figure 6-2 Breakdown of L-Dorsi at various stages of chewing.

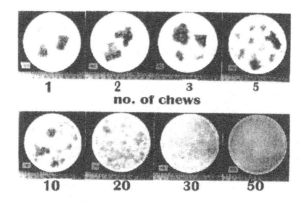

Figure 6-3 Breakdown of an analogue fiber at various stages of chewing.

Also evident is a second interesting phenomenon of self-reassembly prior to swallowing. Meats which do not show this reassembly, such as aged fish tissue or simple fiber analogues (Fig. 6-3), have eating deficiencies usually described as bitty or difficult to clear or form a bolus. This is a further challenge of composite science which non-food systems rarely encounter.

Particulate Composites Exhibiting Brittle Fracture

Typical of this class are biscuits and baked goods whose force-deformation curves show a high modulus value. Strains are small but stresses are high (Fig. 6-4) at fractures, which are usually rapid; their emitted sound is part of the perception of crispness or crunchiness (Fig. 6-5). Geometry plays a major role

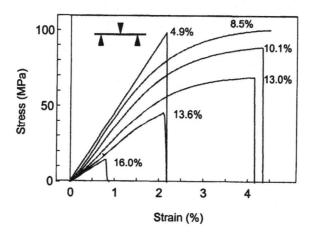

Figure 6-4 Stress-strain curves for bars of heat-set wheat starch at different moisture contents (i.e., plasticization of starch).

in the design of such foods since the forces the jaw can impose are limited. Typically, such foods are aerated so that the forces required are only those to start the crack in each element and several jaw strokes are used in breaking the structure.

Unfortunately, the high modulus of crisp materials is very sensitive to moisture (Attenburrow and Davies, 1993; Levine and Slade, 1993). Brittle, glassy foods become rubbery at relatively low moisture contents (below the level of equilibrium moisture sorption at most ambient conditions), which explains the need for careful packaging.

When placed in the aqueous environment of the mouth, plasticisation commences immediately. For thin-walled porous structures, collapse is rapid and the sensory process is often described as "melting." If the collapsing structure removes saliva faster than it can be replaced, the structure is perceived as "dry." One of the major roles of fat in these products is to waterproof the structure so that crispness is maintained throughout the eating process. The discontinuous phase in these structures is gas, whose flow is not restricted during the deformation of the structure. Those foods where the included phase is liquid can yield brittle fracture by a totally different mechanism.

Liquid Filled Foams

Typical of this class are raw vegetables and fruit (Lucas et al., 1997) with structures that are cellular composites. The cell wall is itself a polymer composite of polysaccharides and the structure is heterogeneous in cell type and cell size. The bonding between cells can also be highly variable.

When raw fruit is eaten, the cell walls are pre-stressed by turgor pressure. Loading causes individual cells to break, releasing juice. The total structure fails as stress is transferred through a row or layer of cells, and the resulting brittle fracture is easily seen since the surface fragments are smooth and can usually be fitted together.

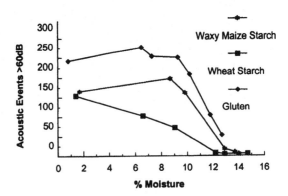

Figure 6-5 Brittle fracture of cereal fractions.

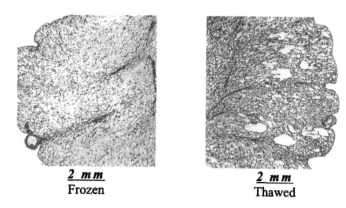

<u>**2 mm**</u>
Frozen

<u>**2 mm**</u>
Thawed

Figure 6-6 Structural damage of strawberry tissue due to freezing, storage, and cooking as seen by a TEM micrograph.

Storage, freezing, and cooking destroy turgor, and even if the cell wall properties remain the same, water flows throughout the structure so the stress on individual walls is lower at the same deformation of the specimen (Warner and Edwards, 1988). The resultant texture is described as "soft and wet" (Fig. 6-6). Cooking, or the ripening of fruit, causes chemical change in structural polymers of cell walls and intercell adhesion. Failure occurs between cells rather than across walls since little or no juice is released, and mealy textures are encountered. Such effects are easily seen by examining the effects of mechanical disruption of fruits in different texture (Fig. 6-7).

A few vegetables have cellular architectures which are resistant to processing, resulting in brittle fractures in the mouth even after cooking. For example, the Chinese water chestnut has walls which are stiff even after heat treatment because of enhanced chemical cross-linking between cells (Fig. 6-8). Turgor pressure plays no part in the crispness of this vegetable (Parker and Waldron, 1995).

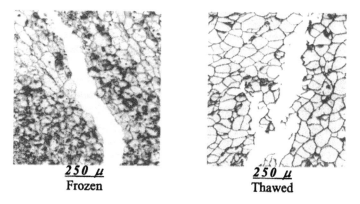

<u>**250 μ**</u>
Frozen

<u>**250 μ**</u>
Thawed

Figure 6-7 Structural damage of carrot tissue due to freezing, storage, and cooking as seen by a TEM micrograph.

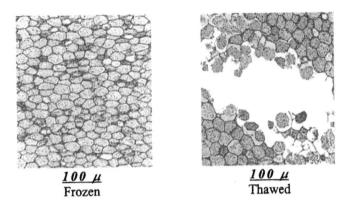

<u>*100 μ*</u>   <u>*100 μ*</u>
Frozen   Thawed

Figure 6-8 Structural damage of water chestnut tissue due to freezing, storage, and cooking as seen by a TEM micrograph.

COLLAPSE UNDER THERMAL TREATMENT

Polymer Networks

In the preceding examples, the mechanical properties of the polymer composites are relatively insensitive to the temperature range from ambient to 37° so that mouth temperature is relatively unimportant. However, this is not always the case. Since gelatin and some carageenan gels melt below 37°, sauces and dressings of this type (i.e., characterized as thick liquids and dispersions of soft particles at ambient temperature) act with lower viscosity in the mouth, which lubricates and adds juiciness or succulence to the foods to which they are applied.

Fat Crystals

The classical example of melting networks is chocolate. At ambient temperature it has all the characteristics of an elastic solid even though liquid oil is present. The crystal network is highly developed and the elastic modulus of individual crystals and network is very high. At ambient temperatures fracture occurs at high speed with low strain but high stress. At 37°, virtually all crystals are melted so that the texture changes from brittle solid to a viscous liquid during the eating process when flavor components are simultaneously released. If chocolate remains cold during eating, none of these expected properties are experienced, which presents a major challenge to ice cream manufacturers (Fig. 6-9).

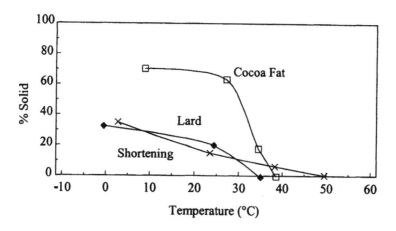

Figure 6-9 Solid fat index profiles of shortening, lard, and cocoa.

Fat Continuous Emulsions

Typical of this class are butter, margarines, and spreads. At high crystal contents these products should have similar properties to chocolate (i.e., elastic solids exhibiting brittle fracture). However, this is not required in materials designed to spread (i.e., exhibiting plastic or pseudoplastic rheology). Though it is not known why, the churning process of butter manufacture solves this problem by cleverly allowing much of the high melting fat to crystallize within the cream. As the churning process occurs, the crystalline droplets are included as a particulate phase and a continuous network is formed only from residual fats which crystallize at lower temperature. The result is a weak, plastic network (Fig. 6-10). In the mouth, the churning process is reversed. Fat crystals melt and an oil-in-water emulsion (cream) is formed via minimal mechanical processing.

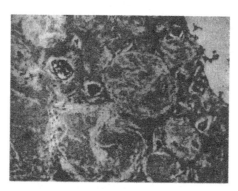

Figure 6-10 Transmission micrograph of butter clusters obtained after churning.

71

With the advent of low-fat spreads, two problems arise. Firstly, how to stabilize the water-in-oil product, and secondly, how to produce a similar cream structure in the mouth. The simple addition of lipophilic emulsifiers results in too stable an emulsion which will not invert (Fig. 6-11) because it is in part stabilized sterically by complete shells of fat crystals (Fig. 6-12). If brittle, these shells fracture both on spreading and eating, giving excessively fast aqueous phase release. However, provided the aqueous phase itself exhibits plastic rheology, the effect is not deleterious.

Considerable research has been done on how to turn aqueous liquids into plastic solids. Plasticity is observed to be a property of concentrated particulate dispersions where weak attractive forces between particles results in a structure that can be rapidly re-established after shearing. To date this has been achieved by creating phase-separated aqueous mixtures (Harding et al., 1995). For example, at equilibrium there is phase separation between polysaccharides and proteins. If the included phase is a gel (elastic) and the continuous phase a viscous liquid or weak gel, then no flow occurs except under shear, and the structure is rapidly re-established when the deformation ceases.

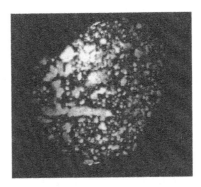

Figure 6-11 Micrograph of stabilized low fat spread showing no inversion of the churning process.

Figure 6-12 Scanning micrograph of a fat crystal shell.

Ice Networks

Frozen desserts depend upon ice structures, and the properties of these foods are determined by the architecture of their networks. Unlike fats, the ice crystal content at any temperature is determined primarily by Raoult's law so that other solutes play a major role in product properties such as hardness and coldness. In the glassy state, the whole product has the properties of an elastic solid. Ice cream in particular is a remarkably complex composite containing a gas phase, fat in globular and dispersed crystals, ice crystals, and a non-frozen matrix (Fig. 6-13). In high sugar concentrations after sheared freezing conditions, ice crystals appear with a low axial ratio (Huang and Platt, 1995). Crystal contacts are minimal, so fresh ice cream has the texture of a concentrated dispersion and is soft and creamy. However, this architecture is not stable except at very low temperatures. Crystal networks normally coarsen via accretion and some Ostwaldt ripening to result in increased hardness and iciness, so it is the control of this process that is critical to ice cream storage. Since water ices (lollies) are frequently frozen quiescently, the thermal gradient from mould to stick is so extreme that it causes dendrites to grow inwards from the wall. The resultant product is anisotropic and fibrous, which accounts for the direction of fracture when biting (Fig. 6-14).

FUTURE TRENDS

This approach to foods as composite materials directs attention to the key processes in product manufacture and changes the focus from formulation to product architecture. In this chapter all of the evidence presented has been qualitative. However, the quantitative models of composite physics are available for application once the key structural features have been identified (Gibson and Ashby, 1988; Vincent and Currey, 1980). This is necessary whenever differentiation of properties within a product type is required and some indication of how this can be done has been given. However, not all of the theories needed are yet available and thus remain a challenge for future food architects.

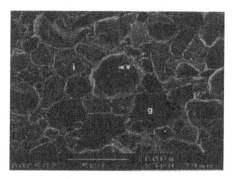

Figure 6-13 Ice cream containing a gas phase (g), fat (v), and ice (i) crystals.

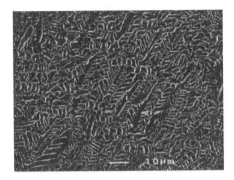

Figure 6-14 Dendrites orientation in a frozen product.

Also to be dealt with are the limitations of the best composite science in building man-made structures compared with the demonstrated capabilities of evolution and biology. The final example of a food composite is the bacterial spore (Fig. 6-15). Not only is this a marvel of sophistication, but it embodies the ideals of most food processors. This structure is ambient stable enough to survive dehydration, freezing, and high pressure, yet with minimum stimulus can regenerate viable cells. The obvious goal is thus to achieve the same with materials less deleterious to the food chain.

CONCLUDING REMARKS

An approach where foods are treated as material composites provides a convenient focus for research attention, but it also exemplifies the complexity of structures, the limitations in construction compared to biological processes, and the opportunities to borrow science from apparently unrelated areas of composite physics. Its further application is recommended to anyone interested in the food they eat.

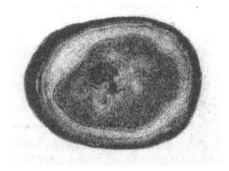

Figure 6-15 Transmission micrograph of a bacterial spore.

ACKNOWLEDGEMENTS

The author acknowledges the contribution of numerous colleagues both past and present. In particular, Mr. J. Judge, S. Ablett, and M. Asquith who have lived through it all.

REFERENCES

Attenburrow G. and Davies A.P. (1993). The Mechanical Properties of Cereal Based Foods in and Around the Glassy State. In: *The Glassy State in Foods*. Blanshard J.M.V. and Lillford P.J., Eds. Nottingham: University Press, pp. 317–331.

Harding, S. E. Hill, S. E. Mitchell, J. R., (1995). *Biopolymer Mixtures*. Nottingham: University Press, pp. 65–83.

Gibson L.J. and Ashby M.F. (1988). *Cellular Solids: Structure and Properties*. Oxford: Pergamon Press.

Huang V. and Platt S. (1995). The latest developments in ice cream technology *Chem. and Ind* 2:51–54.

Hutchings J.B. and Lillford P.J. (1988). The perception of food texture—The philosophy of the breakdown path. *J. Tex. Stud.* 19(2): 103–115.

Levine H, Slade L (1993). The glassy state in applications for the food industry, with an emphasis on cookie and cracker production. In: *The Glassy State in Foods*. Blanshard J.M.V. and Lillford, P.J. Eds. Nottingham: University Press, pp. 333–374.

Lillford P.J. (1991). In: *Feeding and the Texture of Food* Vincent, Julian F. V. Lillford, P. Cambridge: Cambridge University Press, pp. 93–121.

Lucas P.W. et al (1997). In: *Plant Biomechanics*. University of Reading Press, pp. 109–114.

Parker M.L. and Waldron K.W. (1995). Texture of Chinese Water Chestnut: Involvement of Cell Wall Phenolics *J. Sci. Food Agric.* 68(3): 337-346.

Prinz J.F. and Lucas P.W. (1997). An optimization model for mastication and swallowing in mammals *Proc. Roy. Soc. Lond. Biol. Sci..* 264(1389): 1715–1721.

Purslow P.P. (1987). In: *Food Structure and Behaviour*. Blanshard, J. M. V. Lillford, P., Eds. London: Academic Press, pp. 177–197.

Warner M. and Edwards S.F. (1988). *Europhys. Lett.* 5: 623–627.

Vincent J.F.V. and Currey J.D. (1980). *The Mechanical Properties of Biological Materials*. NY: Cambridge University Press.

CHAPTER 7

Thermal Properties and State Diagrams of Fruits and Vegetables by DSC

A.M. Sereno

INTRODUCTION

A recent trend to explain the behavior of food materials during processing and storage is based on an interpretation which considers food materials as systems of water plasticized natural polymers (Slade and Levine, 1991; Roos, 1992). This interpretation is believed to provide an enhanced characterization of the food materials with the help of a phase state diagram where curves showing transition temperatures (e.g., glass transition, melting) are plotted against moisture content. These state diagrams are based on experimental data, most of which may be obtained by differential scanning calorimetry (DSC).

Several food products and model systems have received significant attention during the last ten years, and such information is increasingly available in the literature. Slade and Levine (1991) and Roos (1995) have provided extensive reviews. As an example, results of the application of the DSC methodology to both fresh and processed fruits and vegetables will be presented. In addition, important physical property information can be withdrawn from the thermograms obtained by DSC, namely heat capacities, latent heats from phase changes, reaction heats, and unfrozen water contents. Both state diagrams and thermophysical properties have made the DSC one of the most useful and widespread instruments currently available to the food scientist to characterize food systems.

EXPERIMENTATION

A DSC is an instrument with the capability to measure the difference in heat flow required to maintain both a sample and reference at the same temperature while performing a temperature scan according to a predefined program. Presently, improved differential thermal analyzers (DTA) have been built that allow rigorous monitoring of the individual temperature of a sample and reference inserted in a common compartment with a single heat/cooling source. Under careful calibration and for given ranges of temperature it is possible to obtain quantitative information from these simplified systems with nearly the same accuracy and reproducibility level as with a true DSC. Several authors have produced detailed descriptions of these instruments and their operation (McNaughton and Mortimer, 1975; Pope and Judd, 1977).

To work with food systems, the DSC should be equipped with a low temperature cooling device that can reach −100°C using liquid nitrogen. A material sample with a typical size of 10–20 mg is sealed in 20–60 L aluminum cells. Scanning rates usually range from 2 to 20°C/min, but are most often 5°C/min. To avoid condensation in the compartment containing the cells, this is continually flushed with gaseous dry helium or nitrogen with a flow rate on the order of 30 ml/min.

Most often the material to be studied is first equilibrated under saturated salt solutions, starting either from a fresh product or a freeze-dried portion of it. In order to obtain complete thermal information with high moisture materials like fruits and vegetables, the sample is first cooled (and frozen) to about −100°C and subsequently heated at the chosen scanning program.

All modern DSC instruments are now linked to personal computers, allowing not only the complete on-line registration of the thermogram but also control of the instrument according to any heating program set in advance. These programs range from a simple constant heating rate to complex programs involving different rates in sequence, periods of stabilization (annealing) at constant temperature, and even cycling of the test program.

Thermograms

Figure 7-1 shows typical fruit thermograms obtained with samples equilibrated under different relative humidities. For samples with low water activities, all the water is retained in the solid matrix of the product during the fast cooling process without ice formation. An amorphous glass is obtained which undergoes a glass-rubber transition upon heating, showing in the thermogram as an inflection of the curve. With high moisture content, part of the water is freed from the solid matrix and crystallizes as ice; upon heating after the glass-rubber transition, these crystals melt and the corresponding endothermic peak can be observed in the thermogram. For intermediary moisture content and fast cooling rates, part of the freezable water is trapped inside the vitreous structure. Under slow heating an exothermic crystallization peak [also referred by Flink (1983) as a devitrification peak] is observed after the glass transition and before the ice melts.

Glass Transition, Devitrification, and Melting

As mentioned before, in a typical thermogram for a fruit sample with intermediary moisture content (a_w in the range of 0.8–0.9) scanned between 100°C and 40°C, three phase transitions are normally observed: 1) a glass-rubber second order transition, 2) an ice crystallization (devitrification) exothermic peak, and 3) an ice melting endothermic peak (Fig. 7-2).

The glass-rubber transition produces an increase in the specific heat of the material which occurs in the 10 to 20°C range, corresponding in the thermogram

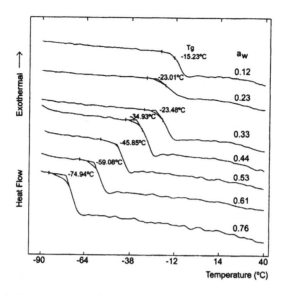

Figure 7-1 Typical thermograms of grapes equilibrated under different a_w (adapted from Sá and Sereno, 1994).

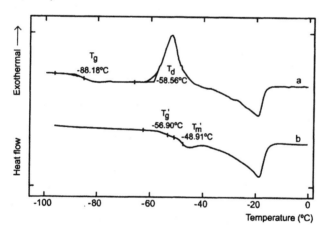

Figure 7-2 Phase transitions during a) the scanning of an intermediate moisture sample and b) the effect of annealing the sample (adapted from Sá and Sereno, 1994).

to an inflection of the curve which stabilizes at a different ordinate. This inflection may be characterized by three points often referred to as the onset, midpoint, and endpoint temperatures of the glass transition (T_g^i, T_g^m, T_g^e) as shown in Figure 7-3. As mentioned before, with samples equilibrated at relative humidities of 80 to 90%, an exothermic peak after the glass transition and before the melting endotherm is often observed. According to Flink (1983), this exothermic behavior is due to devitrification upon rewarming, and results from the crystallization of freezable water trapped in the solid matrix during the fast cooling process. This can be avoided by annealing the sample at a temperature between the glass transition (T_g) and melting (T_m) temperatures (Ablett et al., 1992). After annealing a higher T_g is obtained (corresponding to T_g' as described later) and the devitrification exotherm eliminated; the initial melting temperature decreases with an increase in the size of the melting endotherm (Fig. 7-2).

Foods are complex materials with water as a major component. As a result, both freezing and thawing occurs over a range of temperatures. Different criteria have been proposed to characterize such behavior. Lovric et al. (1987) recommend the specification of two characteristic temperatures: (1) the initial melting temperature (T_m^i) and (2) the peak temperature of the melting endotherm (T_m) (Fig. 7-4). The first is determined as the intersection of the tangent to the inflexion point in the descending branch of the melting endotherm with the base line observed before melting. A final melting temperature can be defined using the same criteria (T_m^f).

Most of the software packages now available for the analysis of thermograms allow automatic identification of T_m^i, T_m, and T_m^f, as well as the calculation of melting heats by numerical integration of the melting endotherm between T_m^i and T_m^f.

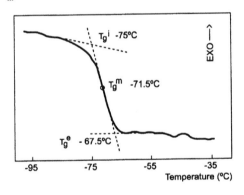

Figure 7-3 Determination of glass transition temperatures (Tg) using DSC.

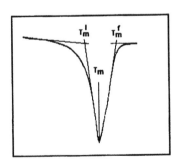

Figure 7-4 Differential scanning calorimetry trace of a typical melting peak.

ANALYSIS OF RESULTS

State Diagram

Transition temperatures obtained from the thermograms may be plotted on a temperature vs. moisture content diagram which defines the different phase areas as shown in Figure 7-5. This phase diagram may be supplemented with lines representing other transitions; for example, in the case of sugar solutions, these lines are typically the solubility and water vaporization curves, usually referred to as the state diagram. Figure 7-5 shows a state diagram for a sucrose solution, which is typical of many food systems and thus useful to understand the path of several food processing operations. Figures 7-6 and 7-7 show the state diagrams for Golden delicious apple and peach jams. These state diagrams are similar to the state diagrams of common sugar solutions like the one presented in Figure 7-5, suggesting that from this point of view the behavior of these natural products is significantly influenced by the sugar they contain.

The glass-transition curves plotted in Figures 7-6 and 7-7 were calculated with the help of a model proposed by Gordon and Taylor (1952) for binary compatible polymer mixtures. The same equation was recommended to predict glass-transition temperatures for water plasticized food systems (Kelley et al., 1987):

$$T_g = \frac{x_s T_{gs} + K.x_w T_{gw}}{x_s + K.x_w} \tag{7-1}$$

where T_g is the glass-transition temperature of the mixture (°C); T_{gs} and T_{gw} the glass-transition temperatures (°C) of the solid matrix and water, respectively; X_s and X_w the corresponding percent contents; and K an empirical constant. A value of $T_{gw} = -135$°C was used for pure water (Johari, 1987). Couchman

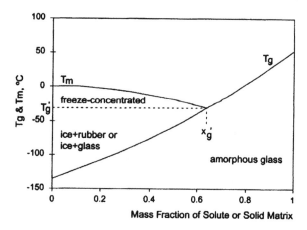

Figure 7-5 General aspect of a state diagram for a food system.

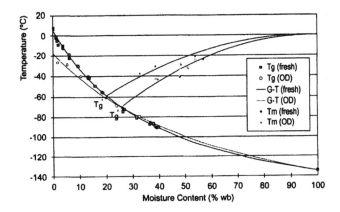

Figure 7-6 State diagram for freeze-dried and osmotic dehydrated Golden D Delicious apple.

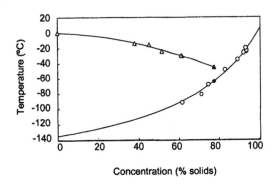

Concentration (% solids)

Figure 7-7 State diagram for a commercial peach jam.

(1978) later proposed an equivalent alternative model.

State diagrams can be used to trace the paths followed along several food processing operations, allowing the technologist to either change operation conditions to reach some desirable state or formulate the product (change the diagram) so as to adapt the material to given processing conditions. As an example, Figure 7-8 presents typical paths followed by foods subject to freeze-drying.

Specific Heat Capacities and Latent Heats

Computer simulation of food processing operations requires detailed information on the physical properties of processed materials. In the case of freezing and thawing operations, specific heat capacities (c_p) and melting enthalpies (ΔH_m) are particularly relevant. For the design and operation of industrial freezing/thawing plants, it has been shown to be very useful to combine both properties in only one, known as apparent heat capacity [$(c_p)_{ap}$].

82

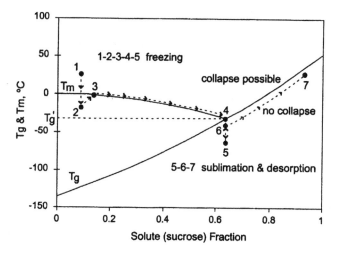

Figure 7-8 Schematic path on the general state diagram of a freeze-drying operation.

With high water content materials such as fruits and vegetables, the mentioned thermal properties are strongly dependent on the fraction of water crystallizing as ice or melting during the process. Differential scanning calorimetry can be used to measure the specific heat capacities and melting enthalpies of foods from a low temperature frozen state to ambient temperature, and thus their combination as an apparent heat capacity value.

According to O'Neil (1966), heat capacities can be calculated from the ratio of heat flow rates determined for each material and a standard reference (usually α-alumina) with respect to a base line obtained with two empty cells. Experimental data on the specific heat capacity of α-alumina has been tabulated by Touloukian and Buyco (1970) in the range of –60 to 40°C.

Figure 7-9 shows a thermogram obtained with a sample of *Golden Delicious* apple. From the four replicated thermograms obtained for each fruit, average values for the apparent specific heat capacity $(c_p)_{ap}$ were calculated and fitted to models proposed by Schwartzberg (1981):

$$(c_p)_{ap} = a' + b'(T_w - T)^2 \quad \text{when} \quad T < T_f \tag{7-2}$$

and Delgado et al. (1990):

$$(c_p)_{ap} = m - n (T_w - T) \quad \text{when} \quad T > T_f \tag{7-3}$$

where a', b', m, and n are adjustable empirical parameters; T_w is the melting point of pure water, and T_f the melting point of the material. The experimental and calculated results obtained are presented in Figure 7-10 and compared to the limited experimental data available in the literature for the same materials.

Unfrozen Water

A fraction of unfrozen water may be calculated from the difference between the total moisture content and amount of frozen water obtained from the ratio of melting heat and melting heat of pure water (333.2 J/kg) (Weast and Astle, 1981). Table 7-1 shows the average values of the initial melting temperature (T_{mi}), melting peak temperature (T_m), melting heats (ΔH_m), and percent of unfrozen water (NFW) for typical products.

STORAGE STABILITY

The physical stability of amorphous foods has been related to the change from a glassy state characterized by very low molecular mobility to a rubbery state. Above Tg the viscosity of the matrix decreases and some physical changes (such as collapse, loss of shape, shrinkage or stickiness) may occur. It is also possible that the state properties of the system (i.e., glassy or rubbery) may contribute to differences in chemical reaction rates in each of the states.

Arrhenius proposed the model now most commonly used to describe the temperature dependence of rate coefficients. However, it has been stated that this model is not the best to represent such dependence in the range from Tg to Tg + 100°C, so an alternative was proposed by Williams, Landel, and Ferry (1955) (Slade and Levine, 1991; Buera and Karel, 1993).

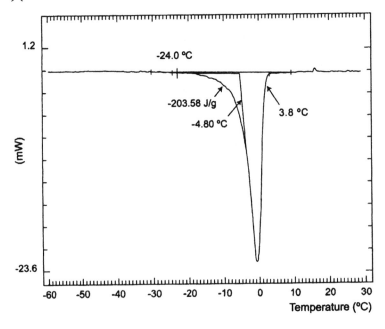

Figure 7-9 Differential scanning calorimetry thermogram of a melting peak of Golden Delicious apple.

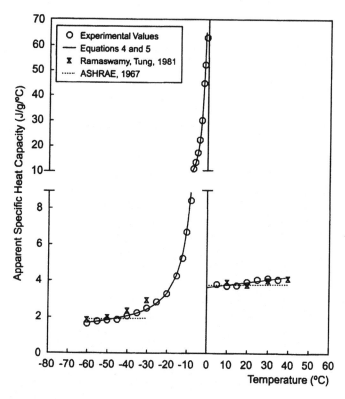

Figure 7-10 Apparent specific heat capacity vs. temperature for apple.

Table 7-1 Measured values of melting points, melting heat, and percent of unfrozen water for typical fruits.

Material	Initial Melting Temperature, °C	Melting Peak, °C	Melting Heat, J/g	UFW, %
Apple	−4.9	−0.7	196.4	27.4
Pear	−5.3	−0.6	191.9	26.3
Tomato	−2.8	0.7	243.9	20.3

According to the Williams-Landel-Ferry (WLF) model, the temperature dependence of a kinetic coefficient would be:

$$\log \frac{K_g}{K} = \frac{-C_3 \left(T - T_g\right)}{C_4 + \left(T - T_g\right)} \qquad (7\text{-}4)$$

where K and K_g represent the browning rate constants at T and T_g, and C_1 and C_2 are constants.

Sá et al., (1998) studied the discoloration of freeze-dried onion stored at a controlled temperature and relative humidity for five months. Three sets of containers conditioned at 33, 44, and 53% relative humidity were stored at 15, 25, 35, and 45°C. The color of the freeze-dried onion powder changed from a uniform cream to yellow brownish and then to brown. Nonenzymatic browning increased with time, temperature, and humidity. A zero-order reaction model produced a good fit with the data, confirming the results of Samaniego-Esguerra et al. (1991).

The temperature dependence of the rate constants was analyzed using the Arrhenius and WLF models. Figure 7-11 is a plot of ln K vs. 1/T. Visual analysis suggests that a better fit could be obtained with the WLF model. An F-test analysis showed that with 90% confidence it is better than the Arrhenius model in the case of onion browning.

FINAL REMARKS

Several practical examples of the use of DSC to measure important properties for the characterization and processing of food materials have been described. Among them, the possibility of plotting state diagrams for products is probably the richest in terms of the information made available to both the food scientist and food technologist. Other relevant properties for the design and optimization of food processing operations are the apparent heat capacity and fraction of unfrozen water on the frozen food product.

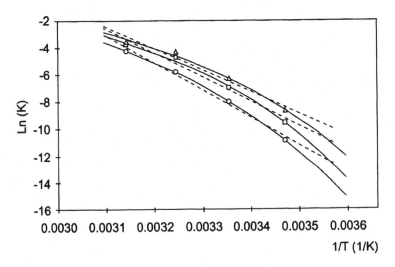

Figure 7-11 Arrhenius and WLF plots of ln K vs. 1/T for the browning of freeze-dried onion (O: a_w = 0.33; □: a_w = 0.44; △: a_w = 0.53; ----- Arrhenius model; —— WLF model).

REFERENCES

ASHRAE (1967). *Handbook of Fundamentals*. New York: American Society of Heating, Refrigerating and Air-Conditioning Engineers.

Ablett S., Izzard M.J., and Lillford P.J. (1992). Differential scanning calorimetric study of frozen sucrose and glycerol solutions. *J. Chem. Soc.* 88(6): 789–794.

Buera M.P. and Karel M. (1993). Application of the WLF equation to describe the combined effects of moisture and temperature on nonenzymatic browning rates in food systems. *J. Food Proc. Pres.* 17(1): 31–45.

Couchman P.R. (1978). Compositional variation of glass transition temperatures. Part 2. Application of the thermodynamic theory compatible polymer blends. *Macromol.* 11: 1156–1161.

Delgado A.E., Rubiolo A.C., and Gribaudo L.M. (1990). Effective heat capacity for strawberry freezing and thawing calculations. *J. Food Eng.* 19: 165–175.

Flink J.M. (1983). Structure and structure transitions in dried carbohydrate materials. In: *Physical Properties of Foods*. Peleg M.and Bagley E.B., Eds.Westport, CT: AVI Publishing, pp. 473–521.

Gordon M. and Taylor J.S. (1952). Ideal copolymers and the second-order transitions of synthetic rubbers. Part I. Non-crystalline copolymers. *J. Appl. Chem.* 2: 493–500.

Johari G.P., Halbrucker A., and Mayer E. (1987). The glass-liquid transition of hyperquenched water. *Nature* 330: 552–553.

Kelley S.S., Rials T.G., and Glasser W.G. (1987). Relaxation behavior of the amorphous components of wood. *J. Mater. Sci.* 22: 617–624.

Lovric T., Pilizota V., and Janekovic A. (1987). DSC study of the thermophysical properties of aqueous liquid and semi-liquid foodstuffs at freezing temperatures. *J. Food Sci.* 52: 772–776.

McNaughton J.L. and Mortimer C.T. (1975). Differential scanning calorimetry. In: *International Reviews in Science: Physical Chemistry Series 2*. Skinner H.A., Ed. London: Butterworths, pp. 1–44.

O'Neill M.J. (1966). Measurement of specific heat functions by differential scanning calorimetry. *Anal. Chem.* 38: 1331–1336.

Pope M.I. and Judd M.D. (1977). *Differential Thermal Analysis. A Guide to the Technique and its Application*. London: Heydon and Sons.

Ramaswamy H.S. and Tung M.A. (1981). Thermophysical properties of apples in relation to freezing. *J. Food Sci.* 46: 724–728.

Roos Y.H. (1992). Phase transitions and transformations in food systems. In: *Handbook of Food Engineering*. Heldman D.R. and Lund D.B., Eds. New York: Marcel Dekker, pp. 437–562.

Roos Y.H. (1995). *Phase Transitions in Foods*. San Diego, CA: Academic Press.

Sá M.M. and Sereno A.M. (1994). Glass transitions and state diagrams for typical natural fruits and vegetables. *Thermochimica Acta* 246: 285–297.

Sá M.M., Figueiredo A.M., Correa A., and Sereno A.M. (1994). Apparent heat capacities, initial melting points and heats of melting of frozen fruits measured by differential scanning calorimetry. *Rev. Esp. Cienc. Tecnol. Alim.* 34(2): 202–209.

Sá M.M., Figueiredo A.M., and Sereno A.M. (1998). Tg and state diagrams for fresh and dehydrated apple by DSC. Presented at the *Seventh International Symposium on Properties of Water*, Helsinki, Finland.

Samaniego-Esguerra C.M. and Boag I.F., and Robertson G.L. (1991). Kinetics of quality deterioration in dried onions and green beans as a function of temperature and water activity. *Lebensm. Wiss. Technol.* 24: 53–58.

Schwartzberg H.G. (1981). Mathematical analysis of the freezing and thawing of foods. Presented at the *AIChE Summer National Meeting*, Detroit, Michigan.

Slade, L. and Levine H. (1991). Beyond water activity: Recent advances based on an alternative approach to the assessment of food quality and safety. *Crit. Rev. Food Sci. Nutr.* 30(2, 3): 115–360.

Touloukian Y.S. and Buyco E.Y. (1970). Specific heat: nonmetallic solids (Vol 5). In: *Thermophysical properties of matter*. Thermophysical Properties Research Center (*TPRC*) Data Series. New York: Plenum Press, pp. 24–29.

Weast R.C., and Astle M.J. (1981). *Handbook of Chemistry and Physics, 63rd ed.* Boca Raton, FL: CRC Press.

Williams M.L., Landel R.F., and Ferry J.D. (1955). The temperature dependence of relaxation mechanisms in amorphous polymers and other glass forming liquids. *J. Am. Chem. Soc.* 77: 3701–3709.

CHAPTER 8

A Thermorheological Model of Corn Starch Dispersion During Gelatinization

W.H. Yang and M.A. Rao

INTRODUCTION

Thermorheological models may be defined as those that have been derived from rheological data obtained as a function of both shear rate and temperature. Such models can be used to calculate the apparent viscosity at different shear rates and temperatures in computer simulation and food engineering experiments. Because the viscosity (η) is independent of shear rate for a simple Newtonian fluid, one may consider only the influence of temperature on the viscosity. For many foods, the Arrhenius equation is suitable for describing the effect of temperature on η:

$$\eta = \eta_{\infty} A \exp(E_a / RT) \qquad (8\text{-}1)$$

where $\eta_{\infty A}$ is the frequency factor, E_a the activation energy (J mol^{-1}), R the gas constant (J mol^{-1} K^{-1}), and T the absolute temperature (K).

For non-Newtonian foods, the simple power law model (Eq. 8-2) can be used to describe shear rate ($\dot{\gamma}$) versus shear stress (σ) data at a fixed temperature:

$$\sigma = K\dot{\gamma}^n \qquad (8\text{-}2)$$

where n is the flow behavior index (dimensionless) and K (Pa s^n) the consistency index. Two very similar thermorheological equations have been obtained by combining the power law and Arrhenius equations. The equation obtained by Christiansen and Craig (1962) was:

$$\sigma = K_{TC} \left(\dot{\gamma} \exp\{E_{ac} / RT\} \right)^{\bar{n}} \qquad (8\text{-}3)$$

An equation commonly encountered in the food engineering literature that has been used in several studies (Harper and El-Sahrigi, 1965; Rao et al., 1981; Vitali and Rao, 1984) is:

$$\sigma = K_{TH} \exp(E_{aH} / RT)\dot{\gamma}^{\bar{n}} \qquad (8\text{-}4)$$

In both Equations 8-3 and 8-4, $\bar{n}$ is the average value of the flow behavior index for data at all the studied temperatures. Vitali and Rao (1984) showed that the activation energy terms in these equations are related:

$$E_{aH} = \bar{n}(E_{aC}) \qquad (8\text{-}5)$$

In many foods such as soups, salad dressings, gravies, and sauces, starch is present in excess water conditions. Because of the drastic increase in magnitude of the apparent viscosity (η_a), the rheological behavior of starch in excess water during the transition from fluid-like to viscoelastic behavior affects heat transfer during food thermal processing (Ball and Olson, 1957). Most models of starch gelatinization were developed under isothermal conditions based on apparent first-order kinetics and the Arrhenius equation to describe the effect of temperature on the gelatinization rate (Kubota et al., 1979; Dolan and Steffe, 1990; Kokini et al., 1992; Okechukwu and Rao, 1996). Because of the temperature history imposed during thermal processing, data obtained under isothermal conditions may not be suitable to describe changes in η_a during gelatinization. In addition to the temperature of the starch sample, shear rate or dynamic frequency has a significant effect on η_a or complex viscosity (η^*), respectively.

A comprehensive thermorheological (TR) model taking into consideration the effect of time (t)-temperature (T) history was developed by Dolan et al. (1989). The model for constant shear rate and starch concentration is:

$$\eta_{dim} = \frac{\eta_a - \eta_{ug}}{\eta_\infty - \eta_{ug}} = [1 - \exp(-k\Psi)]^\alpha \qquad (8\text{-}6)$$

where

$$\Psi = \int T(t) \exp\left(\frac{-E_g}{RT(t)}\right) dt \qquad (8\text{-}7)$$

where η_{dim} is dimensionless apparent viscosity, η_a apparent viscosity at a specific time during heating, η_{ug} apparent viscosity of the ungelatinized dispersion, and η_∞ the highest magnitude of η_a during gelatinization. The model was based on earlier work on protein dough viscosity by Morgan et al. (1989) but extended to include the influence of shear rate, temperature, concentration, and strain history (Dolan and Steffe, 1990). Data obtained using back extrusion and mixer viscometers were used to evaluate the models. The activation energy of gelatinization (E_g) depended on the heating temperature (Dolan et al., 1989), and some of the factors affecting viscosity were negligible (Dolan and Steffe,

1990). The studies of Dolan et al. (1989) and Dolan and Steffe (1990) provide valuable guidelines for identifying the factors affecting the apparent viscosity of starch dispersions (STD) during heating.

Because η^* data are obtained at low strains with minimal alteration of the STD structure, they provide unique opportunities for studying applicable models. Although the empirically obtained frequency shift factor (Ferry, 1980) has been used successfully in time-temperature superposition studies on food polymer dispersions (Lopes da Silva et al., 1994), the applicability of similar (if not identical) scaling of frequency still needs to be explored for STD.

For fluid dynamics and heat transfer investigations related to food processing, the necessary η_a data may be obtained from models developed for η^* data using relationships based on the Cox-Merz rule (Rao, 1992; Rao and Cooley, 1992). Yang (1997) used these procedures to investigate the role of temperature and shear dependent viscosity on heating rates of a canned 3.5% corn STD. In this chapter, the influence of different heating rates and dynamic frequencies on values of η^* during the heating phase (gelatinization) of an 8% corn STD and the relationship between η^* and η_a obtained from steady shear rate data are summarized. These results were described in detail elsewhere (Yang, 1997; Yang and Rao, 1998).

MATERIALS AND METHODS

Corn Starch Dispersions

An unmodified corn starch (American Maize-Product Company) with the commercial specifications: 10.0% moisture, 5.5 pH, and 99.5% granulation through US 200 mesh was used. Starch dispersions at 8% w/w were prepared by mixing with distilled water and held overnight (~16 h) to allow starch hydration.

Dynamic Rheological Data

Dynamic rheological data were obtained using the 4 cm dia parallel plate (gap 500 μm) geometry of a Carri-Med CSL-100 rheometer (TA Instruments). With this type of parallel plate geometry, the drops of paraffin oil placed soon after the storage modulus (G') reached values that were measurable did not penetrate into the corn STD. Magnitudes of the η^* were computed by the rheometer software and the parameters in Equation 8-6 (Dolan et al., 1989) were calculated by nonlinear regression based on a modified Levenberg-Marquardt algorithm (subroutine ZXSSQ, IMSL, Inc.).

The critical factors for obtaining reliable η^* data were: 1) control of the sample volume (0.80 ml) that was loaded carefully on the bottom rheometer plate, 2) removal of excess dispersion from the edge of the top plate, 3) minimizing water evaporation from the sample when the temperature was high, and 4) avoiding paraffin displacement oil by tiny air bubbles from the edge of the top plate.

Heating Rates and Dynamic Frequencies

The effect of different heating rates on the η^* of the 8% corn STD was studied at 1.26 rad s^{-1} and 1% strain as they were heated from 60°C to 92°C at different heating rates (1.6, 2.1, 2.5, 3.0, 4.2 and 6.0°C min^{-1}) and held at 92°C for about 2 min.

In order to examine the effect of ω, η^* data were obtained at: 0.63, 1.88, 5.34, 6.28, 7.85, 12.57, 18.85, 31.41, and 47.12 rad s^{-1} as the 8% corn STD were heated from 60 to 92°C over 15 min (2.1°C min^{-1}).

RESULTS AND DISCUSSION

The rheological properties of a starch paste depend on three factors (Eliasson, 1986): 1) starch granules (dispersed phase): concentration; starch swelling pattern; granule size and size distribution, shape, rigidity, and deformability; 2) amylose/amylopectin matrix (continuous phase): viscoelasticity, amount and type of amylose/amylopectin which has leached from the granules, entanglements, and 3) interactions between the components: starch granule surface, granule-amylose/amylopectin interactions, amylose/amylopectin-granule interactions, and granule-granule contact.

For the η^* versus time data obtained at 0.2 Hz (1.26 rad s^{-1}) on the 8% corn STD, the magnitudes of Eg, k, and α in Equations 8-6 and 8-7 over the range of heating rates studied were: 240 kJ mol^{-1}, 1.95 E + 30, and 0.668, respectively. These values are in reasonable agreement with those of Dolan et al. (1989) who found 210 kJ mol^{-1}, 0.846 E + 26, and 0.494, respectively, considering that η^* instead of η_a and starch from a different source were used. Although the value of R^2 was relatively high ($R^2 = 0.96$), the experimental data showed deviations from the predicted curve for time-temperature history values 5.0×10^{-31} to 1.25×10^{-30} K s (Fig. 8-1). Equations 8-6 and 8-7 with the calculated constants can be used to obtain apparent viscosity from any temperature versus time history during gelatinization provided that the initial and maximum viscosities are known; however, because of granule sedimentation, the former is difficult to determine experimentally. A better fit may be obtained by expressing Eg as a function of temperature and the Arrhenius relationship to be temperature-time history dependent (Dolan et al., 1989; Dolan and Steffe, 1990).

Influence of Heating Rate

The η^* data obtained at a fixed frequency (1.26 rad s^{-1}) but different heating rates followed different curves when plotted against time (Fig. 8-2). However, when the data were plotted against temperature (Fig. 8-3) instead of time, most data collapsed to a single curve. The largest difference in the magnitudes of complex viscosity between the highest (6.0°C min^{-1}) and lowest (1.6°C min^{-1}) heating rates was about 25%. Although heating time plays an important role in determining the extent of starch gelatinization in experiments performed under

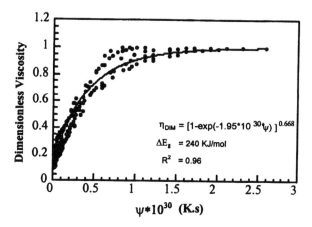

Figure 8-1 Dimensionless viscosity-temperature-time history curve of an 8% corn starch dispersion at different heating rates.

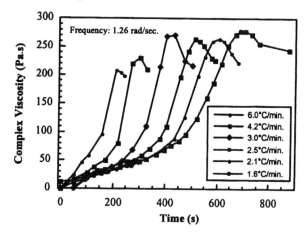

Figure 8-2 Complex viscosity versus the heating time curves of 8% corn STD at a frequency of 1.26 rad s^{-1} and different heating rates.

isothermal heating conditions, the complex viscosity of a corn STD over the range heating rates 1.6°C min^{-1} to 6.0°C min^{-1} was described as a simple function temperature.

Functional Viscosity Model

The influence frequency: At the fixed heating rate 2.1°C min^{-1} and over the range of frequencies employed from 1.26 to 47.12 rad s^{-1}, the η* profiles at a specific frequency versus temperature were similar so that by choosing an arbitrary reference frequency (ωr), all the η* temperature curves at the different frequencies could be reduced to a single curve. Several different frequencies were suitable for use as ωr. The resulting master curve of reduced complex viscosity $\eta_R^* = \eta^*\left(\dfrac{\omega}{\omega_r}\right)$ versus temperature obtained using $\omega r = 6.28$ rad s^{-1} (1.0 Hz) at the heating rate 2.1°C min^{-1} is shown in Figure 8-4. It was fortuitous that a linear frequency ratio was satisfactory over the frequency range 1.26 to 47.12 rad s^{-1} because when data from a limited number of experiments conducted at the high ω values of 62.83 rad s^{-1} (10 Hz) and 78.54 rad s^{-1} (12.5 Hz) at the heating rate of 2.1°C min^{-1} were considered, the frequency scaling could be achieved (Fig. 8-5) by a more general relationship:

$$\eta_R^* = \eta^*\left(\frac{\omega}{\omega_r}\right)^{\beta} \qquad (8\text{-}8)$$

where $\left(\dfrac{\omega}{\omega_r}\right)^{\beta}$ is the frequency shift factor and magnitudes of the exponent β

were determined from experimental data at ω and ωr. For complex viscosity data at 62.83 rad s^{-1} (10 Hz) and 78.54 rad s^{-1} (12.5 Hz) with $\omega r = 6.28$ rad s^{-1}, values of β were 0.913 and 0.922, respectively. It is noteworthy that both increasing and decreasing segments of the viscosity-temperature data at all the frequencies were reduced to a single curve. Complex viscosity data on 8% corn STD from experiments conducted at different heating rates and frequencies were also superposed using 6.28 rad s^{-1} (1.0 Hz) as ωr (Fig. 8-5).

Modified Cox-Merz Rule

Steady shear data on gelatinized 8% corn STD were obtained at 25°C with the Carri-Med CSL-100 rheometer using steady shear rates ($\dot{\gamma}$) in the same range as the frequencies (ω) in dynamic tests. A modified Cox-Merz rule was used to correlate the apparent viscosity-shear rate and complex viscosity-oscillatory frequency data using log-log plots (Rao and Cooley, 1992):

$$\eta^*(\omega) = C\left[\eta_a(\dot{\gamma})\right]^{\alpha c}\big|_{\omega=\dot{\gamma}} \qquad (8\text{-}9)$$

where η^* is complex viscosity (Pa s), η_a apparent viscosity (Pa s), ω frequency (rad s^{-1}), $\dot{\gamma}$ shear rate (s^{-1}), C a constant, and αc the shift factor. For the 8.0% corn STD at 25°C, the constant C and shift factor αc in Equation 8-8 were found to be 5.72 and 0.97, respectively. Yang (1997) successfully utilized the modified Cox-Merz rule together with the complex viscosity master curve in the simulation of natural convection heat transfer to a 3.6% corn starch dispersion being heated in a can.

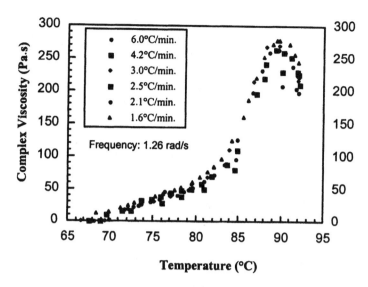

Figure 8-3 Complex viscosity versus the heating temperature curve of an 8% corn STD at a frequency of 1.26 rad s⁻¹ and different heating rates.

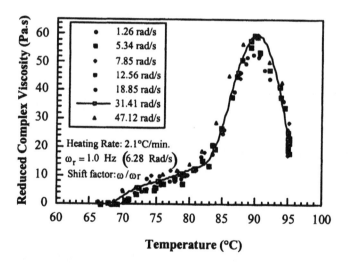

Figure 8-4 Reduced complex viscosity master curve of 8% corn starch dispersion at several oscillatory frequencies and 2.1°C min⁻¹ heating rate as a function of temperature during gelatinization. The data in Figure 8-3 were reduced to a single curve using the frequency shift factor $\left(\dfrac{\omega}{\omega_r}\right)^{\beta}$ with ωr = 6.28 rad s⁻¹.

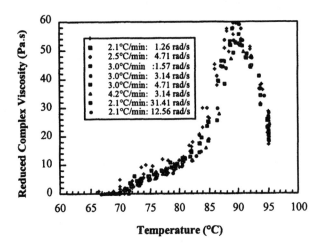

Figure 8-5 Master curve of the reduced complex viscosity of an 8% corn STD at several oscillatory frequencies and heating rates. All data were superposed using the frequency shift factor $\left(\dfrac{\omega}{\omega_r}\right)^\beta$ with $\omega r = 6.28$ rad s^{-1}.

CONCLUSIONS

Using temperature (60–95°C) as the independent variable and shift factors based on a reference frequency, a master curve was developed for reduced η^* during the gelatinization of an 8% corn STD. Similar reduced η^* versus temperature master curves were also developed for 6% and 3.5% corn STD by Yang (1997). The master curves suggest that the studied corn STD exhibited thermorheologically simple behavior (Plazek, 1996). A modified Cox-Merz rule related the complex and apparent viscosity data on the gelatinized 8% corn STD obtained at 25°C.

ACKNOWLEDGMENTS

We are grateful to the Conselho Nacional de Pesquisa e Desenvolvimento (CNPq), Brazil and the Food and Machinery Corporation (FMC) for financial support the American Maize-Product Company for donation of starch; and Terry Heyliger and Rey Elizondo (FMC) for valuable discussion and encouragement.

REFERENCES

Ball C.O. and Olson F.C.W. (1957). *Sterilization in Food Technology*, 1ˢᵗ ed. New York: McGraw Hill Book Company.

Christiansen E.B. and Craig S.E. (1962). Heat transfer to pseudoplastic fluids in laminar flow. *Amer. Inst. Chem. Eng. J.* 8: 154–160.

Dolan K.D., Steffe J.F. (1990). Modeling rheological behavior of gelatinizing starch solutions using mixer viscometry data. *J. Tex. Stud.* 21: 265–294.

Dolan K.D., Steffe J.F., and Morgan R.G. (1989). Back extrusion and simulation of viscosity development during starch gelatinization. *J. Food Proc. Eng.* 11: 79–101.

Eliasson A.C. (1986). Viscoelastic behaviour during the gelatinization of starch: Comparison of wheat, maize, potato and waxy barley starches. *J. Tex. Stud.* 17: 253–265.

Ferry J.D. (1980). *Viscoelastic Properties of Polymers*, 3ʳᵈ ed. New York: John Wiley and Sons.

Harper J.C. and El-Sahrigi A.F. (1965). Viscometric behavior of tomato concentrates. *J. Food Sci.* 30: 470–476.

Kokini J.L., Lai L-S., and Chedid L.L. (1992). Effect of starch structure on starch rheological properties. *Food Technol.* 46(6): 124–139.

Kubota K., Hosokawa Y., Suzuki K., and Hosaka H. (1979). Studies on the gelatinization rate of rice and potato starches. *J. Food Sci.* 44: 1394–1397.

Lopes da Silva J.A., Gonçalves M.P., and Rao M.A. (1994). Influence of temperature on dynamic and steady shear rheology of pectin dispersions. *Carbo. Poly.* 23: 77–87.

Morgan R.G., Steffe J.F., and Ofoli R.Y. (1989). A generalized rheological model for extrusion modeling of protein doughs. *J. Food Proc. Eng.* 11: 55–78.

Okechukwu P.E. and Rao M.A. (1996). Kinetics of cowpea starch gelatinization based on granule swelling. *Starch/Stärke* 48: 43–47.

Plazek D.J. (1996). 1995 Bingham medal address: Oh, thermorheological simplicity, wherefore art thou? *J. Rheol.* 40: 987–1014.

Rao M.A. (1992). Measurement of viscoelastic properties of fluid and semisolid foods. In: *Viscoelastic Properties of Foods*. Rao M.A. and Steffe J.F., Eds. London: Elsevier Applied Science, pp. 207–232.

Rao M.A. and Cooley H.J. (1992). Rheology of tomato pastes in steady and dynamic shear. *J. Tex. Stud.* 23: 415–425.

Rao M.A., Bourne M.C., and Cooley J.J. (1981) Flow properties of tomato concentrates. *J. Tex. Stud.* 12: 521–538.

Vitali A.A. and Rao M.A. (1984). Flow properties of low-pulp concentrated orange juice: Effect of temperature and concentration. *J. Food Sci.* 49: 882–888.

Yang W.H. (1997). *Rheological Behavior and Heat Transfer to a Canned Starch Dispersion: Computer Simulation and Experiment*. Ph.D. Thesis, Cornell University.

Yang W.H. and Rao M.A. (1998). Complex viscosity-temperature master curve of cornstarch dispersion during gelatinization. *J. Food Proc. Eng.* 21(3):191–207.

CHAPTER 9

Integral and Differential Linear and Nonlinear Constitutive Models for the Rheology of Wheat Flour Doughs

J.L. Kokini, M. Dhanasekharan, C-F. Wang, and H. Huang

INTRODUCTION

Food polymers often consist of long chained molecules such as starch, proteins, lipids, and polysaccharides that are viscoelastic. Viscoelasticity generates time dependency and manifests itself in recoil and normal forces. Deformations, which occur during the processing of foods, are generally a combination of shear and extensional flows. Examples include dough sheeting, extrusion, mixing, and protein spinning. Models able to describe the behavior of food materials in all components of stress, strain, and strain rates are called constitutive models (Bird et al., 1987). This chapter will review both linear and nonlinear integral and differential models which have been used with food materials in the authors' laboratory at Rutgers University.

LINEAR VISCOELASTICITY

Linear viscoelasticity is observed when the deformations encountered by food polymers are small enough that the polymeric material is negligibly disturbed from its equilibrium state. One of the most commonly used rheological properties to characterize linear viscoelastic materials is the shear relaxation modulus G (t, γ_o), which is independent of the applied strain in the linear viscoelastic region:

$$G(t, \gamma_o) = \frac{\sigma(t)}{\gamma_o} \qquad (9\text{-}1)$$

where $\sigma(t)$ is the time-dependent shear stress resulting from the applied deformation γ_o. In uniaxial extension the tensile relaxation modulus is given by:

$$E(t, \varepsilon_0) = \frac{\sigma_{11}}{\varepsilon_0} \tag{9-2}$$

where σ_{11} is the extensional stress and ε_0 the extensional strain.

To develop linear constitutive models the "Boltzmann superposition principle" is used. The superposition principle assumes that stresses resulting from strains at different times (and vice versa) can simply be added or superimposed.

$$\sigma(t) = \sum_{i=1}^{N} G(t - t_i) \delta\gamma(t_i) \tag{9-3}$$

where $\delta\gamma(t_i)$ is the incremental strain applied at time t_i and $G(t - t_i)$ is a function which links stress strain behavior. If the strain history is smooth the integral form of this equation can be used. Similar equations can be obtained by making strain the independent variable and stress the dependent variable. The equation is as follows:

$$\sigma(t) = \int_0^t G(t - t') d\gamma(t') \tag{9-4}$$

The measured relaxation modulus as a function of time can then be simulated using a single element or a generalized n element Maxwell model. The shear relaxation modulus for a single element is given by:

$$G(t) = G_0 \exp(-t/\lambda) \tag{9-5}$$

The linear constitutive model is given by:

$$\tau_{ij}(t) = \int_{-\infty}^t G_0 \{\exp[-(t - t')/\lambda]\} \dot{\gamma}_{ij}(t') dt' \tag{9-6}$$

When the data needs more adjustable parameters, the generalized Maxwell model given below is used:

$$\tau_{ij}(t) = \int_{-\infty}^t \sum_{k=11}^n G_k \{\exp[-(t - t')/\lambda_k]\} \dot{\gamma}_{ij}(t') dt' \tag{9-7}$$

where G_k and λ_k are the appropriate moduli and relaxation times of the Maxwell element, respectively. The behavior of the relaxation modulus at sufficiently long times will be dominated by the relaxation time with the largest value, and is called the "longest relaxation time" or "terminal relaxation time." Simulation of the relaxation modulus using the generalized Maxwell model for wheat flour dough is shown in Figure 9-1 (Huang, 1996).

Small amplitude oscillatory measurements are also commonly used to characterize linear viscoelastic properties. The following equations are used for the storage and loss moduli when a generalized Maxwell model is used to simulate linear viscoelastic behavior:

$$G'(\omega) = \sum_{i=1}^{N} \frac{G_i(\omega\lambda_i)^2}{[1+(\omega\lambda_i)^2]} \tag{9-8}$$

$$G''(\omega) = \sum_{i=1}^{N} \frac{G_i\omega\lambda_i}{[1+(\omega\lambda_i)^2]} \tag{9-9}$$

Predictions of G' and G" wheat flour dough are shown in Figure 9-2.

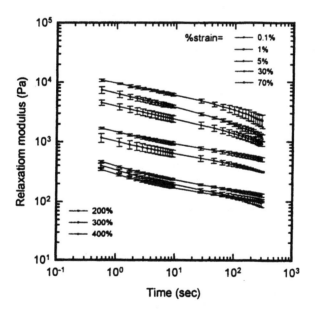

Figure 9-1 Linear and nonlinear shear relaxation moduli for 18% flour dough.

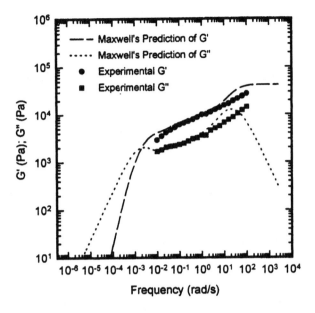

Figure 9-2 Comparison of the experimental and predicted G' and G" using the generalized Maxwell model for 18% protein flour dough at 27°C.

NONLINEAR VISCOELASTICITY

When deformations are rapid or large, the linear viscoelastic theory is no longer valid. Rheological properties like the relaxation modulus G(t) depend not only on time, but also on the magnitude of the deformation. In most food forming operations, deformations are generally both large and rapid. Examples are mixing, sheeting, and extrusion.

The most important nonlinear phenomena are the dependence of the viscosity on shear rate and the appearance of a non-zero first normal stress difference. The transient viscosity at the beginning of steady shear flow depends on the shear rate, and the relaxation modulus depends on the strain magnitude. Oscillatory properties like G' and G" are no longer useful because they are based on the assumption that the stress is sinusoidal, and when nonlinearities occur this is no longer valid.

INTEGRAL VISCOELASTIC MODELS

One approach to analyze nonlinear viscoelasticity is the formulation of nonlinear constitutive equations that make use of empirical equations for quantities such as the rates of creation and loss of entanglements (Carreau et al., 1968). Once the general form of such an equation has been established, selection of the specific nature of the equation is guided by study of the experimental results. In addition, the infinitesimal strain tensor is no longer suitable and must be replaced by the Cauchy and Finger tensors, which are measures of finite strain.

A second-order tensor like the Finger tensor has three scalar invariants for a given deformation. These can be calculated as follows:

$$I_1(B_{ij}) = B_{11} + B_{22} + B_{33} \qquad (9\text{-}10)$$

$$I_2(B_{ij}) = C_{11} + C_{22} + C_{33} \qquad (9\text{-}11)$$

$$I_3(B_{ij}) = 1 \qquad (9\text{-}12)$$

where B_{ij} and C_{ij} are the Finger and Cauchy strain tensors, respectively.

The Boltzmann superposition principle can now be generalized using the Finger tensor to formulate a theory of nonlinear viscoelasticity as follows:

$$\tau_{ij}(t) = \int_{-\infty}^{t} m(t-t')B_{ij}(t,t')dt' \qquad (9\text{-}13)$$

where $m(t - t')$ is the time-dependent memory function and $B_{ij}(t,t')$ the Finger tensor; this is also the equation for a "rubber-like" liquid developed by Lodge (1964).

The constitutive equation that results from Lodge's network theory is:

$$\tau_{ij}(t) = \int_{-\infty}^{t} \frac{G_i}{\lambda_i} \exp\left[-\frac{(t-t')}{\lambda_i} \right] B_{ij}(t,t')dt' \qquad (9\text{-}14)$$

The rubber-like liquid theory is of limited usefulness because it predicts that the viscosity and first normal stress coefficient are independent of shear rate, and there is ample data in the literature to clearly show that this is not the case with most food materials.

The equation proposed by Bernstein, Kearsley, and Zapas (Bernstein et al., 1964) known as the BKZ equation is based on the use of concepts originally used in the development of the theory of rubber viscoelasticity. In this theory the following form of the constitutive equation for viscoelastic material is proposed:

$$\tau_{ij} = \int_{-\infty}^{t} \left[2\frac{\partial \mu}{\partial I_1} C_{ij}(t,t') - 2\frac{\partial \mu}{\partial I_2} B_{ij}(t,t') \right] dt' \qquad (9\text{-}15)$$

where μ is a time-dependent elastic energy potential function given by:

$$\mu = \mu(I_1, I_2, t - t') \qquad (9\text{-}16)$$

This function must be determined experimentally by the study of large rapid deformations, thereby making it impractical.

A form of the BKZ equation that has proven to be more practical is based on the observation that the stress relaxation data for crosslinked rubbers could often be described by a relaxation modulus that is a product of both time-dependent and strain-dependent terms. It is then possible to write the elastic energy function as follows:

$$\mu = \mu(I_1, I_2, t - t') = m(t - t')U(I_1, I_2) \tag{9-17}$$

When the BKZ model is written in this form the Lodge rubber-like liquid becomes a special case of the above equation. Wagner (1976) further simplified the equation and the resulting "factorable" model is given by the following equation:

$$M[(t - t'), I_1, I_2] = m(t - t')h(I_1, I_2) \tag{9-18}$$

where $h(I_1, I_2)$ is called the damping function. This is a form of the memory function, which is separable or factorable and leads to the Wagner constitutive equation:

$$\tau_{ij}(t) = \int_{-\infty}^{t} m(t - t')h(I_1, I_2)B_{ij}(t, t')dt' \tag{9-19}$$

As a first approximation, Wagner (1976) proposed a simple exponential damping function shown below:

$$h(\gamma) = \exp(-n\gamma) \tag{9-20}$$

Osaki (1976) proposed a damping function which is the sum of the following two exponential functions:

$$h(\gamma) = a[\exp(-n_1\gamma)] + (1 - a)\exp(-n_2\gamma) \tag{9-21}$$

Zapas (1966) proposed an alternative form:

$$h(\gamma) = \frac{1}{1 + a\gamma^2} \tag{9-22}$$

Soskey and Winter (1984) proposed a more flexible damping function:

$$h(\gamma) = \frac{1}{1 + a\gamma^b} \tag{9-23}$$

104

These forms have all been shown to be valid in shear flows. Similarly, damping functions are proposed for extensional flows as well. One of those used by Meissner (1971) is:

$$h(\varepsilon) = \left\{ a[\exp(2\varepsilon)] + (1-a)\exp(k\varepsilon) \right\}^{-1} \qquad (9\text{-}24)$$

Damping functions for wheat flour and gluten doughs are shown in Figures 9-3 and 9-4. The form proposed by Osaki was most successful in simulating experimental data for gluten and wheat flour doughs. Figures 9-5 and 9-6 show the comparison of Wagner's model prediction of shear viscosity with the experimental data for 18.8% protein flour and gluten doughs.

The Wagner model showed an under-prediction of steady shear viscosities in the experimental shear rate range of 1.0×10^{-6} to 1.0×10^{-1} sec^{-1}. Higher differences between the experimental and simulated steady shear viscosities were observed in the shear rate region where viscosities were measured using a capillary rheometer. Figures 9-7 and 9-8 show the predictions of the first normal stress coefficient of hard wheat flour and gluten doughs using the Wagner model, which over-predicted the first normal stress coefficient values. The high volume percentage of starch fillers in the doughs was not included in the development of the Wagner model, which may account for the discrepancy between the simulated and experimental results.

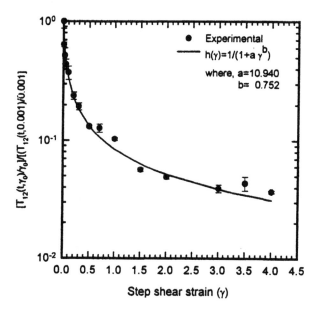

Figure 9-3 Simulation of the shear damping function for 18% protein flour dough.

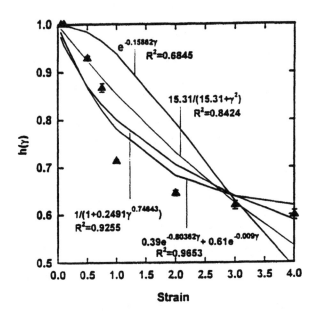

Figure 9-4 Simulation of the damping function h(γ) using four types of mathematical models for 55% moisture gluten dough at 25°C.

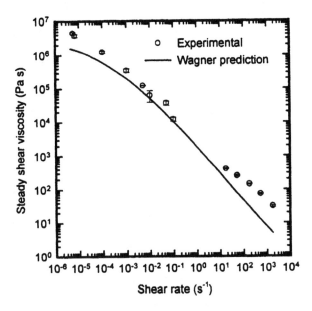

Figure 9-5 Comparison of the Wagner model prediction of the steady shear viscosities with the experimental data for 18% protein flour dough.

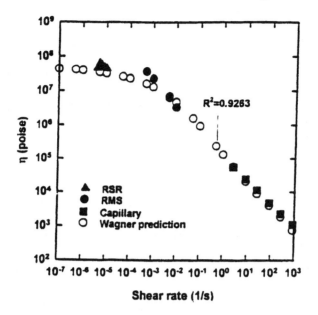

Figure 9-6 Comparison of the experimental data and the Wagner prediction of the steady viscosity for 55% moisture gluten dough at 25°C.

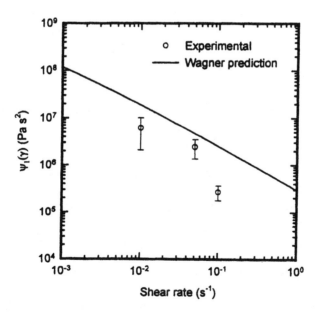

Figure 9-7 Comparison of the Wagner model prediction of the first normal stress coefficients with the experimental data for 18% protein flour dough.

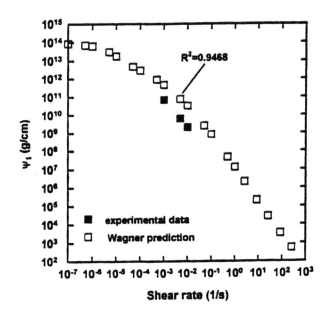

Figure 9-8 Prediction of the primary normal stress coefficient for 55% moisture gluten dough using the Wagner model with two exponential term damping function at temperature of 25°C.

NONLINEAR DIFFERENTIAL CONSTITUTIVE MODELS

Nonlinear differential models are of particular interest in numerical simulations for process design, optimization, and scale-up. This is because integral viscoelastic models are not amenable to use in numerical simulations of complex flows due to the high computational costs involved in tracking the strain history, particularly in three-dimensional flows. Differential models have generally been more popular than integral models in numerical developments (Crochet, 1989) because they have only a few adjustable parameters besides those describing the linear relaxation spectrum, and are therefore most amenable to numerical simulations. Khan and Larson (1987) used the equations of Giesekus (1982), Phan-Thien and Tanner (1977), White and Metzner (1963), Larson (1988), and Acierno et al. (1976) to simulate step-shear, step-biaxial extension, and start-up of a steady uniaxial extension of HDPE without long side branches, and LDPE with long side branches.

In a differential viscoelastic constitutive equation the viscoelastic component of the extra-stress tensor τ^1_{ij} is related to the rate of deformation tensors $\dot{\gamma}_{ij}$ by means of a differential equation. The total stress tensor τ_{ij} is given as the sum of the viscoelastic component τ^1_{ij} and the purely Newtonian component τ^2_{ij} as:

$$\tau_{ij} = \tau_{ij}^1 + \tau_{ij}^2 \qquad (9\text{-}25)$$

where $\tau_{ij}^2 = 2\eta\,\dot{\gamma}_{ij}$. Three differential viscoelastic constitutive equations have been considered for characterizing the rheological properties of gluten dough, and they are described below.

The Giesekus Model

Polymer molecules can be visualized as unbranched or branched chains of structural elements called "beads" joined either by elastic or rigid connections known as "springs" and "rods" which are themselves subjected to thermal Brownian Motion forces. Entanglement loss and regeneration processes cause the relative motion of the beads with respect to the same or neighboring molecules. The relationship between this relative motion and the generating force is described by a configuration-dependent non-isotropic mobility tensor (Giesekus, 1982). The configuration dependence of the mobility is based on the hypothesis of the reptational chain model created by de Gennes (1971). This hypothesis states that the motion of a polymer molecule essentially occurs along a tube around its actual configuration. The final equation obtained is:

$$\left[I_{ij} + \alpha\frac{\lambda}{\eta_1}\tau_{ij}^1 \right]\tau_{ij}^1 + \lambda\,\overset{\triangledown}{\tau}_{ij}^1 = 2\eta_1\dot{\gamma}_{ij} \qquad (9\text{-}26)$$

with a purely Newtonian component $\tau_{ij}^2 = 2\eta\,\dot{\gamma}_{ij}$, where $\dot{\gamma}_{ij}$ is the strain rate tensor and $\overset{\triangledown}{\tau}_{ij}^1$ is the upper-convected (contravariant) derivative. This model has four parameters: η_1, η_2, λ, and α. Parameter α controls the extensional viscosity and the ratio of the second normal stress difference to the first normal stress difference. When $\alpha > 0$, shear thinning behavior is always obtained. The term involving α is the mobility factor that can be associated with anisotropic Brownian motion and anisotropic hydrodynamic drag on the constituent polymer molecules. All the properties of the Giesekus model were obtained numerically due to the highly nonlinear nature of the model equation.

The White-Metzner Model

Lodge (1956) and Yamomoto (1956) developed a molecular model based on the kinetic theory of rubber elasticity (Flory, 1953). In this theory, a flowing polymer system is assumed to consist of long chained molecules connected in a continuously changing network structure with temporary junctions. The White-Metzner model is derived from such a network theory (White and Metzner, 1963). The viscoelastic component of the stress tensor is given by:

$$\overset{1}{\tau_{ij}} + \lambda \overset{\nabla}{\underset{}{\tau}}{}^{1}_{ij} = 2\eta_1 \overset{\bullet}{\gamma}_{ij} \tag{9-27}$$

where η_1 is obtained from the experimental shear viscosity curve, and the function λ is obtained from the experimental first normal stress difference experimental curve. The parameters for the White-Metzner model were found by fitting the experimental shear viscosity curve using the Bird-Carreau dependence as:

$$\eta = \eta_\infty + (\eta_0 - \eta_\infty)(1 + \lambda_v^2 \overset{\bullet}{\gamma}{}^2)^{(n_v-1)/2} \tag{9-28}$$

where λ_v is a time constant for viscosity dependence on shear rate, η_0 zero shear viscosity, and η_∞ infinite shear viscosity. The dependence of the relaxation time on the shear rate is found by fitting the experimental first normal stress difference using a power law type model as:

$$\lambda = \lambda_0 \overset{\bullet}{\gamma}{}^{(n_\lambda-1)} \tag{9-29}$$

so that the first normal stress coefficient ψ_1 is given by:

$$\psi_1 = 2\eta\lambda \tag{9-30}$$

For the White-Metzner model, the transient viscosity η^+ and the transient first normal stress coefficient ψ_1^+ are given by:

$$\eta^+ = \eta(1 - e^{-(t-t_0)/\lambda}) \tag{9-31}$$

$$\psi_1^+ = \psi_1(1 - e^{-(t-t_0)/\lambda} - \frac{t-t_0}{\lambda}e^{-(t-t_0)/\lambda}) \tag{9-32}$$

The Phan-Thien-Tanner Model

Weilgel (1969) proposed an alternative approach to Lodge and Yamomoto's network theory similar to that of Boltzmann's kinetic theory of gases. The stress tensor was shown to assume a Boltzmann integral form, and the kernel of the integral was specified in terms of the creation and destruction rates of network junctions. Phan-Thien and Tanner (1977) used this approach with the modification that the transformations are not affine. They showed that the stress tensor can be explicitly written in terms of an "effective" Finger tensor and assumed specific forms for the creation and destruction rates of the network junctions to derive a constitutive equation containing the free parameters ε and ξ. The final form of the constitutive equation is:

$$\exp\left[\varepsilon\frac{\lambda}{\eta}tr(\tau_{ij}^1)\right]\tau_{ij}^1 + \lambda\left[(1-\frac{\xi}{2})\overset{\triangledown}{\tau_{ij}^1} + \frac{\xi}{2}\overset{\triangle}{\tau_{ij}^1}\right] = 2\eta\dot{\gamma}_{ij} \qquad (9\text{-}33)$$

The parameters η and λ are the partial viscosity and relaxation time, respectively, measured from the equilibrium relaxation spectrum of the fluid. The only adjustable parameters of the model are ε and ξ. The latter can be obtained using the dynamic viscosity-shear viscosity shift according to:

$$\eta'(x) = \eta\left(\frac{x}{\sqrt{\xi(2-\xi)}}\right) \qquad (9\text{-}34)$$

The shear viscosity (η) for the Phan-Thien-Tanner model is given by:

$$\eta = \sum_{i=1}^{n}\frac{G_i\lambda_i}{1+\xi(2-\xi)\lambda_i^2\dot{\gamma}^2} \qquad (9\text{-}35)$$

The first normal stress difference (ψ_1) for this model is given by:

$$\psi_1 = 2\sum_{i=1}^{n}\frac{G_i\lambda_i^2}{1+\xi(2-\xi)\lambda_i^2\dot{\gamma}^2} \qquad (9\text{-}36)$$

The transient shear properties need to be obtained numerically due to the nonlinear nature of the model. The Phan-Thien-Tanner and Giesekus models can be used in multiple modes by choosing a relaxation spectra instead of a single relaxation time and relaxation modulus. This enables good prediction of oscillatory shear properties. Figure 9-9 shows the predictions of the Giesekus, White-Metzner, and Phan-Thien-Tanner models for the shear viscosity of gluten dough. The White-Metzner model resulted in the most accurate estimated values for shear viscosity by using a Bird-Carreau type model which has a power law parameter to predict shear viscosity in the shear thinning regime, as well as a zero viscosity term that enables good prediction in the constant viscosity regime at low shear rates.

Figure 9-10 shows the predictions of the first normal stress coefficient for gluten dough using the Giesekus, White Metzner, and Phan-Thien-Tanner models. Again, the White-Metzner model provided the best fit for the first normal stress coefficient. Figures 9-11 and 9-12 show the predictions of the transient shear properties of gluten dough using the Giesekus, White-Metzner, and Phan-Thien-Tanner models. In this case the White-Metzner model under-predicts the observed transient properties, while the Phan-Thien-Tanner model provided the best fit for all transient shear viscosity and first normal stress coefficient data.

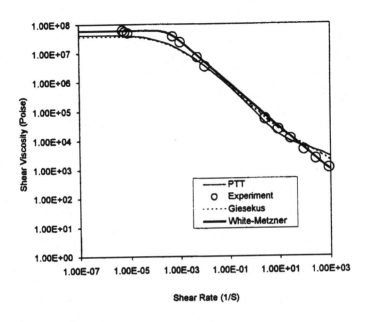

Figure 9-9 Prediction of the steady shear viscosity of gluten dough using Giesekus, Phan-Thien Tanner and White-Metzner model.

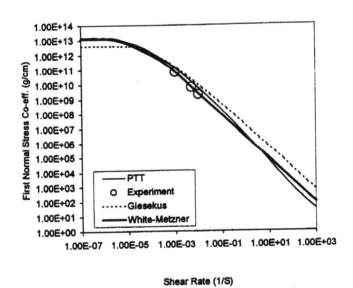

Figure 9-10 Prediction of the first normal stress coefficient of gluten dough using Giesekus, Phan-Thien Tanner and White-Metzner model.

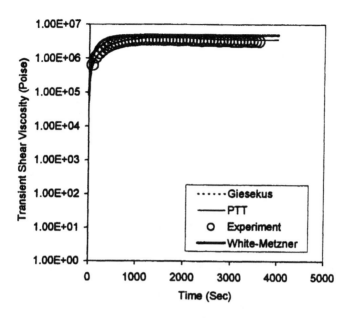

Figure 9-11 Prediction of the transient shear viscosity of gluten dough at a shear rate of 0.01/S using Giesekus, Phan-Thien Tanner and White-Metzner model.

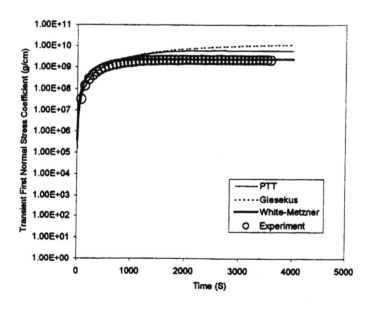

Figure 9-12 Prediction of the transient first normal stress coefficient of gluten dough at a shear rate of 0.01/S using Giesekus, Phan-Thien Tanner and White-Metzner model.

113

CONCLUDING REMARKS

Both differential linear and nonlinear constitutive models clearly demonstrate the same promise in predicting the shear and extensional properties of food materials as they do for synthetic polymers. Constitutive models are a prerequisite for successful simulation of the processing parameters needed to design food processing operations such as extrusion dough sheeting, bubble expansion during baking, and extrusion. One of the reasons why numerical simulation and quantitative predictions for food unit operations have been so limited is because constitutive models have not been available to date. This chapter has attempted to show the basis for the development of such constitutive models to facilitate further building on existing theory, thus leading to the simulation of highly complex food material properties that often are very different from those encountered in synthetic polymers. There is ample evidence that both linear and nonlinear constitutive models are able to predict steady and extensional properties of complex food materials once the assumptions of each constitutive model are fully understood and consistent with the molecular architecture of the food materials selected. Clearly there is much more to be done in examining the behavior of suspensions of highly filled systems of swollen deformable particle suspensions such as starch.

ACKNOWLEDGMENTS

This is Publication No. D10544-13-98 of the New Jersey Agricultural Experiment Station supported by state funds and the Center for Advanced Food Technology (CAFT), a New Jersey Commission for Science and Technology.

REFERENCES

Acierno D., La Mantia P., Marrucci G., and Titomanlio G. (1976). A nonlinear viscoelastic model with structure dependent relaxation times: Basic formulation. *J. Non-Newt. Fluid Mech.* 1: 125–129.

Bernstein B., Kearsley E.A, and Zapas L.J. (1964). Thermodynamics of perfect elastic solids. *J. Res. Nat. Bur. Stand.* 68B:103-113.

Bird R.B., Curtiss C.F., Armstrong R.C., and Hassager O. (1987). Dynamics of Polymer Liquids. Vol. 2, 2nd ed. New York: John Wiley and Sons.

Bueche F. (1952). Viscosity, self-diffusion, and allied effects in solid polymers. *J. Chem. Phys.* 20: 1959–1968.

Carreau P.J., Macdonald I.F., and Bird R.B. (1968). A nonlinear viscoelastic model for polymer solutions and melts. Part II. *Chem. Eng. Sci.* 23: 901.

Crochet M.J. (1989). Numerical simulation of viscoelastic flow: A review. *Rub. Chem. Tech.* 62: 426–455.

deGennes P.G. (1971). Reptation of a polymer chain in the presence of fixed obstacles. *J. Chem. Phys.* 55(2): 572–579.

Flory P.J. (1953). *Principles of Polymer Chemistry.* Ithaca, NY: Cornell University Press.

Giesekus H. (1982). A simple constitutive equation for polymeric fluids based on the concept of deformation dependent tensorial mobility. *J. Non-Newt. Fluid Mech.* 11: 69–109.

Khan S.A. and Larson R.G. (1987). Comparison of simple constitutive equations for polymer melts in shear and biaxial and uniaxial extensions. *J. Rheol.* 31(3): 207–234.

Osaki K. (1976). *Proceedings of the Seventh International Congress on Rheology.* Gothenburg: Tages-Anzeiger/Regina Druck, pp. 104–106.

Larson R.G. (1988). *Constitutive equations for polymer melts and solutions.* Boston: Butterworths.

Lodge A.S. (1964). *Elastic Liquids.* New York: Academic Press.

Meissner J. (1971). Dehnungsverhalten von Polyathylen-Schmelzen. *Rheol. Acta* 10: 230–239.

Phan-Thien N. and Tanner R.I. (1977). A new constitutive equation derived from network theory. *J. Non-Newt. Fluid Mech.* 2: 353–365.

Soskey P.R. and Winter H.H. (1984). Large step shear strain experiments with parallel-disk rotational rheometers. *J. Rheol.* 28(5): 625–634.

Wagner M.H. (1976). Analysis of time-dependent nonlinear stress-growth data for shear and elongational flow of a low-density branched polyethylene melt. *Rheol. Acta* 15(2): 136–145.

White J.L. and Metzner A.B. (1963). Development of constitutive equations for polymeric melts and solutions. *J. Appl. Polym. Sci.* 8: 1867.

Yamomoto M. (1956). The viscoelastic properties of network structure. 1. General Formalisms. *J. Phys. Soc. Jap.* 11(4): 413–421.

Zapas L.J. (1966). Viscoelastic behavior under large deformations. *J. Res. Nat. Bur. Stds.* 70A: 525–532.

CHAPTER 10

Ultrafiltration of Apple Juice

J.E. Lozano, D.T. Constenla, and M.E. Carrín

INTRODUCTION

The application of ultrafiltration (UF) as an alternative to conventional processes for the clarification of apple juice was demonstrated by Heatherbell et al. (1977). However, the acceptance of UF in the fruit processing industry is not yet complete because there are problems with the operation and fouling of membranes. During UF two fluid streams are generated: the ultrafiltered solid free juice (permeate) and the retentate with variable insoluble solids which, in the case of apple juice, are mainly remains of cellular walls and pectin. Permeate flux (J) results from the difference between a convective flux from the bulk of the juice to the membrane and a counter diffusive flux or outflow by which solute is transferred back into the bulk of the fluid (Fig. 10-1). The value of J is strongly dependent on hydrodynamic conditions, membrane properties, and the operating parameters. The main driving force of UF is the transmembrane pressure (ΔPTM), which in the case of hollow fiber ultrafiltration systems (HFUF) can be defined as:

$$\Delta PTM = (P_i + P_o)/2 - P_{ext} \qquad (10\text{-}1)$$

where P_i is the pressure at the inlet of the fiber, P_o the outlet pressure, and P_{ext} the pressure on the permeate side. In practice, the J values obtained with apple juice are much lower than those obtained with water only. This phenomenon is attributable to various causes, including resistance of the gel

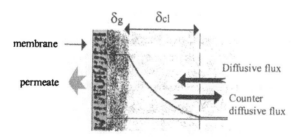

Figure 10-1 Concentration profile during ultrafiltration.

layer, the concentration polarization boundary layer [defined as a localized increase in concentration of rejected solutes at the membrane surface due to convective transport of solutes (Porter, 1972)], and plugging of pores due to fouling. Some of these phenomena are reversible and disappear after cleaning of the UF membranes and others are definitively irreversible.

With respect to the above-described situation, the objectives of this chapter are:

1. review the stationary permeate flux (J_{lim}) of apple juice through a HFUF module and verify the application of accepted predictive models
2. study the permeate flux with time considering the effect of operating parameters such as pressure and recirculating flow rate (Qr)
3. microscopically study the gel-polarized layer (Cg) formation due to pectins, including the effect of pectinolitic enzymes

Experimental data was obtained in a pilot scale UF unit (Amicon model DCSOP) with a single hollow fiber cartridge (Polysulfone, MWCO in the 30,000–100,000 range). The unit had a 40 L storage tank, sanitary stainless steel pump and pipes, a flow reversing valve, pressure control, and permeate and retentate stream flow meters.

STATIONARY PERMEATE FLUX

It is well known (Iritani et al., 1991) that the transmembrane pressure-permeate flux characteristic for ultrafiltration shows a linear dependence of J with ΔPTM at lower values of pressure (1st region), while at higher pressures (2nd region) the permeate flux approaches a limiting value (J_{lim}) independent of further increase in pressure. The last situation is assumed to be controlled by mass transfer.

Figures 10-2 and 10-3 show the variation of J with ΔPTM as a function of the volume concentration ratio (VCR, defined as the initial volume divided by retentate volume at any time), which is a measurement of the retentate concentration and Qr, respectively (Constenla and Lozano, 1996). Pressure independence (2nd region) was observed to occur at a higher pressure at higher Qr. The point at which the pressure independence is evident is called optimum transmembrane pressure ($\Delta PTMo$). The reduction of J_{lim} with Qr can thus be associated with a reduction in the boundary layer due to an increase in turbulence.

On the other hand, the optimal ΔPTM values were nearly independent of VCR when $Qr > 10$ L/min. A hysteresis effect in the permeate flux attributable to the consolidation of the gel layer (Omosaiye, 1978) was also observed, and the area enclosed by the hysteresis loop was found to be greater at lower Qr and VCR values. Traditionally, correlations of J with ΔPTM and VCR have been determined by parameter fitting of the experimental data, but since these polynomial functions have no physical basis, a large number of experimental data is needed for determination of J. Therefore, other theoretical and semi-empirical approaches should be considered.

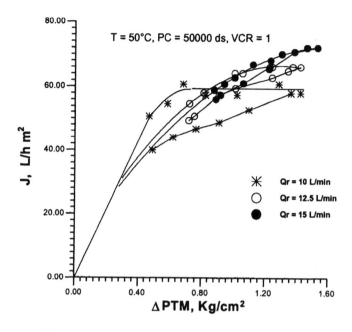

Figure 10-2 Effect of *ΔPTM* and *Qr* on *J* at 50°C.

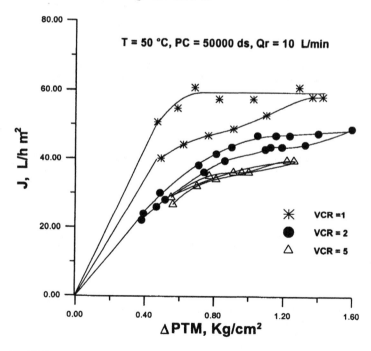

Figure 10-3 Effect of *ΔPTM* and VCR on *J* at 50°C.

Film Theory

One of the simplest theories for modeling flux in the 2nd region is the Film Theory (FT), which requires less experimental data for determination of the involved parameters than the several other models for mass transport in membranes. Solving the differential equation for the boundary layer produces the following equation for total permeate flux (Rautembach and Albrech, 1989):

$$J = k \ln [(C_g - C_{per}) / (C_b - C_{per})] \tag{10-2}$$

where J is the flux, $k = D/\delta_{bl}$ the mass transfer coefficient, D the diffusion coefficient, δ_{bl} the boundary layer thickness, C_b the bulk concentration of rejected macromolecules (mainly pectin), C_g the gel concentration, and C_{per} the solute concentration in permeate. To determine k some empirical correlations based on Bird and Steward's heat mass transfer analogy can be used (Chiang and Cheryan, 1986):

$$Sh = A' Re^{\alpha} Sc^{\beta} \tag{10-3}$$

where Sh, Re, and Sc are the Sherwood, Reynolds, and Schmidt numbers, respectively, and the parameters A', α, and β will depend on the flux conditions and geometry of the system. It is well known that undepectinized fruit juices are pseudoplastic in nature and the (apparent) viscosity included in dimensionless numbers may be described by the well-known power law equation. The numbers in Equation 10-3 must take into account this rheological behavior (Constenla and Lozano, 1996). During the ultrafiltration of apple juice, C_{per} is many times lower than C_b and C_g. Equation 10-2 can thus be simplified to:

$$J = k \ln (C_g/C_b) \tag{10-4}$$

In the hollow fiber system used in this study, the length of the concentration profile entrance region was much greater than the channel length, indicating the concentration profile was still developing during flow down the channel (Chiang and Cheryan, 1986). To evaluate FT application, the dimensionless number in Equation 10-3 must be calculated. Solutions like those proposed by Leveque or Gröver, depend on the degree of development of the concentration profile. For laminar flow systems, if the velocity profile is fully developed but the concentration boundary layer is developing along the fiber, the Leveque solution of Equation 10-3 can be used with $A' = 1.86$, $a = 1/3$, and $\beta = 1/3$ (Cheryan, 1986). The Leveque solution fits the results quite well (predicted values are slightly over the experimental fluxes) compared with Gröver's (Fig. 10-4), suggesting that the assumed velocity and concentration profile were appropriate (Constenla and Lozano, 1996).

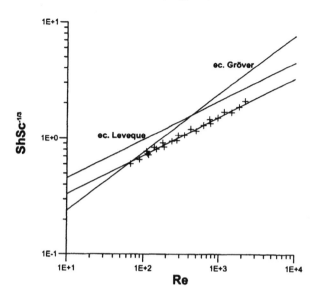

Figure 10-4 Mass transfer correlations, with the dotted line representing Equation 10-5.

Surface Renewal Theory

Because the film model frequently underestimates the values of the permeate flux, Koltuniewicz (1992) added a correction factor that takes into account the lateral migration of macromolecules from the membrane to the bulk of the feed stream (Fig. 10-5). This lateral migration is responsible for the instability of the concentration polarization layer and increases in the total permeate flux. According to this theory, J may be written as:

$$J = (J_i - J_f)\left(\frac{s}{s+A}\right) + J_f \tag{10-5}$$

where J_i and J_f are the initial and final flux, respectively, A the rate of flux decline, and s the rate of surface renewal. The A parameter can be considered as proportional to the transmembrane pressure ($A = a^* \Delta PTM$) and J_{lim} results (Koltuniewicz, 1992) in:

$$J_{lim} = \frac{s}{a^* R_m} + J_f \tag{10-6}$$

where R_m is the membrane resistance to flux. Equation 10-6 may be used to determine s experimentally, as can R_m using water as solvent. It is also possible to estimate flux J for a given rate of surface renewal s, constant Qr, and VCR as:

121

$$J = J_f + \left(\frac{\Delta PTM}{R_m} - J_f \right) \left(\frac{s/a*}{s/a*+\Delta PTM} \right) \qquad (10\text{-}7)$$

In the case of apple juice ultrafiltration, the characteristic resistance of the membrane was calculated from the water flux variation with ΔPTM. With this R_m value ($R_m = 6.366 \times 10^5$ Pa $\cdot$h/m) and the final flux J_f (determined from the flux decline with time at constant Q and VCR), both the ratio $s/a*$ and J_{lim} were calculated (Constenla and Lozano, 1996). These experimental data were finally fitted to Equation 10-7 and plotted in Figure 10-6, which shows excellent agreement between the experimental data and SRM predictions.

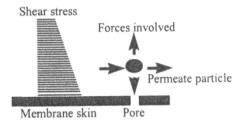

Figure 10-5 Sketch of forces involved during apple juice ultrafiltration.

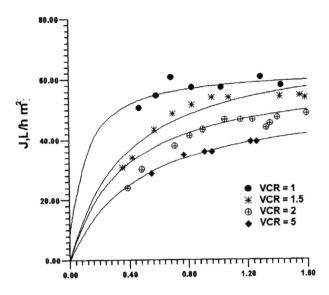

Figure 10-6 Comparison between experimental data and those predicted by the SR model (Eq. 10-7).

Resistance-in-Series Model (RIS)

Another approach to predict the flux of permeate is the resistance-in-series model. Theoretically this model could predict J throughout the range of ΔPTM as:

$$J = \frac{\Delta PTM}{R_m^{\cdot} + R_g + R_{bl}} \qquad (10\text{-}8)$$

(10-9)where R'_m is the intrinsic resistance of the membrane which is a property of the mechanical and chemical structure of the membrane material; R_g is the gel resistance which depends upon the extent of polarization and physical properties of both the gel layer and permeate which must pass across it; and R_{bl} is the boundary layer resistance. Chiang and Cheryan (1986) considered that $R_p = R_g + R_{bl}$ is a function of pressure $(R_p = \phi \cdot \Delta PTM)$ so that Equation 10-8 can be rewritten as:

$$J = \frac{\Delta PTM}{R_m^{\cdot} + \phi \Delta PTM}$$

where ϕ is the resistance index and a constant for a particular membrane-solute combination. When ΔPTM is low, the value of J is controlled by the membrane resistance. However, when ΔPTM is large the permeate flux approaches $1/\phi$. To illustrate, some of the R_m, ϕ, and $\phi \cdot \Delta PTM$ values calculated during the HFUF of apple juice are listed in Table 10-1.

Table 10-1 Values of intrinsic resistance (R'_m), with the resistance index (ϕ) and $\phi \cdot \Delta PTM$ products of the resistance-in-series model.

VCR	$R'_m \times 10^6$ (Pa h/m)	ϕ (m2 ·h/L)	$\phi \cdot \Delta PTM \times 10^6$ (Pa h/m)
1.0	1.079	0.0172	3.037
2.0	1.609	0.0227	3.620
3.0	1.815	0.0205	4.008
5.0	2.105	0.0254	4.486

Clearly R'_m and $\phi \Delta PTM$ are of the same order of magnitude. Therefore, the simplification proposed by Chiang and Cheryan (1986) to the RSM represented by Equation 10-8 resulted in a poor fitting with the experimental data. It was also found that there was a definite correlation between the resistance R'_m and VCR. The increasing nature of R'_m with regard to VCR suggests continuous formation of fouling with ΔPTM.

PERMEATE FLUX AS A FUNCTION OF TIME

Most of the many models proposed in the literature for representing J are semi-empirical and practical equations (Table 10-2). However, membrane fouling mechanisms may be studied through the classical laws of filtration under constant pressure (Table 10-3). During the UF process (Iritani et al., 1991) J behaves as in cake filtration only at the very beginning, which is attributable to the formation of the gel layer with minor counter-diffusion flux.

The variation of J with a normal molecular weight cut-off of 50,000 as a function of time and recirculating flow rate for both constant ΔPTM (73.5 kPa) and concentration (VCR = 1) is shown in Figure 10-7. As previously indicated, pectin and other large solutes like starch that are normally found when unripened apples are processed tend to form a fairly viscous and gelatinous-type layer on the skin of the asymmetric fiber. Flux decline due to this phenomenon can be reduced by increasing the flow velocity on the membrane. Traditionally correlations of J with ΔPTM and VCR have been determined by parameter fitting of the experimental data. It was found that the following exponential equation proposed in the SRT model (Constenla and Lozano, 1996) fit appropriately:

$$J = J_F + (J_O - J_F) \exp(-At) \tag{10-10}$$

where J_0, J_F, and A values can be obtained at different Qr and constant values of VCR and ΔPTM. The experimental data were finally fitted to Equation 10-9 and are plotted in Figure 10-7. As presumed, an increase in Qr significantly increased the permeate flux, which was reflected as an extensive increase in the A parameter.

Table 10-2 Some equations representing permeate flux as a function of time.

Representative Equation	Author
$J = J_F + (J_0 - J_F) \exp(-A\,t)$	Koltuniewicz, 1992
$J = [(J_0)^{-2} + 2\,K_4\,t]^{-1/2}$	Mietton-Peuchot et al., 1984
$4\,J = J_0 - B \ln(VCR)$	Heatherbell et al., 1977
$J - J_F = (J_0 - J_F)\,[1 - \exp(-t/B)]$	Probstein et al., 1979

0, 1, and F are zero, initial, and final time; Vp is the permeate volume; and A, B, and K_4 are constants

Table 10-3 Classic filtration models (pseudoplastic fluids).

Mechanism	Scheme	Representative Equations
(1) Total pore blocking	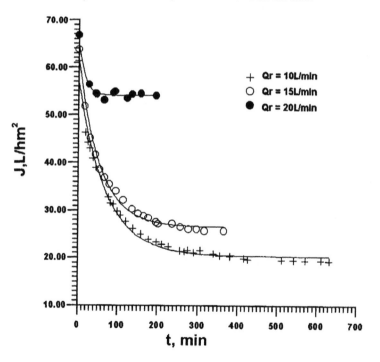	$J_0 - J = K_1\ V_p \ln (J/J_0) = -K_1$
(2) Partial pore blocking		$\ln (J/J_0) = -K_2\ V_{p1}/J - 1/J_0 = K_2\ t$
(3) Progressive pore blocking		$J = J_0\ (1-K_3\ V_p/2)(3n+1)/2nJ = J_0\{1+[(n+1)/n]\ J_0\ K_3\ t\}^{-(3n+1)/(n+1)}$
(4) Cake filtration		$(1/J)^n = (1/J_0)^n + K_4\ V_p\ (1/J)^{n+1} = (1/J_0)^{n+1}+[(n+1)/n]\ K_4\ t$

K_1, K_2, K_3, and K_4 are experimental constants; and n is the flow behavior index

Figure 10-7 Variation of J with time at VCR = 1 (the full line represents Eq. 10-10).

Influence of VCR on the Permeate Flux

The authors found that in the case of pseudo-plastics fluids such as the apple juice retentate (Constenla and Lozano, 1996), the following equation was satisfactory for describing the decrease in J with VCR during the UF of apple juice in hollow fiber based on classical cross-flow filtration laws at constant pressure (Mietton-Peuchot, 1984):

$$J = (Jo^{-(n+1)}) + (n + 1/n)\ K_4 t^{-1/(n+1)} \qquad (10\text{-}11)$$

where n is the flow behavior index and K_4 an experimental constant. Theoretically this equation could also predict J throughout the range of VCR. However, different operative conditions restrain the VCR up to a maximum of 14. The effect of the operative conditions may be summarized as follows (Constenla and Lozano, 1996).

Molecular weight cut-off of membranes: The permeate flux became independent of the solute rejection characteristic of the hollow fibers after a few minutes of operation. This effect is commonly attributed to the build-up of the concentration polarization/gel layer.

Transmembrane pressure (ΔPTM): As Figure 10-8 shows during the UF of apple juice in the mass transfer region, a 60% increase in ΔPTM was only reflected as a 5% increase in J.

Recycling flow (Qr): Acceleration of the apple juice retentate near the membrane surface removed the accumulated macromolecules by reducing the effect of the concentrated polarization gel layer. Due to the low diameter of the hollow fibers, high tangential velocities were obtained at laminar rates. Figure 10-9 shows the drastic effect of Qr on J maintaining the ΔPTM constant.

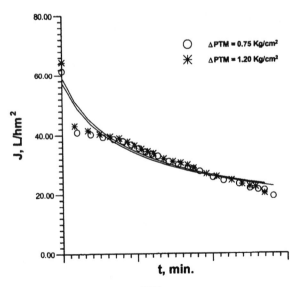

Figure 10-8 Variation of *J* with *ΔPTM* (the full lines represent Eq. 10-11).

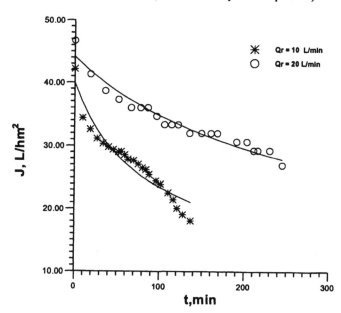

Figure 10-9 Influence of *Qr* on *J* (the full lines represent Eq. 10-11).

Permeate Flux and Volume Concentration Ratio

Some of the proposed exponential models that empirically relate *J* to VCR (Table 10-2) fit reasonably well. For example:

$$J = Jo - B \ln (VCR) \qquad (10\text{-}12)$$

where J_0 is the initial permeate flux and B a constant which depends on the system, operating conditions, and juice properties. Decrease of flux with concentration was found to be nonlinear and changes in the rate of permeation were better followed when plotted against ln VCR (Fig. 10-10). The rate of flux decrease in *J* was divided into three periods, the first of which was characterized by a rapid decrease in *J* within just a few minutes. During the second period (up to VCR ≈ 3) the variation of *J* was unstable depending on the fiber cut-off, but then *J* approached a linear and steady logarithmic decrease with VCR. This behavior was explained by considering the resistance to flux as two separate additive resistances in series: (I) the membrane resistance (R'_m) and (II) the concentration polarization/gel layer resistance (R_p). During the first period R_p increased very fast, reaching a value equivalent to that of R'_m. In the second region, the R'_m value was still an important component of the total resistance and *J* was not completely independent of the fiber properties. Finally, during the last period R_p was dominant and the cut-off of the hollow fiber became irrelevant.

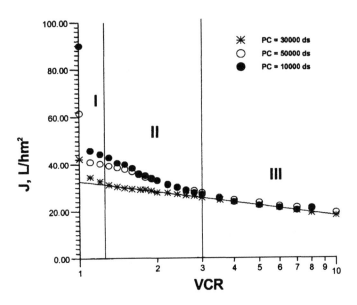

Figure 10-10 Decrease of permeate flux with ln VCR for hollow fibers with different MWCO (the full line represents Eq. 10-12).

HOLLOW FIBER ULTRAFILTRATION OF APPLE JUICE

As previously mentioned, during the HFUF of apple juice, pectin and other large solutes are brought to the membrane surface by convective transport. Consolidation of this gel layer on the HFUF membrane has a drastic effect on the performance of the operation. In order to understand the phenomena that govern the gel layer, VCR as well as Cg concentration and width evolution with time were studied using transmission electron microscopy (TEM) and a lab HFUF unit with a single hollow fiber (cut-off 50,000) operated in batch mode (Fig. 10-11). Assayed fibers were cut in rings at determined distances from the inlet, fixed with OsO_4 and ruthenium red, and then dehydrated and embedded with Spurr resin. Fiber rings were finally cut with a LKB microtome and observed in a TEM JEOL 100 CXL microscope. The effect of VCR was determined over the 0–1.6 Kg/cm^2 range of ΔPTM.

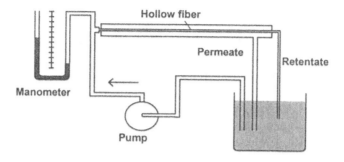

Figure 10-11 Single fiber UF system.

Micrographic Study Results

Figure 10-12 shows a scanning electron micrograph (SEM) of a HFUF fiber clearly revealing the polysulfone skin (approx. 5 μm width) where filtration occurs. Obtained TEM micrographs show a cut of the inside surface of the filtration skin under different running conditions such as with clean fiber, after pectin filtration, and after enzyme immobilization. Some interesting conclusions from this study are that: 1) during the HFUF of apple juice a pectin gel layer was effectively developed for VCR ≥ 3 and any other increase with VCR (Figs. 10-13a–d), and 2) the solute concentration in the C_g was in the range predicted by the FT (0.02–0.2 g/ml) for Q_r values in the 10–15 L/min range and gel layer width in the 300–700 nm range.

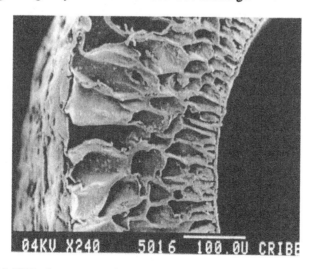

Figure 10-12 SEM of an asymmetric polysulfone membrane cross section (ROMICON MWCO = 50,000 ds. magn. = 240 ×).

129

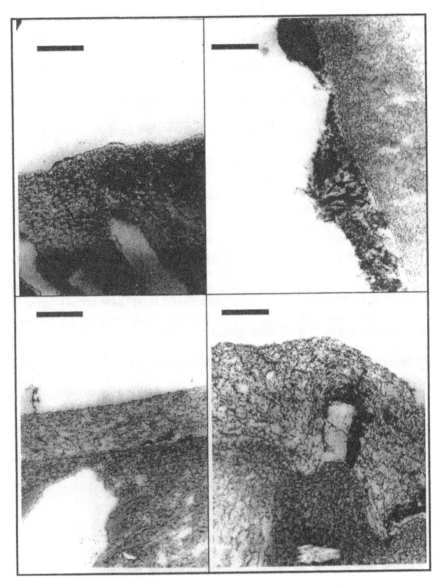

Figure 10-13 Increase in gel layer thickness with (a) VCR = 1, (b) VCR= 2, (c) VCR = 3, and (d) VCR = 4 (δg = 7000 Å, bar = 0.2 μm).

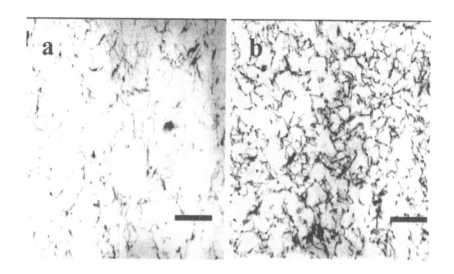

Figure 10-14 TEM of different pectin gels: (a) $Cg = 0.017$ g/mL and (b) $Cg = 0.20$ g/mL (bar = 0.2 μm).

Cheryan (1986) listed C_g values obtained by UF of different systems (gelatin, blood, milk, etc.) via calculation by extrapolation of J (linear zone) as a function of the retentate concentration. These values were in the 0.07 g/mL range for protein complex, and up to 0.6 g/mL in the case of human blood serum. The gelatin gel layer concentration was reported at $C_g = 0.25$ g/mL.

Though a pectinase treatment completely removed the gel, the permeability of the membranes was not fully recovered. Direct flow-through cleaning resulted in a compaction and permeability reduction of the asymmetric membrane (Fig. 10-15). However, reverse flow-through cleaning (back-flushing) removed the gel layer and recovered the porous structure of the fiber, (Fig. 10-16) which showed a clear correlation with the macroscopic structure of the membrane.

Additional results were achieved by physically immobilizing the commercial pectinase (Rohapect D5S) on the membrane in order to hydrolyze the gel layer. A primary enzyme adsorption layer was produced (Fig. 10-17) which led to another resistance due to immobilization of the enzyme on the membrane skin. After several hours of UF, the total resistance due to both the enzymes and pectin made the gel layer thicken by solute transport from the diffusive layer and diminish from the pectin digested away by the immobilized enzyme (Fig. 10-18).

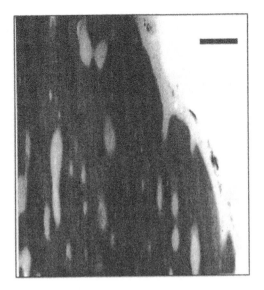

Figure 10-15 Membrane after direct flow-through cleaning (bar = 1 µm).

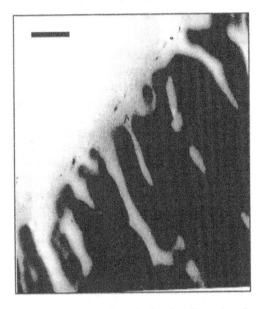

Figure 10-16 Membrane after reverse-through cleaning (bar = 1 µm).

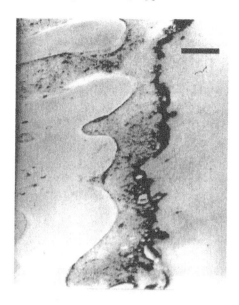

Figure 10-17 Membrane skin after enzyme immobilization (bar = 1 μm).

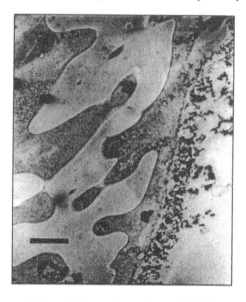

Figure 10-18 Membrane with immobilized pectinase after pectin UF (bar = 1 μm).

REFERENCES

Cheryan M. (1986). *Ultrafiltration Handbook.* Lancaster, PA: Technomic Publishers.

Chiang B. and Cheryan M. (1986). Ultrafiltration of skim milk in hollow fibers. *J. Food Sci.* 51(2): 340–344.

Chiang B.H. and Cheryan M. (1987). Modeling of hollow fiber ultrafiltration of skim milk under mass-transfer limiting conditions. *J. Food Eng.* 6: 241–255.

Constenla D.T. and Lozano J.E. (1996). Predicting stationary permeate flux in the ultrafiltration of apple juice. *Lebens.-Wissen. Technol.* 29: 587–592.

Heatherbell D.A., Short J.L., and Stauebi P. (1977). Apple juice clarification by ultrafiltration. *Confructa.* 22: 157–169.

Iritani E., Hayashi T., and Murase T. (1991) Analysis of filtration mechanism of crossflow upward and downward ultrafiltration. *J. Chem. Eng. Japan* 24(1): 39–44.

Koltuniewicz A. (1992) Predicting permeate flux in ultrafiltration on the basis of surface renewal concept. *J. Memb. Sci.* 68: 107–118.

Mietton-Peuchot M., Milisic V., and Ben Aim, R. (1984) Microfiltration tangentielle des boissons. *Le Lait.* 64: 121–128.

Omosaiye O., Cheryan M., and Matthews M. (1978). Removal of oligosaccharides from soybean water extracts by ultrafiltration. *J. Food Sci.* 43: 35–358.

Porter M. (1972) Concentration polarization with membrane ultrafiltration. *Indus. Eng. Chem. Prod. Res. Dev.* 11(3): 234–248.

Probstein R., Leung W., and Alliance Y. (1979). Determination of diffusivity and gel concentration in macromolecular solutions by ultrafiltration. *J. Phys. Chem.* 83(9): 1228–1236.

Rautembach R. and Albrecht R. (1989). *Membrane Processes, 2nd ed.* New York: John Wiley & Sons, p. 78.

CHAPTER 11

Simulation of Drying Rates and Quality Changes During the Dehydration of Foodstuffs

G.H. Crapiste

INTRODUCTION

Air drying has been used worldwide for centuries to preserve food and agricultural products. At present, there is an increasing demand for high-quality shelf-stable dried fruits and vegetables. Drying equipment and operating conditions for these products must be selected and designed on the basis of raw material characteristics, quality requirements on the final dry product, and economic analysis. Despite these considerations, application of mathematical modeling and computer simulation in dryer design and optimization is still rather scarce because of the complex nature of the materials and difficulty in quantitatively representing food quality changes during drying.

Food drying is a complex process involving simultaneous mass and energy transport phenomena. Understanding of different mass and heat transfer mechanisms inside and between the targeted food and drying air, as well as knowledge of thermophysical, equilibrium, and transport properties of both systems are required to model and simulate the process (Bruin and Luyben, 1980; Okos et al., 1992; Crapiste and Rotstein, 1997). During drying, food materials undergo physical, chemical, and biological changes that can affect quality attributes like texture, color, flavor, and nutritional value. The kinetics of these phenomena which are complex functions of temperature, moisture content, food composition, and several other factors, have not been sufficiently studied. Consequently, there is a lack of simulation and optimization in food drying from a quality standpoint, which in most cases is restricted to individual pieces or a simple thin layer bed (Mishkin et al., 1983; Lee and Pyun, 1993; Coonce et al., 1993; Pezzutti and Crapiste, 1996). The objective of this chapter is thus to review some fundamental topics in modeling and simulation of fruit and vegetable drying in packed-beds, with particular emphasis on changes in product quality during the process.

QUALITY CHANGES IN FOODS DURING DRYING

The quality of dried foods depends not only on the drying process but the quality of raw materials, pretreatment processes (i.e., blanching, chemical treatment, freezing) and storage conditions of the final product. Some comprehensive overviews of the influence of processing on food quality and the most common degradation reactions are given by Priestley (1979), Karmas and Harris (1988), Aguilera and Stanley (1990) and Villota and Hawkes (1992). Changes in food materials due to dehydration have been reviewed by Bruin and Luyben (1980) and Okos et al. (1992). The major effects of dehydration that cause quality degradation in fruits and vegetables are summarized in Table 11-1.

Physical and Structural Changes

The acceptability of dried products depends mainly on their structural properties since texture is one of the attributes used in judging their quality. Shrinkage, destruction of the cellular system, and loss of volatile aroma components are the most important physical and structural changes that occur during the drying processes. Rehydration is maximized when structural disruption at the cellular level and shrinkage are minimized, as in freeze-drying.

Shrinkage of food particles during drying has been extensively measured and modeled (Lozano et al., 1983; Ratti, 1994; Zogzas et al., 1994), but information on food shrinkage in packed beds is rather scarce (Ratti and Mujumdar, 1995). Changes in the shrinkage and porosity of apple and potato slices for both individual particles and packed beds are shown in Figure 11-1, where shrinkage is related to water loss, geometry, and the drying rate. A nearly linear correlation with moisture content is obtained except at very low moistures.

Table 11-1 Quality changes in foods during drying.

Type	Factor	Quality Effect
Physical and structural	Shrinkage	Volume, texture, rehydration ability
	Cell structure damage	Texture, rehydration ability, solute loss
	Volatile retention	Aroma loss
Chemical and organoleptic	Browning reactions	Darkening, off-flavor development
	Lipid oxidation	Rancidity, off-flavor development
	Pigment degradation	Color loss
	Enzyme inactivation	Flavor and pungency loss
Nutritional	Microbial death	Microbial survival
	Protein denaturation	Loss of biological value
	Vitamin degradation	Loss of nutritive value

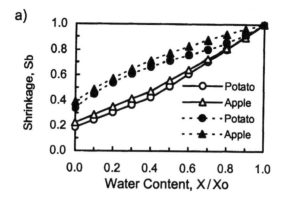

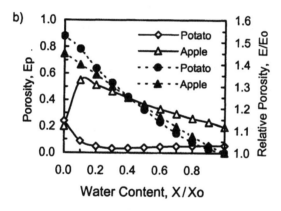

Figure 11-1 Shrinkage (a) and porosity changes (b) of apple and potato during drying.

A total volume reduction of 75–85% is usually obtained when drying fruits and vegetables. With the exception of high porous tissues such as apples, porosity changes smoothly with drying, indicating that the structural shrinkage is more important than the formation of air-filled pores. At low moisture contents where shrinkage decreases or the structure collapses, a significant increase or decrease in porosity is observed (Lozano et al., 1983). Regardless of any of these conditions, bed porosity increases continuously with drying due to shrinkage and deformation of particles.

Some major effects of heating and dehydration are breakdown and degradation of the middle lamella, separation and rupture of cell walls, lysis of cytoplasm, loss of membrane functionality, and structural collapse of cells. Most of these are associated with physicochemical changes such as cellulose crystallization, pectin depolymerization and solubilization, protein denaturation, and starch gelatinization which cause loss of water-holding capacity, rehydration ability, soluble solids, and textural and mechanical properties (particularly flexibility and strength). These phenomena are very difficult to model and so few attempts have been made to represent their rates in terms of drying conditions. However, Wenz and Crapiste (1996) were able to demonstrate the effect of heating and drying on the degradation of apple and potato tissue membrane semipermeability by measuring with lixiviation of potassium (Fig. 11-2). They determined that the destruction of cells by heat treatment in these products follows a pseudo first-order kinetics.

Chemical and Organoleptic Changes

Organoleptic quality is severely affected during drying since color and flavor are the most important sensory attributes of dried foods. The development of color due to enzymatic and non-enzymatic browning is one of the major problems that occurs during the processing and storage of dehydrated foods. Browning reactions also produce off-flavor development and loss of nutritional value. Enzymatic browning is due to the effect of enzymes that catalyze the hydroxylation and oxidation of phenolic compounds. Nonenzymatic browning is a Maillard reaction involving amino groups and reducing sugars that result in brown polymeric pigments. The rates of both types are strongly influenced by composition, temperature, water activity, and pH.

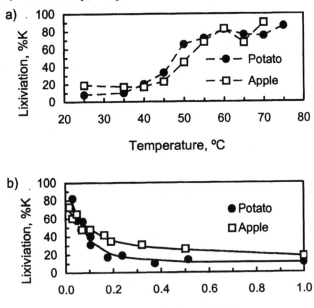

Figure 11-2 Effect of heating and drying on cell structure damage.

The development of color in different foods has been extensively studied. Although several mathematical models have been proposed, in most cases a zero-order reaction with or without an induction period or a first-order reaction has been found to describe the kinetics of browning. The effects of temperature and moisture content on the reaction rate constant for color changes in onion and garlic that follow pseudo first-order kinetics (Pezzutti and Crapiste, 1997a) are shown in Figure 11-3.

Chemical and thermal treatments such as sulfating or blanching are used to control enzymatic browning by enzyme inactivation in fruits and vegetables. However, thermal treatment is not recommended in products such as onion and garlic where some desirable enzymatic reactions involved in the production of the characteristic flavor and pungency are required. The enzyme inactivation reactions and loss of pungency can be modeled by first-order kinetics (Bruin and Luyben, 1980; Pezzutti and Crapiste, 1996).

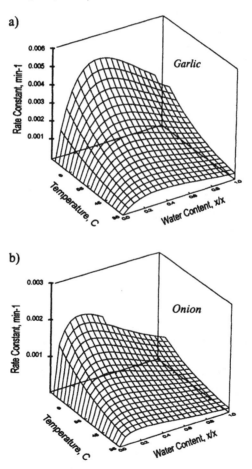

Figure 11-3 Reaction rate constants for color change during heating and drying.

Rancidity and off-flavor development due to lipid oxidation and color loss due to pigment degradation are also important causes of quality deterioration. However, several studies indicate that these changes occur mainly during storage (especially in dehydrated foods) rather than during the drying process. Lipid oxidation is difficult to model since the reaction is very complex (i.e., in steps and autocatalytic) and depends on temperature, water activity, food and lipid composition, ultraviolet light, oxygen concentration, presence of metals, natural antioxidants, and chelating agents.

Since the stability of pigments in foods is affected by many factors, their degradation follows complex mechanisms, which do not follow the same kinetics. However, most important degradation reactions such as chlorophyll, anthocyanins, and carotenoids have been reported to follow pseudo first-order kinetics (Bruin and Luyben, 1980; Villota and Hawkes, 1992).

Nutritional Changes

Microbiological aspects are of great importance to food dehydration from a quality and stability standpoint. The basic objective of the process is to reduce water activity to a level at which microbial spoilage is minimized in order to extend the shelf-life of the product. Although dehydration inhibits the growth of microorganisms, part of the initial load may survive the process and cause problems during storage or after rehydration. The mathematical modeling of microorganism inactivation during drying is usually described with a thermal inactivation mechanism since the rate of thermal death follows a first-order kinetics with a temperature and moisture-dependent reaction rate constant (Bruin and Luyben, 1980).

Loss of nutritional quality is mainly due to the effect of temperature and dehydration on vitamins and proteins, but most of the research has focused only on the former. The stability and degradation rates of vitamins such as ascorbic acid, thiamin, and riboflavin are highly affected by temperature, water content, pH, ionic strength, and metal traces. Because it is particularly heat sensitive at high moisture contents, heating and oxidation may destroy considerable amounts of ascorbic acid during the drying of fruits and vegetables. Various models and mechanisms have been used to describe the degradation of vitamins during the processing and storage of foods, but the destruction rates are often considered to be of first order. Kinetic parameters for vitamin destruction in food products have been reviewed by Villota and Hawkes (1992), though it is often assumed that the loss of biological value due to protein degradation during drying is not a major problem (Okos et al., 1992).

MATHEMATICAL MODELING

A schematic view of the drying problem and governing differential equations that represent mass and energy transfer in packed beds are presented in Table 11-2 for batch drying and Table 11-3 for continuous through-flow drying.

Table 11-2 Model for packed beds (batch drying).

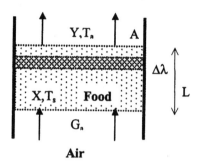

Mass Balances

Food
$$\left[\frac{\partial X}{\partial t}\right]_\lambda = -\frac{n_w a_v}{\rho_s(1-\varepsilon)}$$

Air
$$\left[\frac{\partial Y}{\partial t}\right]_\lambda = \frac{n_w a_v}{\rho_a \varepsilon} - \frac{1}{AL_0}\frac{G_a}{\rho_a \varepsilon}\frac{\rho_s(1-\varepsilon)}{\rho_{so}(1-\varepsilon_o)}\frac{\partial Y}{\partial \lambda}$$

Energy Balances

Food
$$\left[\frac{\partial T_s}{\partial t}\right]_\lambda = \frac{a_v}{\rho_s(1-\varepsilon)Cp_s}\left[h_g(T_a - T_s) - n_w \Delta H_s\right]$$

Air
$$\left[\frac{\partial T_a}{\partial t}\right]_\lambda = -\frac{h_g a_v}{\rho_a \varepsilon Cp_a}(T_a - T_s) - \frac{1}{AL_0}\frac{G_a}{\rho_a \varepsilon}\frac{\rho_s(1-\varepsilon)}{\rho_{so}(1-\varepsilon_o)}\frac{\partial T_a}{\partial \lambda}$$

Initial and Boundary Conditions

The mathematical models involve mass and energy balances for air and product streams along with the kinetics of the coupled mass and heat transfer phenomena taking place in the bed. The governing differential balance equations are expressed in the moving coordinate λ, with the term $\rho_s (1 - \varepsilon) / \rho_{so} (1 - \varepsilon_o)$ accounting for bed shrinkage during drying.

Table 11-3 Model for packed beds (continuous through-flow drying).

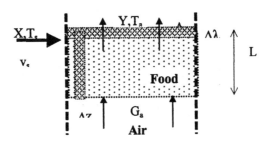

Mass Balances

Food $\qquad \left[\dfrac{\partial X}{\partial z}\right]_\lambda = -\dfrac{n_w a_v}{\rho_s (1-\varepsilon) v_s}$

Air $\qquad \left[\dfrac{\partial Y}{\partial z}\right]_\lambda = \dfrac{n_w a_v (1-\varepsilon)}{\rho_a \varepsilon v_s} - \dfrac{1}{AL_0}\dfrac{G_a}{\rho_a \varepsilon v_s}\dfrac{\rho_s(1-\varepsilon)}{\rho_{so}(1-\varepsilon_0)}\dfrac{\partial Y}{\partial \lambda}$

Energy Balances

Food $\qquad \left[\dfrac{\partial T_s}{\partial z}\right]_\lambda = \dfrac{a_v}{\rho_s(1-\varepsilon) v_s Cp_s}\left[h_g(T_a - T_s) - n_w \Delta H_s\right]$

Air $\qquad \left[\dfrac{\partial T_a}{\partial z}\right]_\lambda = -\dfrac{h_g a_v}{\rho_a \varepsilon v_s Cp_a}(T_a - T_s) - \dfrac{1}{AL_0}\dfrac{G_a}{\rho_a \varepsilon v_s}\dfrac{\rho_s(1-\varepsilon)}{\rho_{so}(1-\varepsilon_0)}\dfrac{\partial T_a}{\partial \lambda}$

Boundary Conditions

The most important information needed for dryer modeling and simulation—including thermophysical and equilibrium properties of moist air and food products as well as heat and mass transport coefficients—is summarized in Table 11-4. The evaluation of thermophysical and equilibrium properties as a function of temperature, water content, and drying conditions has been discussed extensively in previous works (Ratti and Crapiste, 1995; Ratti and Mujumdar, 1995; Pezzutti and Crapiste, 1996, 1997b; Crapiste and Rotstein, 1997).

Drying calculations are based on knowledge of the drying kinetics of food particles. Based on material behavior, one of the methods given in Table 11-4 have been used to estimate the drying kinetics of fruits and vegetables:

Table 11-4 Properties and kinetics used in drying models for packed beds.

Food and Air Properties

Sorptional equilibrium	$a_w, \Delta H_s$
Thermophysical	$\rho_s, S_p, Cp_s, \rho_a, Cp_a$
Bed	$a_v, \varepsilon, S_{bed}$
Convective transfer coefficients	h_g, k_g

Drying Kinetics

Diffusive model

$$\frac{\partial X}{\partial t} = \nabla(D_{eff} \nabla X);$$

$$D_{eff} = D_o(X) \exp\left[-\frac{\Delta E_d(X)}{RT}\right]$$

Characteristic drying curve

$$n_w = \frac{k_g\left[a_w p_{w\infty}(T_s) - p_{w\infty}\right]}{\left[1 + (\Phi / X_0)Bi_m\right]}; \quad \Phi = \Phi(X)$$

Kinetics of Quality Deterioration

Cell damage (T effect)

$$\frac{dN_c}{dt} = -k N_C; \quad k = k_o \exp(-\Delta E / RT)$$

Color change

$$\frac{d DE}{dt} = K_c(DE_f - DE);$$

$$K_c = K_0(X) \exp\left[-\Delta E_c(X) / RT\right]$$

Flavor (pungency) loss

$$\frac{dP_E}{dt} = -k_E(T) P_E;$$

$$\frac{dP_T}{dt} = \left[-k_E(T) + k_C(T)\right] P_E$$

1. A diffusive model in terms of an effective transport coefficient depending on temperature, water content, and shrinkage (Pezzutti and Crapiste, 1997a)
2. A characteristic drying curve model in terms of a generalized drying parameter depending on water content (Ratti and Crapiste, 1992)

The overall model is completed with the kinetics of quality deterioration. In selecting quality factors to evaluate the drying process, consideration must be given to the importance of the attribute and magnitude of changes for the particular food under study. As is shown in Table 11-4, changes in color and cell damage or loss of flavor strength assessed in terms of total and enzymatically produced pyruvic acid were considered in this case (Wenz and Crapiste, 1996; Pezzutti and Crapiste, 1996, 1997b).

SIMULATION RESULTS

When batch and continuous through-flow packed bed dryers were simulated, the initial conditions for batch drying and boundary conditions for continuous drying were assumed to be constant temperature, water content, and quality attributes for the whole bed. A uniform distribution of air in the bed cross-sectional area was also assumed. The mathematical model for the process consists of a set of partial differential and algebraic equations that must be solved simultaneously. This was achieved by using finite differences for the spatial derivatives in the direction of the airflow (λ) and the method described by Gear to integrate the resulting ordinary differential equations with respect to time (batch drying) or z (continuous drying).

The simulation of apple, potato, onion, and garlic drying allowed the effects of batch or continuous drying, air flow rate, air temperature, relative humidity, particle size, and bed depth to be studied. Figure 11-4 shows some results of the simulation for continuous drying of a packed bed of potato slices at the bottom and top of the bed where the maximum and minimum of the variables are usually located. Prediction of moisture content, temperature, color, and flavor strength profiles developed during the batch drying of garlic in packed beds is shown in Figure 11-5. The same modeling technique can be used to estimate changes in other quality factors during drying if their kinetics are known.

The results of such experiments imply a need to optimize the drying process for quality factors. As such simulation models can be useful tools for predicting the optimum air conditions needed to dry products to a required moisture content in a reasonable processing time while still maintaining acceptable quality attributes. The models can also be used to optimize the process to achieve minimum deterioration given the constraints imposed by product specifications, equipment characteristics, and economic considerations.

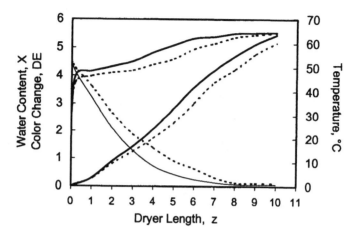

Figure 11-4 Drying curves, temperature profiles, and color changes at two bed locations for the drying of potato slices (G_a = 0.015 kg/s, T_a = 65°C).

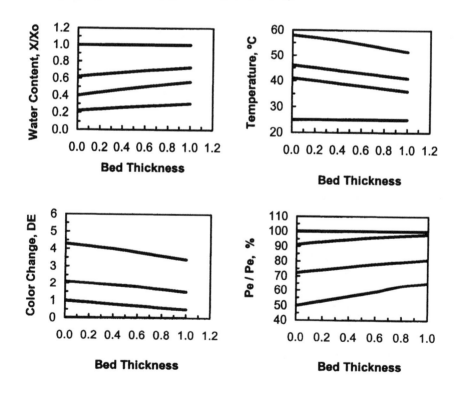

Figure 11-5 Local profiles during the batch drying of garlic in packed beds.

CONCLUSIONS

A review of the major causes of quality degradation during fruit and vegetable dehydration reveals that the complex nature of these materials makes it very difficult to represent physical and structural changes in a quantitative form. This is not surprising since most reactions causing organoleptic and nutritional changes in foods follow complex mechanisms and depend on several factors. In many cases such changes can be represented by pseudo zero or first order kinetics since the reaction rate constant depends mainly on temperature and water content, but specific experimental information for the product under study is required.

Complete models to simulate the dehydration of food pieces in packed beds were presented for batch and continuous drying. These models were based on differential mass and energy balances in the dryer along with kinetic equations describing transport phenomena in individual particles and variations in product quality. The models were solved numerically in order to obtain the profiles of moisture content, temperature, and quality attributes as a function of time or dryer position for different food products and drying conditions. It was thus shown that modeling and simulation can be used to select, design, and optimize dryers for fruits and vegetables since they are able to take into account major quality losses during the process.

NOTATION

a_w	water activity
A	transversal area of the dryer section, m^2
A_s	external area, m^2
a_v	area per unit volume, m^2/m^3
Bi_m	mass transfer Biot number
Cp_a, Cp_s	specific heat of air and product on dry basis, J/kg K
D_{eff}	effective diffusion coefficient, m^2/s
DE, DE_f	color change
G_a	dry air mass flow rate, kg/s
h_g	heat transfer coefficients, J/m^2 s K
k	pseudo first order reaction rate constant, s^{-1}
k_C, k_E	reaction rate constants for flavor loss, s^{-1}
k_g	mass transfer coefficients, kg/m^2 s kPa
K_c	reaction rate constant for color change, s^{-1}
L, L_o	bed depth, m
m_s	mass of dry solid, kg
N_C	number of undamaged cells
n_w	drying rate, kg/m^2s
P_E, P_T	enzymatic and total pyruvic acid content, g/kg
p_v	water vapor partial pressure, kPa
R	gas constant, J / kmol K
S_p, S_{bed}	particle and bed shrinkage coefficients

t	time, s
T_a, T_s	air and product temperature, °K
v_s	product velocity, m/s
X	product water content, kg/kg
Y	air absolute humidity, kg/kg
Z	coordinate, m
ΔE, ΔE_c	activation energy for reaction, J/kmol
ΔEd	activation energy for the diffusion process, J/kmol
ΔHs	heat of sorption, J/kg K
ε	bed porosity
Φ	generalized drying parameter
λ	dimensionless coordinate following shrinkage
ρ_a, ρ_s	air and dry solid density, kg/m^3

REFERENCES

Aguilera J.M., Chirife J., Flink J.M., and Karel M. (1990). Computer simulation of non-enzymatic browning during potato dehydration. *Lebensm.-Wiss. U.-Technol.* 8: 128–133.

Aguilera J.M. and Stanley D.W. (1990). *Microstructural Principles of Food Processing and Engineering.* Essex, UK: Elsevier Applied Science.

Bruin S. and Luyben K. (1980). Drying of food materials: A review of recent developments. In: *Advances in Drying, Vol. 1.* Mujumdar A.S., Ed. Washington, DC: Hemisphere Publishing, pp. 155–215.

Coonce V.M., Levien K.L., and Torres J.A. (1993). Mathematical models for drying rates and quality indicator changes during barley malt kilning. *Cereal Food World* 38(11): 822–830.

Crapiste G.H. and Rotstein E. (1997). Design and performance evaluation of dryers. In: *Handbook of Food Engineering Practice.* Valentas K.J., Rotstein E., and Singh R.P., Eds. Boca Ratón: CRC Press, pp. 125–166.

Karmas E. and Harris R.S. (1988). *Nutritional Evaluation of Food Processing, 3rd ed.* New York: Van Nostrand Reinhold.

Lee D.S. and Pyun Y.R. (1993). Optimization of operating conditions in tunnel drying of food. *Drying Technol.* 11(5): 1025–1052.

Lozano J.E., Rotstein E., and Urbicain M.J. (1983). Shrinkage, porosity and bulk density of foodstuffs at changing moisture contents. *J. Food Sci.* 48: 1497–1502.

Mishkin M., Saguy I., and Karel M. (1983) Dynamic optimization of dehydration processes: minimizing browning in dehydration of potatoes. *J. Food Sci.* 48: 1617–1621.

Okos M.R., Narsimhan G., Singh R.K., and Weitnauer A.C. (1992). Food dehydration. In: *Handbook of Food Engineering.* Heldman D.R. and Lund D.B., Eds. New York: Marcel Dekker Inc., pp. 437–562.

Pezzutti A. and Crapiste G.H. (1996). Modelado del deterioro organoléptico en la deshidratación de ajo y cebolla. In: *Proceedings of the First Latinoamerican Congress on Food Engineering, Vol. II.* Valencia, Spain: Universidad Politécnica de Valencia. pp. 290–300.

Pezzutti A. and Crapiste G.H. (1997a). Color change during dehydration of onion (Allium cepa L) and garlic (Allium sativum L). *Acta Hortic.* 433: 455–462.

Pezzutti A. and Crapiste G.H. (1997b). Sorptional equilibrium and drying characteristic of garlic. *J. Food Eng.* 31: 113–123.

Priestley R.J. (1979). *Effects of Heating on Foodstuffs.* London: Applied Science Publishers.

Ratti C. (1994). Shrinkage during drying of foodstuffs. *J Food Eng.* 23: 91–105.

Ratti C. and Crapiste G.H. (1992). A generalized drying curve for shrinking food materials. In: *Drying'92.* Mujumdar A.S., Ed. Amsterdam: Elsevier Science, pp. 864–873.

Ratti C. and Crapiste G.H. (1995). Determination of heat transfer coefficients during drying of foodstuffs. *J. Food Proc Eng* 18: 41–53.

Ratti C. and Mujumdar A.S. (1995). Simulation of packed bed drying of foodstuffs with airflow reversal. *J. Food Eng.* 26: 259–271.

Villota R. and Hawkes J.G. (1992). Reaction kinetics in food systems. In: *Handbook of Food Engineering.* Heldman D.R. and Lund D.B., Eds. New York: Marcel Dekker, pp. 39–144.

Wenz J.J. and Crapiste G.H. (1996). Deterioro celular en el procesamiento de productos frutihortícolas. In: *Proceedings of the First Latinoamerican Congress on Food Engineering, Vol. II.* Valencia, Spain: Universidad Politécnica de Valencia. pp. 327–336.

Zogzas N.P., Maroulis Z.B., and Marinos-Kouris D. (1994). Densities, shrinkage and porosity of some vegetables during air drying. *Drying Technol.* 12: 1653–1666.

CHAPTER 12

Vacuum Impregnation in Fruit Processing

P. Fito, A. Chiralt, J.M. Barat, and J. Martinez-Monzó

INTRODUCTION

Food technology frequently deals with fruit-fluid systems (FFS). Table 12-1 shows some examples where a fruit tissue is put in contact with a fluid (liquid or gas) phase in fruit preservation processes such as drying, fermentation, thermal processing, packaging, and storage. The heat and mass transfer processes in such systems have usually been modeled considering the food solid (fruit) as a continuous phase. Nevertheless, tissue cellular structure also plays an important role in the definition of mechanisms involved in the process (Aguilera and Stanley, 1990).

In relation to the fruit structure role, a fast mass transfer mechanism (hydrodynamic mechanism: HDM) has been described as occurring in process operations where a porous solid is immersed in a liquid phase and changes in temperature or pressure take place (Fito, 1994; Fito and Pastor, 1994; Fito et al., 1994a). The occluded gas inside the product pores is compressed or expanded according to the pressure or temperature changes, while the external liquid is pumped into the pores in line with the gas compression. An effective exchange of the product's internal gas for the external liquid was found to be promoted in vacuum impregnation (VI) operations, where a vacuum pressure ($p_1 \sim 50\text{--}100$ mbar) is imposed on the system for a short time (t_1), and the product remains immersed in the liquid for a time t_2 after the atmospheric pressure (p_2) is reestablished (Fito et al., 1996; Salvatori et al., 1998a). The volume fraction of a sample (X) impregnated by external liquid when mechanical equilibrium was

Table 12-1 Some fruit-fluid systems (FFS) in fruit processing.

Fruit-Liquid Systems	Fruit-Gas Systems
Canned fruits	Air dehydration
Fruits in dairy products	Chilling
Osmotic dehydration (OD)	Freezing-Thawing
Re-hydration	Modified atmosphere packaging
Boiling and cooking	Edible film applications
Candied fruits	Smoking

achieved has been modeled as a function of the compression ratio ($r \sim p_2/p_1$), sample effective porosity (ε_e), and sample volume deformations at the end of the process (γ) and the vacuum step (γ_1) as described in Equation 12-1 (Fito et al., 1996). If $\gamma = \gamma_1 = 0$, Equation 12-1 gives the VI relationship for stiff products.

$$\varepsilon_e = \frac{(X - \gamma)r + \gamma_1}{r - 1} \qquad (12\text{-}1)$$

The possibility of introducing an external solution with specific/selected solutes inside product pores makes VI a viable tool in fruit processing. The addition of preservatives (anti-microbials, anti-browning agents, pH reducers), fast water activity depression, and modifying thermal properties, are some of the possible applications of VI (Chiralt et al., 1999). Impregnation of fruit pores has been seen to occur without vacuum action when fruit remains immersed in a liquid phase for a long time (e.g., syrup, canned, and candied fruits) due to the capillary forces and pressure and temperature fluctuations in the system (Barat et al., 1998). Gas-liquid exchange promotes a structural change in the fruit tissue at the same time as chemical and physical modifications may also occur. Therefore, fruit behavior in different processes may be greatly affected by a previous VI with a specific solution.

The aim of this chapter is to analyze the effect of VI on the physical and structural properties of fruit as a function of the solute concentration of the external solution. The influence of VI pretreatment on fruit development throughout osmotic processes for short (such as minimally processed) and long (candies) treatments is also discussed.

VI CHANGES IN FRUIT

The effect of VI on the physical and structural properties of fruits has been studied in several products (Salvatori et al., 1998a; Martínez-Monzó et al., 1996, 1998) by introducing hypotonic, isotonic, and hypertonic solutions in fruit tissue. The greater the fruit porosity, the greater the effectiveness of VI to promote the required change, so apples were chosen as a useful model to analyze the changes provoked by VI.

Composition Changes

Figure 12-1a shows the development in composition (water and soluble solids) water activity and freezable water (x_{fw}) of Granny Smith apples as a function of an external solution's sugar concentration when the fruit was VI. These values corresponded to VI experiments ($t_1 = 15$ in, $t_2 = 15$ min) using rectified grape most as external solution in apple samples. Nevertheless, the mass fraction of any component i (water or solutes) reached in a VI product (x^{VI}_i) can be estimated from Equation 12-2, which is deduced from a mass balance in a system in terms of the initial composition (x_i^0) of a sample, the mass fraction of the impregnated solution (x_{HDM}), and its composition (y_i). The x_{HDM} value can be obtained from the product's VI response (impregnated volume fraction: X) and initial product (ρ^0) and solution (ρ^{IS}) densities (Eq. 12-3). From these equations, the required solution concentration of a determined component (water, sugar, acid, additive) can be calculated to achieve the desired final level in the product.

$$x_i^{VI} = \frac{x_i^0 + x_{HDM} \, y_i}{1 + x_{HDM}} \qquad (12\text{-}2)$$

$$x_{HDM} = X \frac{\rho^{IS}}{\rho^0} \qquad (12\text{-}3)$$

Concentration changes promoted by concentration gradients were not taken into account in Equation 12-2 due to the short time used in VI. Nevertheless, when driven forces dependent on concentration are high (or VI times are long), deviations of the predicted final concentrations can be found (Fito and Chiralt, 1997; Chiralt et al., 1999).

Physical Property Changes in Fruit

Relative changes in the mechanical and thermal properties, as well as density and color of apple samples as promoted by VI were evaluated (Barat et al., 1997; Martínez Monzó et al., 1996, 1997, 1998) as a function of their external solution (hypo-, iso- and hypertonic) concentrations. The relative change in physical properties (RCPP) was defined (in percentage) as the value in the impregnated sample minus the value in the fresh one, divided by the latter.

Stress relaxation tests were performed to analyze mechanical behavior. Figure 12-1b shows the RCPP values for the apparent elasticity modulus, velocity, and total relative level of the stress relaxation. When isotonic solutions were used (cell turgor unaltered), no significant differences in the initial (E) and asymptotic (E_a) moduli between fresh and VI apples were found. Nevertheless, the relaxation rate (B) and total relaxation level (A) increased in the VI samples in relation to the impregnation degree (X). These results led to the changes in the viscoelastic behavior of the isotonic VI samples principally being attributed to the exchange of gas (compressible during the mechanical test) for liquid which flowed out from the pores throughout compression (Martínez Monzó et al., 1996). When hypertonic solutions are used for VI, sample osmotic dehydration (OD) simultaneously occurs. This contributes to changes in the chemical and physical properties of a product, promoting turgor losses and complete loss of cell elasticity after plasmolysis. The apparent elastic modulus thus decreases sharply, increasing the viscous character. VI with hypotonic solutions only implies a greater level of stress relaxation, as can be explained by an outflow of the intracellular liquid corresponding to the cell rupture promoted by excessive turgor (Pitt, 1992).

Optical properties also change due to VI. The gas-liquid exchange in a fruit implies a more homogeneous refraction index through a sample, and consequently an increase in product transparency. So, when color was measured by diffuse reflection, a decrease in the reflection coefficients was obtained for VI samples as compared with fresh samples, thus implying lower values of the clarity and chrome color coordinates (L*, C_{ab}*) and small changes in hue (h_{ab}*) (Martínez-Monzó et al., 1997). The optical properties of highly porous products such as apples are the most affected by VI (Fig. 12-1c).

VI with isotonic solutions increases the thermal conductivity of porous fruits due to gas substitution, although slight modifications are produced in thermal diffusivity because of the simultaneous density (ρ_a) increase (Fig. 12-1d). Changes are greatly dependent on the total porosity and pore distribution in relation to the heat flow direction (Barat et al., 1994). An increase in the solution concentration promotes an expected reduction in all thermal properties as compared to impregnation without compositional changes.

Changes in Fruit Tissue Structure

CryoSEM observations (Bomben and King, 1982) of VI samples with hypo-, iso-, and hypertonic solutions were carried out in several fruits (Martínez-Monzó et al., 1998; Barat, 1998; Salvatori, 1997). When hypo- or isotonic

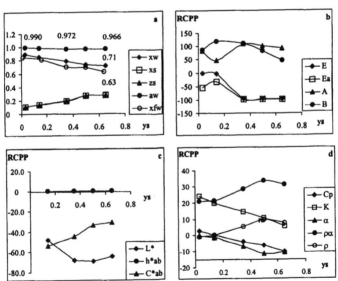

Figure 12-1 Changes in composition, freezable water, water activity, and physical properties (RCPP) promoted by VI as a function of the solute content of an impregnation solution.

solutions were used, no cellular alterations or debonding was observed in micrographs, but the extracellular spaces in the sample appeared completely flooded by the external solution. The most relevant structural change promoted by VI was thus concluded to be the filling of the product pores with the external solution. In many cases some native liquid was expelled during the vacuum step and replaced by the external solution (Salvatori, 1997). In VI hypertonic treatments, OD of the tissue promotes plasmolysis but much less cell shrinkage of the cellular wall is observed as compared to that which occurs in osmosed tissue at normal pressure (Fig. 12-2). Moreover, cell wall observations of osmosed tissue by TEM show a much better preserved cell wall ultrastructure (i.e., similar to fresh fruit texture) when VI is used to depress water activity in the minimal processing of fruits (Alzamora and Gerschenson, 1997).

HYPERTONIC SOLUTIONS

In fruit processing the more frequent fruit-liquid systems have hypertonic solutions in the liquid phase. For example, concentrated sugar solutions in OD processes are normally used to decrease a_w in minimally processed fruits and vegetables and some deep processed fruits such as candies or jam. Vacuum treatment leads to faster osmotic processes due to the coupled action of HDM and pseudo-diffusion mechanisms (PDM) (Fito, 1994; Shi et al., 1995). When VI of products with OS was carried out at the beginning of the process by Fito et al. (1994b), OD was called pulsed vacuum osmotic dehydration (PVOD). In PVOD, VI takes place during the first 5–10 min by applying a vacuum pulse in the tank. This implies a fast compositional change in the product that will affect the osmotic driving force and mass transfer kinetics (Barat et al., 1997; Fito and Chiralt, 1997). Barat et al. (1998) found that VI also affected the development of osmosed samples until the final equilibrium was reached in terms of mass, volume, density, and structure changes.

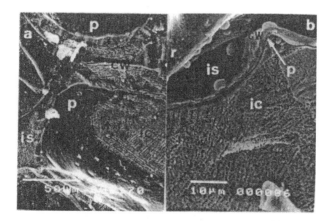

Figure 12-2 CryoSEM micrographs of apple tissue treated with hypertonic solutions (65 Brix) at 30°C for 30 min, with (a) and without (b) VI with the solution.

Figure 12-3 shows the influence of VI on the general pathway of a sample's mass and volume for a long-term osmotic process. Two-step behaviour was observed in all cases. In Step 1 at short times (24–48 h depending on process conditions), weight and volume losses occurred (principally due to the relatively fast water loss of the product) until a minimum value was achieved at a critical time (t_c). The t_c values were affected by the OS concentration, temperature, and kind of osmotic treatment (OD and PVOD); the higher the OS concentration and temperature the lower the t_c value, whereas the vacuum pulse greatly decreased t_c (Barat et al., 1998). Sample compositional changes occurred throughout Step 1, and at t_c the fruit liquid phase (FLP) had the same overall soluble solid composition as the external OS in all cases. A compositional pseudoequilibrium (Nicolis and Prigogine, 1977) situation can thus be assumed at t_c.

In Step 2 after t_c there were gains in mass and volume until asymptotic values were reached (50–100 days depending on the process conditions). It is remarkable that the asymptotic value was achieved more quickly for volume than weight in the OD processes, whereas this difference was not observed in the PVOD treatments (Fig. 12-3). In the case of PVOD the volume recovery was also slightly higher. Sample mass overcame its initial value in all experiments, which indicates that the void volume of the pores was filled by the external solution when the true equilibrium was achieved.

In agreement with the above described behavior, the development of porosity and density in apples throughout an OD process follows the pathway shown in Figure 12-4. PVOD treatments lead to near zero porosity after vacuum pulses due to the filling of sample pores with OS, while bulk density reaches a constant value when compositional equilibrium occurs at $t = t_c$, this being almost equal to the density of the OS. In OD treatments porosity greatly increases during the first period ($t < t_c$), but after that tends toward zero, due to the progressive filling of sample pores (Barat et al., 1998).

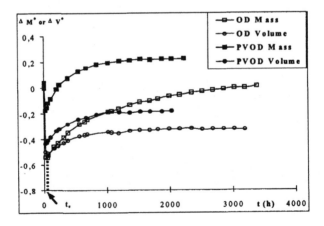

Figure 12-3 Development of volume and mass relative change in apple samples throughout long-term OD and PVOD treatments with 55 Brix OS at 40°C.

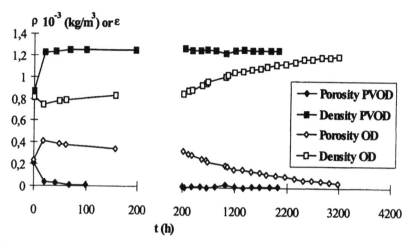

Figure 12-4 Development of density and porosity in apple tissue throughout long-term OD and PVOD treatments with 55 Brix OS at 40°C. (Please note the change of the *y* axes scale).

In the first step the sample pore impregnation in PVOD and the great water fluxes may explain the macrostructural features observed for sample mass, volume, and porosity. Nevertheless, in the second longer step other hydrodynamic mechanisms occurring because of generated pressure gradients must be taken into account to explain sample development and differences between OD and PVOD treatments.

THE WAY TO EQUILIBRIUM

The pressure gradients responsible for the weight and volume recovery in long-term osmotic processes ($t > t_c$) has been attributed to the deformation-relaxation phenomena of the shrunken cell wall matrix. In fact, shrinkage of cell walls during the first step ($t < t_c$) that is associated with water loss, implies an accumulation of free energy in the system as mechanical stress (Barat et al., 1998). Relaxation of stressed cell walls promotes bulk fluxes of OS and pressure gradients if a product remains immersed in the OS while the sample volume recovers. The consequences of these structural changes greatly affect some engineering aspects of osmotic process development such as the equilibrium criteria and final liquid retention capability of the samples.

Short Time Processes

Different cell wall behaviour was observed at the microstructural level between OD and PVOD treated samples. Figure 12-2 shows CryoSEM micrographs of a few osmosed cells (30 min treatment) located near (2–3 mm) the sample surface, where the different cell wall behaviour for the OD and PVOD treatments can be observed. Cell wall (CW) deforms, remaining bonded to plasmalemma (P), during the cell shrinkage caused by water loss in OD processes (Salvatori et al., 1998b). However, the CW separates from the plasmalemma throughout shrinkage in PVOD, and the liquid phase from the intercellular spaces (is) flows into the cell cavity (CC) through the cell wall (Martínez-Monzó et al., 1998). This different behaviour can be explained in terms of the varying friction forces (viscosity) of the fluid phase in the is when flow is promoted.

Figure 12-5 schematically shows the different pathways of the initially bonded layers CW and P as a function of the module of forces acting on each surface (S) of this double layer throughout the water loss process. In Figure 12-5 the part of CW and P beside the intercellular space [from one bonding zone (BZ) A to another B in the cell] was drawn for the cases where the layers may be considered an elastic structure. The reaction forces to the shrinkage-associated force (F_S) are the deformation resistance of the layers (F_R) plus that which is associated with the fluid pressure drop (SΔP) when it flows towards the generated volume in the is. The module of action-reaction forces on the elastic double layer in dynamic equilibrium increases in line with the overall water loss (ΔM_w) (degree of layer deformation) and water loss rate (dΔM_w/dt) (deformation rate). When is is occupied by gas (in OD), the force component SΔP is negligible and CW bonded to P deforms while the gas phase flows (flux J_{gas}) in the is to the generated void (intercellular space-generated volume: ISGV, Fig. 12-5c). On the contrary, when a liquid phase occupies the is (in PVOD), the liquid pressure drop is much greater and the F_S value quickly overcomes the cell wall-plasmalemma adherence force (F_A). In this situation the cell wall-plasmalemma separates, thus promoting the liquid flow (flux J_{liquid}) through the permeable CW towards the generated void into the cell cavity (**ccgv**) and CW withstands the deformation. The water loss rate (usually higher in PVOD than in OD) which defines the cell deformation rate will also affect the critical value of ΔM_w at which CW-P separation occurs due to the viscoelastic response of the bonding points.

Figure 12-6a,b shows cryo-SEM micrographs of osmosed cells (55 Brix OS and 40°C) at t = t_c. From the different structural response to osmotic stress in very short OD and PVOD treatments, the differences observed in the tissue microstructure may be understood. In the OD processes a shape inversion occurred in the cellular tissue at the compositional equilibrium time because of the CW shrinkage: the is became more cylindrical, whereas cells took on an extremely shrunken, irregular shape that maintained the cell bonding zones (BZ). A reduction of the cell size-pore size ratio because of the cell shrinkage can also be seen, which is coherent with the sample porosity increase in the first step of the OD treatments (Fig. 12-4). In the PVOD treatments the cell wall did not shrink with plasmalemma but did deform to some extent because of the total volume loss, and the intercellular spaces appeared partially collapsed due to polyhedral packaging of the non-turgid cells.

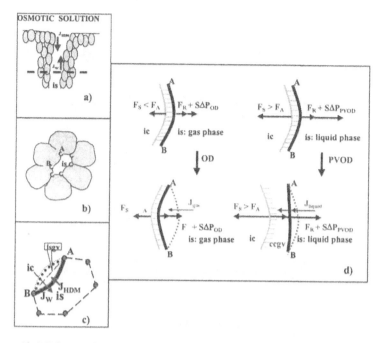

Figure 12-5 Scheme of the CW-plasmalemma ensemble deformation and debonding in OD and PVOD processes:

a) Idealized **is** space beside the interface
b) Cross-section of the **is** showing cell bonding points (A, B)
c) Deformation pathway of CW-plasmalemma ensemble between the A and B bonding points
d) Forces acting on the CW-plasmalemma ensemble leading to CW deformation (OD) or debonding (PVOD)

Long Time Processes

When water loss is stopped, the relaxation of a shrunken cell matrix is the prevailing phenomenon that provokes mass transfer in a food system, which defines its development towards minimum free energy over a wide time range (several weeks/months). In OD processes, the relaxation of cell walls implies the separation of the CW-P layers at a determined time and the suction effect on the external solution, the latter of which seems to provoke not only cell filling through the permeable cell wall (transmitted cell to cell from the interface), but also the progressive filling of the void intercellular spaces (porosity decrease) while mechanical equilibrium is being achieved. This simultaneous flow into the pores could be promoted by small temperature fluctuations and those pressure fluctuations associated with the relaxation phenomena occurring in the cells. In this sense it is remarkable that small gas expansions produced by pressure or temperature changes represent irreversible phenomena in fruit tissue, as gas

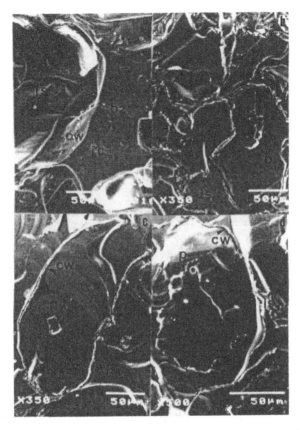

Figure 12-6 CryoSEM micrographs of apple tissue treated with OD (a and c), PVOD (b and d), and 55 °Brix OS at 40°C, t = t$_c$ (above), and t > t$_c$ (below) when the weight recovery was maximum.

immediately leaves the system due to density differences. However, in the case of PVOD treatments, sample weight gain in the second period must be attributed principally to cell refilling because of the relaxation and subsequent increase in total volume, although the rebuilding of the cellular arrangement after relaxation may also increase the extracellular volume.

Figure 12-6 (c and d) shows osmosed apple tissue at the time of maximum sample weight recovery for OD and PVOD treatments. The cell walls have recovered their initial roundness and the intercellular spaces their typical shape to quite an extent for both treatments. Likewise, in the OD samples the is appears almost completely filled with OS, and the plasmalemma and cell walls are completely separated. A greater degree of irreversibility is observed in the cell shape recovery (and therefore the sample volume recovery) in the OD osmotic treatments with highly concentrated OS. The lower the OS concentration, the higher the structural recovery (Barat, 1998). Some void is and the still deformed CW appeared in the more internal part of the OD-treated samples when high OS concentrations were used. On the contrary, PVOD-treated samples with high OS concentrations had better cell wall roundness, showing a reduced plasmalemma volume in the middle of the recovered cell volume (Fig. 12-6d).

MASS TRANSFER MECHANISMS AND DRIVING FORCES

From the above discussion the complexity of mass transfer phenomena in structured (cellular) tissues is evident. Not only are mechanisms dependent on concentration gradients responsible for water or sugar gain in OD, but so are mechanisms dependent on pressure gradients. These pressure gradients can be imposed on the food tissue system (vacuum pulses), but can also be generated by capillary forces, mechanical relaxation of deformed cell matrices, or depressions occurring inside the sample caused by volume (water) losses. These mechanisms act together in the tissue during the osmotic process, but the intensity of each one varies according to the different process times. However, in a macroscopic sample the osmotic processes develop to different degrees as a function of the distance to the sample surface as has been demonstrated in previous compositions and structural profile analyses (Salvatori et al., 1998b). Therefore, macroscopic properties of samples (such as mean composition) will contain the overall effects of several mechanisms acting in different intensity. Nevertheless, the observation of clearly separated steps in sample development throughout long-term osmotic processes reveals the different time scales in which each mass transfer mechanism acts in these cases. The above discussed micro- and macrostructural features analyzed throughout a long-term osmotic process in relation to the prevailing mass transfer mechanisms at each process time have been schematized in Figure 12-7.

The Liquid Retention Capability of a Solid Matrix

In PVOD processes, the capability of a tissue to retain its fruit liquid phase (FLP) inside a solid cellular matrix (SM) after osmotic treatments is also affected by previous VI of the sample. By using a ternary diagram the changes in system composition may be represented by considering a ternary system constituted of water, soluble solids, and insoluble compounds of the solid matrix (inert components). Figure 12-8 shows the product pathway throughout an OD process with the same active solutes in fruit and OS. Axes show water and solute concentrations while the inert concentration is the distance of any point to the hypotenuse. In Figure 12-8, Point 1 represents the raw material, and Point 2 the first pseudoequilibrium step at $t = t_e$ when the equality of concentration of the FLP and OS is achieved. Any point between 1 and 2 may be the final situation in an OD treatment. In the case of the candied fruits process, the fruits will not only reach Point 2, but will also increase the FLP/SM ratio, going from Point 2 to Point 3. The line from the origin to point OS indicates the points have the same composition in the FLP as the OS ($y_s = z_s$). Point 5 represents the

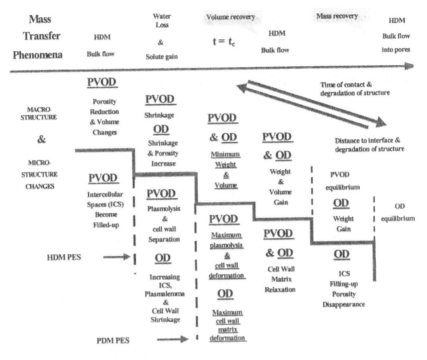

Figure 12-7 Prevailing mass transfer mechanisms as related to macro- and microstructural changes in line with process time or distance to the sample surface (interface) (PES: pseudo-equilibrium situation).

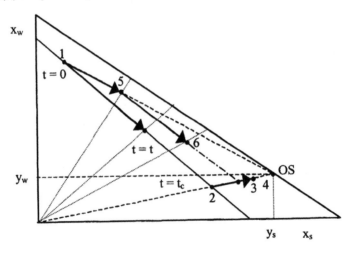

Figure 12-8 Development of product composition (mass fractions) in a ternary diagram (water-solutes-inert) throughout OD and PVOD processes. The point OS represents the OS concentration.

concentration changes produced due to a VI of Sample 1 with the OS, and Point 6 would be the result of a PVOD treatment for a relatively short time. Comparison of the trajectories 1-2-3 and 1-5-6-3 in a ternary diagram reveals the greater FLP/SM ratio achieved in the PVOD treatments for a determined FLP concentration.

PRACTICAL CRITERIA FOR PSEUDOEQUILIBRIUM SITUATION DEFINITION

The analysis of mass transfer kinetics in fruit immersed in a hypertonic solution requires the definition of driving force and therefore the equilibrium criteria. As different mechanisms are involved in mass transfer phenomena on a wide time scale, a practical approach will be useful to establish the equilibrium criteria depending on the actual length of the process and prevailing action of the different mechanisms in each case. Although coupling may occur in big samples, a practical analysis will allow the grouping of different processes on the basis of their length and the prevailing driving force acting at the end time.

Table 12-2 shows classification of different osmotic processes that take this into account. A summary of criteria for a practical definition of equilibrium is also given. Accordingly, the equality of a_w between FLP and OS is a good measure of OD, but with very long treatment times (as in the case of candied fruits) the matrix relaxation and gas release are the best determinants of the final FLP/SM ratio.

Table 12-2 Criteria for a practical definition of equilibrium for different processes in FFS.

Time Scale	Process	Operation	Controlling Mechanism(s)	Equilibrium Condition
Minutes	Very short	Vacuum impregnation	HDM	$\Delta P_{int-ext} = 0$
Hours	Short	Osmotic dehydration	PDM and CDM	$\Delta a_i = 0$ $z_s = y_s$
Days	Long	Osmotic dehydration	PDM and CDM	$\Delta P_{DRP} = >0$ $\Delta V = 0$
Weeks/ Months	Very long	Fruit candy	Gas releases & pore filling	$\Delta P_{DRP} = 0$ $\Delta M = 0$

CMD: Cell matrix deformation
$\Delta P_{int-ext}$: Pressure differences between the exterior and interior of the product
ΔP_{DRP}: Pressure differences associated with the deformation-relaxation phenomena of the cell matrix

NOTATION

A: Relative relaxation level of the stress
B: Stress relaxation rate (s^{-1})
E: Apparent elasticity modulus (Pa)
E_a: Asymptotic elasticity modulus (Pa)
α: Thermal diffusivity (m^2/s)
K: Thermal conductivity (W/mK)
Cp: Specific heat at constant pressure ($JK^{-1}g^{-1}$)
L^*:   Clarity in the CIELab color scale
C_{ab}^*: Chrome in the CIELab color scale
h_{ab}^*: Hue in the CIELab color scale
F_R: CW mechanical resistance associated force (N)
F_S: Cell shrinkage associated force (N)
ic: Intracellular content
is: Intercellular space
J_{gas}: Gas flux (kg/m^2h)
J_{liquid}: Liquid flux (kg/m^2h)
p: Plasmalemma
ΔM_w: Water loss (kg water/kg initial sample)
F_A: CW-plasmalemma adherence force (N)
$S\Delta P$: Associated force to fluid pressure drop (N)
t: Time (h)
t_c: Critical time (minimum sample weight and volume) (h)
x_{fw}:   Freezable water (kg ice/kg sample)
x_i: Mass fraction of component i (w: water, s: solutes) in the fruit
a_i: Thermodynamic activity of component i (w or s) in the fruit
x_{HDM}: Mass fraction of external solution impregnated by HDM
X: Volume fraction of external solution impregnated by HDM
x_i^{VI}: Mass fraction of component i reached in a VI process
γ: Volume relative deformation of the sample after VI
y_s: Mass fraction of solutes of the external solution
z_s: Mass fraction of soluble solids of the fruit liquid phase (FLP)
ΔM^o: Weight loss referred to the initial mass sample (kg/kg)
ΔV^o: Volume loss referred to the initial volume sample (m^3/m^3)
ε_e: Effective porosity
ρ_a: Fruit apparent density (kg/m^3)
ρ: Fruit real density (kg/m^3)
r: Compression ratio ($\sim p_2/p_1$) in VI operations

ACKNOWLEDGMENTS
The authors wish to thank the Comisión Interministerial de Ciencia y Tecnología (CICYT), the U.E. (STD3 program), and the CYTED Program for the financial support of this work.

REFERENCES

Aguilera J.M. and Stanley D.W. (1990). *Microstructural Principles of Food Processing and Engineering*. Essex, UK: Elsevier Applied Science.

Alzamora S.M. and Gerschenson L.N. (1997). Effect of water activity depression on textural characteristics of minimally processed fruits. In: *New Frontiers in Food Engineering. Proceedings of the Fifth Conference of Food Engineering*. Barbosa-Cánovas G.V., Lombardo S., Narsimhan G., and Okos M.R., Eds. New York: AICHE, pp. 72–75.

Barat J.M., Martínez-Monzó J., Alvarruiz A., Chiralt A., and Fito P. (1994). Changes in thermal properties due to vacuum impregnation. In: *Proceedings of ISOPOW. Practicum II*. Argaiz A., López-Malo A., Palou E., and Corte P., Eds. Puebla: Universidad de las Américas, pp. 117–120.

Barat J.M., Alvarruiz A., Chiralt A., and Fito P. (1997). A mass transfer modelling in osmotic dehydration. In: *Engineering and Food at ICEF 7*. Jowitt R., Ed. Sheffield: Academic Press, pp. G:81–84.

Barat J.M. (1998). *Desarrollo de un Modelo de la Deshidratación Osmótica Como Operación Básica*. PhD Thesis. Valencia: Polytechnical University.

Barat J.M., Chiralt A, and Fito P. (1998). Engineering/Processing—Equilibrium in cellular food osmotic solution systems as related to structure *J. Food Sci.* 63(5):836–847.

Bomben J.L. and King C.J. (1982). Heat and mass transport in the freezing of apple tissue. *J. Food Technol.* 17: 615–632.

Chiralt A., Fito P., Andrés A., Barat J.M., Martínez-Monzó J., and Martínez-Navarrete N. (1999). Vacuum impregnation: A tool in minimally processing of foods. In: *Processing foods: quality optimization and process assessment*. Oliveira F.A.R. and Oliveira J.C., Eds.. Boca Raton, Fla: CRC Press.

Fito P. (1994). Modelling of vacuum osmotic dehydration of foods. In: *Water in Foods. Fundamental Aspects and their Significance in Relation to Processing of Foods*. P. Fito, A. Mulet, and B. Mckenna, Eds. London: Elsevier Applied Science, pp. 313–328.

Fito P. and Pastor R. (1994). On some non-diffusional mechanism occurring during vacuum osmotic dehydration. *J. Food Eng.* 21: 513–519.

Fito P., Chiralt A., Serra J., Mata M., Pastor R., André A., and Shi X.Q. (1994a). *An Alternating Flow Procedure to Improve Liquid Exchanges in Food Products and Equipment for Carrying out Said Procedure*. European Patent 94500071.9.

Fito P., Andrés A., Pasto R., and Chiralt A. (1994b). Vacuum osmotic dehydration of fruits. In: *Process Optimization and Minimal Processing of Foods*. Singh P. and Oliveira F., Eds. Boca Ratón: CRC Press, pp. 107–121.

Fito P., Andrés A., Chiralt A., and Pardo P. (1996). Coupling of hydrodynamic mechanism and deformation-relaxation phenomena during vacuum treatments in solid porous food-liquid systems. *J. Food Eng.* 27: 229–240.

Fito P. and Chiralt A. (1997). Osmotic dehydration: An approach to the modelling of solid food-liquid operations. In: *Food Engineering 2000*. Fito P., Ortega-Rodríguez E., and Barbosa-Cánovas G.V., Eds. New York: Chapman & Hall, pp. 231–252.

Martínez-Monzó J., Martínez-Navarrete N., Fito P., and Chiralt A. (1996). Cambios en las propiedades viscoelásticas de manzana (*Granny Smith*) por tratamientos de impregnación a vacío. In: *Equipos y Procesos para la Industria de Alimentos. Vol. II. Análisis cinético, Termodinámico y Estructural de los Cambios Producidos durante el Procesamiento de Alimentos*. Ortega E., Parada E. and Fito P., Eds. Valencia: Servicio de Publicaciones de la Univ. Politécnica de Valencia, pp. 234–243.

Martínez-Monzó J., Martínez-Navarrete N., Fito P., and Chiralt A. (1997). Effect of vacuum osmotic dehydration on physicochemical properties and texture of apple. In: *Engineering and Food at ICEF 7*. Jowitt, R. Ed. Sheffield: Academic Press, pp. SG 17–20.

Martínez-Monzó J., Martínez-Navarrete N., Fito P., and Chiralt A. (1998). Mechanical and structural changes in apple (var. Granny Smith) due to vacuum impregnation with cryoprotectants. *J. Food Sci.* 63(3):499–503

Nicolis G. and Prigogine I. (1977). *Self-Organisation in Nonequilibrium Systems. From Dissipative Structures to Order Through Fluctuations*. New York: John Wiley & Sons.

Pitt R.E. (1992). Viscoelastic properties of fruits and vegetables. In: *Viscoelastic Properties of Foods*. Rao M.A. and Steffe J.F., Eds. London: Elsevier Applied Science, pp. 49–76.

Salvatori D. (1997). *Deshidratación Osmótica de Frutas: Cambios Composicionales y Estructurales a Temperaturas Moderadas*. Ph.D. Thesis. Valencia: Polytechnical University.

Salvatori D., Andrés A., Chiralt A., and Fito P. (1998a). The response of some properties of fruits to vacuum impregnation. *J. Food Proc. Eng.* 21: 59–73.

Salvatori D., Albors A., Andrés A., Chiralt A., and Fito P. (1998b). Analysis of the structural and compositional profiles in osmotically dehydrated apple tissue. *J. Food Sci.* 63(4):606–610.

Shi X.Q., Fito P., and Chiralt A. (1995). Influence of vacuum treatment on mass transfer during the osmotic dehydration of fruits. *Food Res. Int.* 28: 445–454.

CHAPTER 13

Engineering Trends in Food Freezing

R.H. Mascheroni

INTRODUCTION

Food freezing is an industrial activity whose worldwide magnitude increases continuously and at a higher rate compared with the average food process. For medium to large outputs and apart from other technical features, food-freezing equipment is always of the continuous type. This is logical because continuous operation lowers costs and yields more uniform, better quality products. The design and construction of freezing equipment have reached high quality standards in recent years, so there are always one or more systems suitable for any type of raw or preprocessed food. Table 13-1 presents a concise summary of current options in the continuous equipment market.

From this wide range of alternatives, those that are important for the still reduced Latin-American market are used for freezing small to medium outputs of hamburgers, raw and processed meats (fillets, breaded, scallops and other), meat-based products, poultry, fish, vegetables, ice cream, and in lower volumes, fruit, prepared meals, and bakery products. These circumstances limit the market to belt and fluidized bed freezers, cryogenic equipment, and two-stage freezing (mainly cryogenic-mechanical).

From a historical perspective, the first priority of equipment designers and manufacturers has been to develop machines capable of ensuring frozen foods of the highest possible quality. However, this tendency was restricted by the cost factor of freezing, which for each product and commercial condition, establishes an equilibrium between quality and costs (Löndahl and Eek, 1991).

Especially in Europe, important concerns have been raised very recently on sanitary aspects of the design and operation of freezing equipment, and this has shifted such cost factors as investment and operation to a third order of importance. It has also exerted a strong influence on the production of new designs (Löndahl and Eek, 1991; Lelieveld, 1995; Eriksson Delsing and Löndahl, 1995).

Table 13-1 Options in continuous food freezing equipment.

Mechanical Refrigeration
Air Freezing:
•   Belt freezers
•   Fluidized bed freezers
•   Trolley tunnels
•   Container tunnels
•   Boxed foods tunnels
Contact freezing:
•   Horizontal plate freezers
Immersion freezing in liquids:
•   Immersion only
•   Immersion plus aspersion
Cryogenic Refrigeration
With CO_2:
•   By aspersion
With N_2:
•   By aspersion
•   By immersion
Two-Stage Freezing
Cryogenic-mechanical (combined):
•   Pre-freezer: Standard cryogenic equipment
•   Pre-freezer: Specific cryogenic equipment
Mechanical-mechanical
•   Plate plus belt

PRESENT EQUIPMENT DESIGN OBJECTIVES

These are established for the equipment to meet the following targets (in descending priority):

- To obtain a frozen product of maximum final quality
- To comply with strict sanitary conditions, including ease of cleaning
- To reduce fixed and operating costs to a minimum compatible with the first two targets

Final Quality

As indicated before, this is the first factor to be considered. The trend is towards using the equipment and operating conditions required to obtain a product of the best possible quality so long as the increase in processing costs is justified from an economic point of view (this includes both the several different stages of the cold chain that can affect the product quality and willingness of the market to pay higher prices for better quality). Generally such improved quality is obtained when using higher freezing rates and constant controlled operating conditions. Factors determining the freezing rate are as follows:

- Temperature of the refrigerating medium
- Effective heat transfer coefficient
- Product size and shape
- Physical properties of the system
- Others (initial and final temperatures, relative humidity)

Temperature of the Refrigerating Medium

Freezing time is approximately proportional to the reciprocal of the difference of initial (*Ti*) and refrigerating medium (*Tr*) temperatures, or $1/(Ti - Tr)$ (Cleland and Earle, 1984; Salvadori and Mascheroni, 1991). This means that provided a refrigerant temperature is lowered, the processing time will always decrease proportionally.

There are two main regions in industrial freezing with respect to refrigerating medium temperature: 1) mechanical refrigeration, where the minimum temperature that can be achieved in economically feasible conditions is about –40°C; and 2) cryogenic refrigeration, where the temperature is determined by the evaporation of the refrigerant used (about –79°C for CO_2 and –196°C for N_2). That is to say, the choice of a particular refrigerant will in practice establish the working temperature so it constitutes a factor that is not prone to admit any important further influence.

Effective Heat Transfer Coefficient

The heat transfer coefficient *h* is a measure of the ability of an interface placed between a system (food product) and refrigerating medium to exchange heat. Its value is determined by refrigerant fluid-dynamics (velocity and flow regime), type of contact, existence of packaging, air space between product and packaging, and size and spatial distribution of the product (Tocci and Mascheroni, 1995). As a general rule, high refrigerant velocity and turbulence increase *h* significantly, as does the absence of packaging.

The influence of *h* on the freezing time is strong for low *h* values (e.g., low air velocities or presence of highly insulating packaging), but this effect becomes rapidly weaker for high *h* values (high velocities and turbulence, absence of packaging). In this regard, when *h* values are sufficiently high the freezing time remains almost constant (Cleland and Earle, 1984; Salvadori and Mascheroni, 1991).

Product Size and Shape

Freezing time usually depends on a product's thickness raised to a power between 1 and 2 (i.e., comparing a product with a thickness A and a product with a thickness 2A, this second one will have a freezing time of about four times longer than the first one, under the same operating conditions). From a different point of view, assuming that the freezing equipment has enough refrigerating capacity, it can be said that if the size of a product were reduced in half it would become possible to freeze almost four times the original product weight in the same time period. In many cases it is simply not possible to change a product's size, but to handle it to a certain extent (in single layers of thin thicknesses, unlike layers or containers of greater thickness) equivalent of a size reduction, so allowing heat transfer to be optimized.

Concerning shape, it is known that upon transferring heat through several product surfaces, the freezing time decreases proportionally (Cleland et al., 1987; Salvadori et al., 1997). For example, freezing is approximately three times longer in a slab than a sphere whose diameter is equal to the slab thickness, and two times longer than in a cylinder of the same diameter. The same type of ratios apply for packages or boxes transferring heat across several or all of their surfaces.

Other Factors

The following are variables over which little control or handling can be done:

- Thermal properties: Because freezing times are proportional to the reciprocal of thermal diffusivity (α, the ratio of thermal conductivity to the product of density and specific heat capacity) (Salvadori and Mascheroni, 1991) it is only possible to influence such property in particular cases; for example, by packing the product better, removal of air voids, or dehulling.
- Initial temperature: This can be modified in cooked or fried products by natural or forced refrigeration before freezing.
- Final temperature: In some cases cooling can be stopped before reaching the target final temperature in the thermal center, leaving it to get uniform in cold storage and thus saving up to 15% of the freezing time without introducing further heat load.
- Relative humidity: In refrigeration by air, a low ambient relative humidity decreases the freezing time due to the heat absorbed by surface ice to sublimate. However, sublimation does not seem to be a suitable solution because it produces negative effects on quality (weight loss, spottiness, rancidity) and increases frost formation on the evaporators, implying longer defrosting periods.

At present most modifications or new designs take advantage of some of the possible ways of increasing freezing rates to speed up the initial freezing stages of sensitive products (generally attained by using lowered temperatures or increased heat transfer coefficients). Sensitive products are those that may deform, break into pieces, stick together or to the freezer, or dehydrate during freezing. In this faster initial stage, the aim is to rapidly freeze the surface layer of the food so as to protect its structure by preventing risks of damage or excessive weight losses. A list of sensitive products (not intended to be exhaustive) is presented in Table 13-2.

Table 13-2 Sensitive Foods Requiring Special Freezing Conditions.

Soft:
• Strawberries (tendency to crush)
• Raspberries and other berries (tendency to break into pieces)
• Meats as hamburgers, fillets, and scallops (deform in belt freezers)
• Ice creams, dough, pies, pastry
• Semiliquids, cream pies (deform and drip)
Sticky:
• Shrimp, prawns, and other peeled shellfish
• Strawberries in fluidized beds
• Meat pieces in belt freezers
• Fillets in belt freezers
• Ravioli and other pasta
• Potatoes (diced and uncooked French fries)
Tendency to dehydrate:
• Hamburgers and meatballs (because they are made of minced meat)
• Fish fillets
• Peeled shellfish
• Mushrooms
• Cooked products
Tendency to brown:
• Mushrooms
• Cut fruits

Methods to Reduce the Freezing Rate

To deal with soft, wet, sticky, and fragile products, two procedures are commonly followed, but others not addressed here may also be possible.

Design Modifications in Circulation Conditions

Some equipment manufacturers have modified the air circulation patterns through the belts in their freezers to avoid uni-directional air flow and enable most of the food surface to receive the refrigerating airflow. As such, York International has developed what they call "Double Impingement Airflow" (York International, 1992). Figure 13-1 presents a scheme of the air flow pattern in a freezer designed with that purpose. This flow pattern makes air go upwards in the lower half of the spiral freezer and downwards in the upper half, producing a much more even freezing with a parallel increase in the freezing rate, shorter freezing times, and lower sublimation weight losses. Other manufacturers (Ross Industries, 1995) have introduced the concept of boundary layer control (impingement freezing) or equivalent developments which consist of mechanical refrigeration equipment with yield results comparable to those of cryogenic type, but at a substantially lower cost. This is achieved by using air jets collimated from special nozzles which literally eliminate the stagnant boundary layer that surrounds the product and cause simultaneous impingement to the upper and lower product surfaces. Figure 13-2 presents a scheme of the air nozzles and flow pattern around the food. This technology reduces the duration of freezing up to six times (depending on the ratio of area to volume of the product) and weight losses to about one third compared with the conventional belt freezer. A comparative example is given in Table 13-3 for a 133 g beef hamburger frozen from an initial temperature of 0°C to a final equilibrium temperature of −18°C (Ross Industries, 1995).

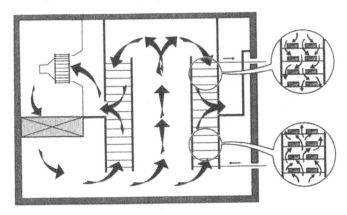

Figure 13-1 Scheme of air flow pattern in a double impingement airflow freezer.

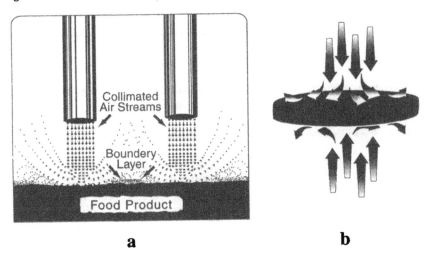

Figure 13-2 Operational scheme of an impingement freezer: a) air nozzles and b) flow around the product to be frozen.

Table 13-3 Effect of freezing equipment on freezing time and weight loss for a hamburger patty.

Freezing Method	Freezing Time (min)	Weight Loss (%)
Belt (conventional)	17.0	1.6
Cryogenic	3.2	0.5
Impingement	3.2	0.5

Application of a Fast Initial Freezing Stage (crusting)

As expressed previously, the idea here is to freeze the surface layer of the product using a rapid method that ensures quality but does not increase the total expense of freezing by cryogenic methods (about 10 times as expensive as the mechanical technique). The proportion of the food frozen by this initial stage is about 20–30% by volume, which is considered sufficient to secure a good surface quality. After this pre-freezing stage, products are automatically transferred via a continuous belt to the finalizing freezer (also continuous, normally a belt freezer). This can be carried out either via a mechanical-mechanical or cryogenic-mechanical combined method. An example of the first is the so-called Super Contact Product Surface Freezing Tunnel (Food Systems Europe, 1994), of which a scheme is shown in Figure 13-3. It consists of a metallic plate with a built-in extruded refrigerating coil in which temperatures ranging from –40 to –60°C are used. A 10 μm thick polyethylene film slides over the plate and is continuously loaded with the product causing the lower surface of the food to freeze from contact with the cold plate. The technique is appropriate for flat unwrapped solid foods and liquid products in aluminum foil trays, and can be complemented with an air impingement system to simultaneously pre-freeze the upper product surface or even completely freeze the food (especially for low production rates).

The cryogenic-mechanical combined method (two-stage freezing) can be achieved either with standard cryogenic or purpose-designed equipment. The first type is generally an aspersion or liquid N_2 immersion freezer whose advantage lies in the possibility of using already existing equipment that may also serve as a normal freezer when necessary (i.e., during production peaks or failure of the mechanical freezer). The advantage of purpose-designed equipment is to have a machine with the performance required to meet specific objectives. An example is the CRUSToFREEZE (AGA, 1995) which uses a continuous belt over which the product is fed. A food is initially immersed in liquid N_2 and then receives a spray of the same refrigerant, leaving the pre-freezer as a firm, non-agglomerated individual product. This system is recommended for shellfish and any type of fruit or vegetables. The retention time ranges from 5 to 60 s according to product type and size. Another example of purpose-designed equipment is the AGA Freeze F (AGA, 1993) which consists of a continuous stainless steel belt over which a liquid N_2 spray is spread which causes the surface temperature to reach –80 to –100°C. Immediately thereafter products are loaded on the very cold belt so that their undersides freeze instantaneously (especially suitable for flat products). The retention time varies from 5 to 25 s according to product size.

171

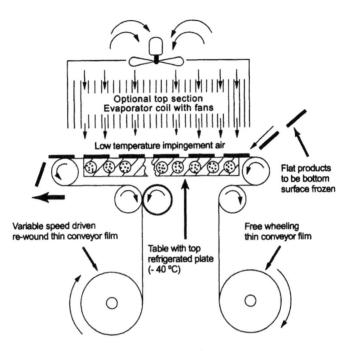

Figure 13-3 Scheme of a contact product surface freezer.

Commonly Used Freezing Equipment

To summarize the items addressed up to here, Table 13-4 is presented which lists the most frequently used equipment to freeze common products. As can be seen, in most cases a certain food product can be adequately frozen in two or more different types of equipment. The choice to buy a determined type of freezer is typically based on freezing costs, but also to be considered is the availability of idle equipment (if the need is mainly for seasonal crops where the same freezer would be used for several different products throughout the year or even during the same week or day). In some cases portable equipment is moved to different locations to optimize its use.

HYGIENIC AND SANITARY TOPICS

Sanitary topics have become of such great importance that they influence both design and economic modifications to the freezing process. Since prevention of microbiological problems is key, the existence of several fatal cases of contamination and microbial poisoning has pushed Europe and the USA to identify weak points in the design and operation of freezing equipment and prescribe directions for improving both aspects.

Table 13-4 Foods and equipment commonly used in freezing.

	Belt Tunnel	Cryogenic Tunnel	Trolley or Box Tunnel	Fluidized Bed	Combined Equipment
Hamburger patties	X	X			XX
Meats (diced, balls)	X	X		X	XX
Meat and fish blocks			X		
Meat and fish fillets	X	X	X		X
Meat, fish & poultry pieces	X	XX	X		XX
Fish (whole)		X	X		
Poultry (whole)			X		XX
Prepared foods (i.e., TV dinners)	X	X	X (boxed)		XX
Shrimps, scallops	X	X		X	X
Peas				X	
Potatoes and carrots (diced, sliced, fried)				X	
Green beans				X	XX
Asparagus	X	X			X
Artichoke hearts				X	
Brussel sprouts				X	
Cauliflower and broccoli florets		XX		X	
Corn on the cob	X	X		X (sliced)	
Berries		X		X	X
Cherries		X		X	X
Fruit pieces				X	X
Fruit or vegetable puree		X			X
Apples and pears (diced, sliced)				X	
Parsley and celery	X	X		X	
Spices				X	
Spinach			X		
Mushrooms	X	X	X (packaged)	X	X
Ice creams	X	X			X
Pasta products	X	XX	X (hardening)		
Bread, croissants, other pastries	X	XX	X		X
Pizzas, quiches	X	XX			XX
Pocket sandwiches	X	XX	X		
Breaded nuggets	X	XX			
Fruit tarts and pies	X	XX	X		
	X	XX			

X: common use
XX: random use

The principal defects found in the evaluations were:

- Operating conditions with equipment zones and/or time periods with temperatures above 0°C (including cleaning periods and time the equipment stands idle during the weekend)
- Zones (non-curved corners with recessed surfaces or joints, floor joints, open or concave frameworks, and service conduits) where food residues accumulate with juices or defrosting or cleaning water
- Zones not accessible to cleaning systems (normally corners and parts of frames, engines, and service conduits)

This encouraged (mainly in the European Union) the development of the following directives.

Specifications for Food Processing Machines

- Must allow all materials contacting the food to be cleaned before each use
- Must be smooth, without crevices or edges that may retain organic materials
- Must keep protrusions, edges, and openings to a minimum (screws, screwheads, and rivets are allowed only if imperative)
- Must be easy to clean and disinfect, and inner surfaces must be curved
- Must be able to drain all liquids without obstructions
- Must prevent liquid accumulation and avoid living organisms and organic material from reaching zones where cleaning is not possible
- Must prevent ancillary substances (e.g., lubricants) from contacting the food

Criteria for Hygienic Design

- Materials contacting the food must be non-toxic, non-absorbent, and resistant to the product, cleaning, and disinfecting agents
- Reinforced plastics and elastomers must not allow the product to penetrate
- Elastomer compression must be controlled
- Surfaces must allow draining (slope greater than 3%)
- Surface roughness must be less than 0.8 μm
- Features that must be avoided include:
    —metal-to-metal joints
    —misaligned connections of equipment and pipelines
    —crevices in seals and joints
    —o-rings
    —screw threads
    —sharp edges (radii greater than 6 mm are recommended)
    —dead zones

Up to a few years ago the design of food freezers centered on optimizing heat transfer and lowering operation costs. In the same sense most hygienic concerns only referred to the automatic cleaning of moving belts, removing debris and product particles during the defrosting period, and deep cleaning during periods of nonuse. However, the impact of the previously cited directives was so strong that major manufacturers redesigned their equipment almost completely. For example, Frigoscandia reformed its FLoFREEZE (Eriksson Delsing and Löndhal, 1995) by lowering the position of the product-carrying tray to floor level (the refrigeration equipment was placed beside the tray rather than below it as in previous models) and thus reducing the height of the freezer. Also, all platforms and structures were eliminated, thus lowering the heat capacity of the equipment and shortening the cooling down and warming up periods (Fig. 13-4). In addition, the cleaning process was redesigned with a first stage involving removal of product debris and snow via manual rinsing of walls and floor, and a second stage with an automatic cleaning that includes an initial rinsing, use of a detergent (as a foam), and a final rinsing to eliminate cleaning materials and debris. These modifications mean higher designing and construction costs and are not necessarily compensated by shorter stoppage periods for cleaning and disinfection or lower needs of cleaning materials and water.

REDUCTION OF FIXED AND OPERATION COSTS

As was previously expressed, costs are usually relegated to a third order of importance. In most cases fixed costs increase when equipment needs to be redesigned, but the same is not applied to operation costs. As a result, most of the improvements achieved in recent years (including automatic operation control, reduced weight losses, less defrosting and cleaning periods, and lower consumption of chemicals, cleaning water, and stabilization periods) clearly favor a higher final product quality and equipment output as well as lower operating costs. In many cases these favorable variations in operation costs balance the increase in fixed costs.

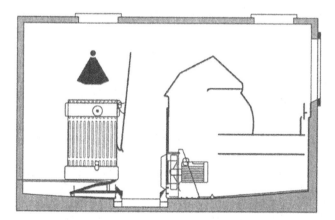

Figure 13-4 Cross-sectional scheme of a fluidized bed freezer FLoFREEZE.

FINAL REMARKS

The constant competition for obtaining small improvements in the final quality of frozen foods (particularly in those more sensitive to mechanical damage during the process) must be balanced against market pressure towards ensuring the best sanitary conditions of frozen products when design modifications in continuous food freezers are considered. Another factor, which is valid throughout the refrigeration sector, is the need to adapt refrigeration equipment to replace refrigerants since the legal period for use of CFCs is rapidly coming to an end. The release of very low output continuous freezing equipment (from 100 kg/h on) is yet another fact of interest for countries where low outputs are common, as this will enable many current freezing operations that are done batchwise (with difficult design and control) to be conducted in standardized conditions, thus ensuring better final quality and simpler handling.

REFERENCES

AGA (1993). AGA Freeze F Cryogenic pre-freezer brochure.

AGA (1995). CRUSToFREEZE brochure.

Cleland A.C. and Earle R.L. (1984). Freezing time predictions for different final product temperatures. *J. Food Sci.* 49: 1230–1232.

Cleland D.J., Cleland A.C., and Earle R.L. (1987). Prediction of freezing and thawing times for multi-dimensional shapes by simple formulae. Part 1: Regular shapes. *Int. J. Refrig.* 10: 156–164.

Eriksson D. K. and Löndhal G. (1995). Hygienic considerations in food freezing. In: *Proceedings of the 19th International Congress of Refrigeration. Vol. I.* Paris: International Institute of Refrigeration, pp. 382–391.

Food Systems Europe (1994). Super contact product surface freezing tunnel brochure.

Lelieveld H.L.M. (1995). Refrigeration and hygienic considerations in food freezing. In: *Proceedings of the 19th International Congress of Refrigeration. Vol. IIIa.* Paris: International Institute of Refrigeration, pp. 15–20.

Löndhal G. and Eek L. (1991). Innovations in food freezing technology. In: *Proceedings of the 18th International Congress of Refrigeration. Vol. IV.* Montréal: International Institute of Refrigeration, pp. 1956–1961.

Ross Industries (1995). BLC 200. Freezing, cooling and tempering tunnels brochure.

Salvadori V.O. and Mascheroni R.H. (1991). Prediction of freezing and thawing times of foods by means of a simplified analytical method. *J. Food Eng.* 13: 67–78.

Salvadori V.O., De Michelis A., and Mascheroni R.H. (1997). Prediction of freezing times for regular multi-dimensional foods using simple formulae. *Lebens-Wiss u. Technol.* 30: 30–35.

Tocci A.M. and Mascheroni R.H. (1995). Heat and mass transfer coefficients during the refrigeration, freezing and storage of meats, meat products and analogues. *J. Food Eng.* 26: 147–160.

York International (1992). York International: Meeting the challenge of global integration. *Quick Frozen Foods Int.* (7): 67–78.

CHAPTER 14

Evaluation of Freezing and Thawing Processes Using Experimental and Mathematical Determinations

A.C. Rubiolo

INTRODUCTION

Food preservation implies the control of certain factors that reduce or inhibit the rate of undesirable reactions in a food product. A simple method to obtain this effect is decreasing the temperature of the product. When a biological system is cooled to below 0°C, the water portion solidifies, thus freezing the food. Since most products after being stored at low temperature are thawed for consumption, both such operations are important in food processing where they are often carried out by placing a food in an air stream to achieve a heat exchange.

For a process analysis it is always necessary to know how much heat is required to take a product from an initial to a final state and how product temperature is rapidly modified during this time. Furthermore the heat exchange rate is another factor that influences product quality and stability. Such knowledge is important for designing equipment and optimizing its performance.

The thermal response of objects immersed in an air stream during freezing or thawing processes depends on the Biot number which provides a measure of the relative magnitude of external (air velocity and temperature) and internal (product size and geometry) heat transfer resistances for evaluating the effect of operating conditions. Although the basic heat-transfer theory that involves process variables is well known, the mathematical development for obtaining the solution of the governing partial differential equations with the initial and boundary conditions of interest only considers homogeneous situations for specific geometries (slab, prism, cylinder, sphere). The heat exchange between the air and solid, in addition to the simultaneous phase change of water into ice in the product are the principal problems generally solved with unsteady-state heat equations which are used to calculate process times. Temperature solutions to predict times are obtained by considering the continuous depression of the freezing point in an unfrozen food portion. This procedure assumes the behavior of an ideal solution, which is reasonable since water—the major component in the food—crystallizes over a range of temperatures due to the freezing point depression caused by dissolved solutes.

179

Another problem during this process is the mass transfer between the solid surface and surrounding medium which is mainly studied to determine the water loss caused by evaporation or sublimation. Therefore, many variables in the solid should be evaluated to relate process conditions to the crystallization rate which is often responsible for most of the changes during food freezing. These and other methods are essential to minimizing damage to the product.

The damage undergone during the freezing process has been attributed to a number of factors, but one of the most important is the rate of ice growth. The nature of the ice formation affects the crystal size in that larger ice particles have a stronger influence on the tissue and its components than smaller crystals. The simplest thermodynamic equation indicates that ice is formed when the addition of free energy from the phase change (negative) and water-ice interface (positive) is a negative final result. An ice nucleus persists if the temperature change is large enough to produce a lower final energy value. After nucleation occurs, less energy is needed for increasing crystal size. Therefore fast cooling is preferred over slow cooling since the former can produce more ice crystals reducing the possibility of crystal growth.

The location of ice is another source of irreversible damage in food tissues. When intercellular ice crystals are initially formed it indicates that freezing outside the cell took place earlier, which means that this extracellular accumulation of ice and membrane shrinkage due to cell water loss occurs at a low freezing rate. However, intracellular ice crystals predominate when the freezing rate is high. This behavior can mainly be explained by considering the difference between the solution compositions located inside and outside the cell. For example, larger molecular weight compounds inside the cell increase the freezing point depression so that more energy must be removed to produce solidification.

Some studies have shown that the extent and type of injury to a given tissue caused by ice crystal location and size are governed not only by the freezing rate but also frozen storage temperatures and conditions. Ice crystals in frozen plant tissues grow in size at the expense of smaller ones during this stage and early thawing since larger crystals increase stability and reduce the interfacial energy. This recrystallization can also have a negative effect on cells. Such structural damage leads to the loss of firmness, texture changes, and modifications to water binding and fluid retaining capabilities after thawing.

In addition to the factors cited above, chemical changes in frozen foods are also important to quality preservation. Alterations to tissue integrity can cause or accelerate chemical reactions which are responsible for the loss of compounds required in a healthy diet. Diffusion coefficients and intercellular composition can be different after thawing due to modifications of cell membrane permeability and integrity which affect reaction rates since originally separated compounds such as enzymes and cations are mixed. This is the case with the ascorbic acid and enzyme ascorbate oxidase that are present in many plant tissues. Therefore an oxidation reaction of the ascorbic acid can be catalyzed when the structural barriers of the plant tissue are disrupted after the freezing-thawing process. Furthermore the process can take place at different rates according to the presence of hydrogen, metal, or oxygen mixed with the enzyme and substrate.

Taking this into consideration, different freezing and thawing rates were used to analyze the influence of process air velocities and temperatures in the exudate formation and vitamin C loss of frozen and thawed strawberries. Experimental temperatures were compared with predicted values by using heat and heat-mass transfer models in order to determine the behavior of each corresponding solution in process rate calculations.

MATHEMATICAL MODELS

Applying the energy change equation for unsteady state heat transfer to a solid spherical particle, the partial differential equation 14-1 is obtained, such equation is known as Fourier's second law of heat conduction:

$$\rho Cp(T)\frac{\partial T}{\partial t} = \frac{1}{r^2}\frac{\partial}{\partial r}\left(K(T)r^2\frac{\partial T}{\partial r}\right) \qquad (14\text{-}1)$$

where the initial (I.C.) and boundary (B.C.) conditions used are:

$$\text{I.C.} \quad T = T_0 \qquad\qquad\qquad r \text{ and } t = 0 \qquad (14\text{-}2)$$

$$\text{B.C.} \quad -K(T)\frac{\partial T}{\partial t}\bigg|_{r=R} = h(T - T_a) \qquad r = R \text{ and } t \geq 0 \qquad (14\text{-}3)$$

$$\frac{\partial T}{\partial r} = 0 \qquad\qquad\qquad r = 0 \text{ and } t \geq 0 \qquad (14\text{-}4)$$

where ρ is the density, $Cp(T)$ the heat capacity, $K(T)$ the thermal conductivity, h the heat transfer coefficient, r the radial distance, T the solid temperature, t the process time, T_a the ambient air temperature, and R the sphere radius.

Solutions to these equations for constant thermal properties were integrated in the corresponding solid volume to define its average temperature, and the analytical solutions of unsteady state heat transfer problems at low Biot numbers without phase changes were modified to obtain simple algebraic freezing and thawing solutions. First a temperature-dependent correlation for the effective heat capacity was derived to account for changes in the sensible and latent heat of solidification due to ice formation. The effective heat capacity was then combined with the average temperature solutions to estimate the time during the process in which a temperature based on the product's average enthalpy would reach its final value when the product was removed from the system (Schwartzberg, 1981; Rubiolo de Reinick, 1985). The effective heat capacity equation for temperatures lower than the initial freezing point was thus derived as (Schwartzberg, 1976):

$$Cp(T) = C_f + \frac{\Delta H_0(n_{w0} - bn_s)(T_0 - T_i)}{(T - T_i)^2} \qquad (14\text{-}5)$$

The enthalpy equation was defined as:

$$\Delta H = (T - T_i)\left[C_f + \frac{(n_{w0} - bn_s)\Delta H_0}{T_0 - T} \right] \qquad (14\text{-}6)$$

where C_f is the heat capacity of the frozen food, ΔH_0 the latent heat of water crystallization at 0°C, n_{w0} the total water content in the food, bn_s the unfreezable water bound to solutes, T_0 the temperature of water solidification, and T_i the initial freezing temperature of the food.

The equation to calculate the freezing time for temperature below T_i (when $\overline{T} < T_1$) is:

$$\theta = \frac{V\rho(1+fBi)}{hA}\left[C_f \ln\left(\frac{L_1(T_a - T_1)}{T_a - \overline{T}}\right) + \frac{(n_{w0} - bn_s)\Delta H_0(T_0 - T_i)}{(T_0 - T_a)^2} \right.$$
$$\left. \times\left(\ln\left(\frac{L_1(T_0 - \overline{T})(T_a - T_1)}{(T_0 - T_1)(T_a - \overline{T})}\right) + \left(\frac{(T_1 - \overline{T})}{(T_0 - T_1)}\right)\left(\frac{(T_0 - T_a)}{(T_0 - \overline{T})}\right)\right)\right] \qquad (14\text{-}7)$$

where fBi is the Biot number modified by the geometric factor f, V/A is R/3, T_1 the initial solid temperature below T_i, $\overline{T}$ the average temperature, and h the heat transfer coefficient between the air and solid product that is calculated considering the other adjacent objects (Rubiolo de Reinick and Schwartzberg, 1986).

The thermal conductivity to obtain the Biot number can be assumed as an average value for the frozen and thawed states (Rubiolo de Reinick, 1992). However, an equation in which these changes with temperature are taken into account can be introduced in heat transfer solutions with non-phase changes, but obtaining freezing time predictions that are more difficult to compute. A simple equation for the thermal conductivity variation with temperature below T_i is:

$$K(T) = K_f + (K_o - K_f)\frac{T_i - T}{T - T_0} \qquad (14\text{-}8)$$

where K_0 and K_f are the thermal conductivities of unfrozen and frozen food, respectively.

When other effects at the surface are considered in the heat transfer process such as water evaporation or sublimation, the boundary conditions at the surface (Eq. 14-3) can be changed as follows:

$$-K(T)\frac{\partial T}{\partial t}\bigg|_{r=R} = h(T-T_a) + K_m L_v (C-C_a)_{r=R} \qquad (14\text{-}9)$$

where k_m is the mass transfer coefficient, L_v the latent heat of evaporation or sublimation, C the water concentration at the food surface, and C_a the water concentration just adjacent to the surface of the solid for the water vapor in the air.

A mass transfer partial differential equation is necessary for calculating the water concentration in the solid (Tocci and Mascheroni, 1991):

$$\frac{\partial C}{\partial t} = \frac{1}{r^2}\frac{\partial}{\partial r}\left(Dr^2\frac{\partial C}{\partial r}\right) \qquad (14\text{-}10)$$

with the following initial and boundary conditions:

$$C = C_i \qquad\qquad r \text{ and } t = 0 \qquad (14\text{-}11)$$

$$\frac{\partial C}{\partial r} = 0 \qquad\qquad r = R \text{ and } t \geq 0 \qquad (14\text{-}12)$$

$$-D\frac{\partial C}{\partial r}\bigg|_{r=R} = k_m (C - C_a)_{r=R} \qquad r = 0 \text{ and } t \geq 0 \qquad (14\text{-}13)$$

where D is the water diffusion coefficient in the solid.

These mass transfer equations have analytical concentration solutions for constant temperature but are not for cases where the moisture content is a function of the product surface temperature and environmental conditions. Therefore, the prediction of simultaneous temperature and concentration changes during this process were carried out using numerical methods (Hayakawa and Succar, 1982).

Coupled heat and mass transfer differential equations can be solved by taking into account the phase change during freezing and the water lost by evaporation or sublimation because of the different moisture content in the air and fluid just adjacent to the product surface. After different numerical methods are considered for the prediction of the product temperatures and concentrations during the process, heat and mass balances can then be used to calculate the overall energy transfer and total water loss at the surface. Furthermore, an average temperature can be obtained from the average enthalpy of the product.

Table 14-1 presents the time it took to reach a center temperature of $-18°C$ in strawberries. It was obtained by solving partial differential equations with a difference finite method and considering both heat transfer and coupled heat and mass transfer in order to involve the water loss in the process (Delgado, 1997).

183

Table 14-1 Operating conditions and times required during strawberry freezing.

Sample N°	Air Velocity m/s	Ambient Temp. °C	Initial Temp. °C	Exp. Time (min)	Calculated Time (min)			
					Heat eq.	Error	H. + m. eq.	Error
1	2.58	−20	14.5	39.9	40.6	−1.86	38.5	3.4
2	3.56	−20	13.5	31.8	34. 7	9.20	30.0	−5.5
3	4.54	−20	18.3	31.9	34.8	9.08	30.3	−5.0
4	0.51	−20	13.5	91.3	86.7	−5.05	84.7	−7.2
5	2.58	−30	14.0	20.8	20.3	−2.02	19.8	−7.6
6	3.56	−30	14.4	15.4	18.3	18.87	17.2	11.4
7	4.54	−30	17.1	13.7	16.2	17.60	15.8	15.1
8	0.51	−30	16.2	40.3	38.0	−5.66	37.5	−6.9

Average errors between experimental and predicted times were 8.66% when heat transfer equations were used and 7.76% for coupled heat and mass transfer. Consequently, in those cases where the difference obtained between the predicted and experimental values was small and errors were attributed not only to models but also experimental uncertainty, the freezing times were reasonably predicted with any of these models. However, the strawberry geometry is a source of error in both experimental and calculated values since it is difficult to guarantee that the thermocouple for measuring the central temperature was properly located in the product and an equivalent radius (determined by assuming a sphere volume equal to the strawberry volume evaluated as a right circular cone) was used as the sphere dimension for the finite difference method. Figure 14-1 shows the average temperature calculated from the average enthalpy and predicted central temperature versus time. As demonstrated, the times required to obtain average temperature changes were smaller than those for obtaining the same change in central temperatures, and the difference between these times was larger when the temperature range change included the freezing point where the phase change heat was involved.

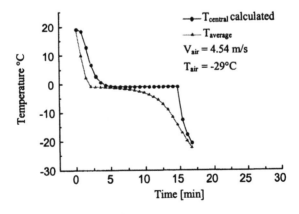

Figure 14-1 Average and central temperatures versus time for strawberry freezing.

Because a product would reach equilibrium temperature without changing its total enthalpy after being located in an insulated place, the average temperature based on the average enthalpy change by the product predicts the shortest period of time to remove the product from the freezing stage to a storage chamber without modifying the room temperature and using the least energy to obtain appropriate preservation conditions in the food.

The rate and extent of the water transfer were dependent on the convective mass transfer coefficient as well as the moisture content difference in the air stream and near the solid surface. This last value was obtained from the remaining water on the food surface by assuming a vapor-liquid equilibrium. As long as ice was not present, a decrease on the water concentration was determined as a function of the diffusion coefficient in the food which is affected by the internal and peripheral product structure and temperature.

The water loss experimentally determined by the differences in strawberry weight before and after freezing and the value calculated by the mass balance between the initial and final water concentrations in the volume of the solid are presented in Table 14-2. Final concentrations in the solid were the values in each volume fraction obtained by the finite difference method used for solving the mass transfer partial differential equation when the solid central temperature was $-18°C$. The greatest experimental and calculated values were obtained when freezing times were larger (i.e., for lower air velocities), but the differences between these values might be reduced by improving the accuracy of either the experimental data or parameter predictions.

Many factors in the product affected the water loss in the strawberries during the process. Especially influential were the high initial temperature and large exposed area (related to the dry weight) which increased the mass transfer. However, this comparison between water loss as a function of temperature and exposed area is not always very accurate due to the random variation of these variables. Furthermore the small degree of water loss in this case also makes such variations difficult to analyze. The times calculated with the heat transfer model were therefore similar to those values obtained with the coupled system of heat and mass transfer, but this last method also provided a base for measuring the influence of environmental conditions in the food water variation and resulting deviations.

Table 14-3 provides the thawing time and water loss when the strawberries reached a central temperature of $-1°C$ after being frozen under homogeneous air conditions ($-30°C$, 2.58 m/s) to control the cooling rate and minimize the difference in fluid crystallization and dehydration. Results show that changes in food water binding and fluid retaining capability produced irreversible or metastable conditions after thawing that facilitated the water separation. In addition, the effect of a high air temperature was more easily detected than a high air velocity during thawing, but both variations increased the water loss and decreased the process time. Differences between inter- and intracellular local compositions and the water transported between these domains were particularly observable after freezing and thawing from the food exudate.

185

Table 14-2 Water loss in strawberries after freezing.

Sample N°	Radius m 10^{-2}	Strawberry Dry Weight (So) kg 10^{-3}	Area/So m^2/kg	Water Loss/Dry Matter kg/kg DM	
				Experimental	Calculated
1	1.03	0.826	210	0.058	0.049
2	1.16	1.180	180	0.052	0.049
3	1.23	1.320	180	0.068	0.056
4	1.19	1.039	197	0.109	0.046
5	1.06	0.917	202	0.077	0.049
6	1.10	0.957	183	0.050	0.044
7	1.08	0.672	206	0.072	0.049
8	0.98	0.633	226	0.105	0.047

Table 14-3 Operating conditions and determined values for thawing strawberries.

Sample N°	Air Vel.	Ambient Temp.	Radius m 10^{-2}	Initial Temp. °C	Time (min) heat h. + m.		Water Loss/ Dry Matter kg/kg DM	
	(m/s)	°C			Eq.	Eq.	Exp.	Calc.
9	2.580	6	1.15	-12.10	58.66	59.00	0.061	0.057
10	3.560	6	1.14	-12.56	50.33	50.84	0.051	0.064
11	4.540	6	1.30	-14.10	57.66	58.50	------	0.077
12	0.508	22	1.11	-12.16	43.17	45.16	0.782	0.133

EXPERIMENTAL DETERMINATION

Exudate was experimentally assessed by considering the water drip loss from strawberries placed in a box and left three hours in a chamber to reach a given ambient temperature. Thawed strawberries were removed from the box, blotted on paper towels, and then weighed to determine differences with initial weights. Exudate was calculated considering that the air reached saturation moisture at the strawberry surface temperature. The water loss by diffusion was then determined as water exudate assuming that it separated from the strawberry and remained in a liquid state.

Consequently, aspects related to water release due to membrane denaturation and structure reorganization after freezing were not included, the nonthermal processes were very poorly characterized, and those mathematical equations associated with the water equilibrium change after crystallization were not indicated (Fennema et al., 1973). Although the magnitude of these values may provide a basis for comparing product behavior after the application of said processes, any resulting errors still cannot indicate the influence of process conditions.

As can be observed in Table 14-4, experimental exudate values were very different and did not show any correlation with process conditions. However, the similar calculated values in all the cases indicated a connection to the product area. Water loss results were obtained after adding exudate values to those determined in Table 14-3 corresponding to the freezing-thawing process.

The total weight loss of thawed and wiped strawberries was also different, meaning that the water release was not only dependent on process conditions but also the food tissue and structure and equations used which could not adequately take into account the water equilibrium. Experimental values were larger than those calculated since not all the thermodynamic and kinetic processes were involved in the mathematical equations used to define this freezing-thawing phenomenon.

In contrast to the difficulty in determination, the amount of water loss was not significant to the total weight of the fruit but could be important in relation to the physical or textural properties of the thawed product. The exudate can contain soluble solutes, however the quantities involved in the little solution released are smaller than the loss of food nutritive compounds mostly modified by chemical reactions.

An example of interest is vitamin C, an organic acid soluble in water that is present in fruits at different ranges of composition whose concentration rapidly decreases. The loss of this vitamin can be attributed to chemical degradation, although the variation can also occur by washing during processing (Laing et al., 1978). As this vitamin is one of the most labile nutrients present in foods, its retention ensures that other biocomponents remain without alteration during conservation and processing. Therefore in food freezing vitamin C is determined for its quantity and as a tool to assess the effectiveness of the process in terms of nutrient conservation. The initial value in strawberries varies from 0.05 to 0.1 kg/100 kg, but this quantity rapidly decreases when the fruits are not preserved adequately.

Table 14-5 shows the loss rate of vitamin C in strawberries that were frozen under different processing conditions and stored one day at different temperatures. Initial and final vitamin C values were determined in the same sample by cutting the strawberries in halves. One part was used immediately after thawing and the other half after storing one day at different temperatures (Amer, 1997). The retention values obtained from the ratio of final and initial values in the samples were higher than 0.7 for any freezing process after storing one day at $-24°C$. Values higher than 1 were attributed to the experimental error

Table 14-4 Water loss during strawberry freezing and thawing.

Sample N°	Dry Matter kg 10^{-3}	Exudate kg/kg DM		Water Loss kg/kg DM		Weight Loss kg/kg DM
		Exp.	Calc.	Exp.	Calc.	Exp.
9	1.045	0.009	0.026	0.070	0.083	0.070
10	0.966	0.059	0.027	0.110	0.091	0.304
11	1.413	0.026	0.021	0.026	0.098	0.159
12	0.979	0.048	0.022	0.230	0.155	0.318

in the determination of vitamin C by high performance liquid chromatography because of the low stability of this substrate during sample preparation. These results indicate that during freezing, the degradation of vitamin C was very small even at lower air velocities. Because the retention values in the samples frozen using the same process conditions but stored one day at –4°C were lower than those stored at –24°C, the storage temperature was determined to have a great influence on the degradation of vitamin C. The final substrate concentration not only depended on the kinetic constant modified by the temperature but also the reaction time. The freezing times and difference between them were not important compared to the storage time of 24 h even at lower air velocity when the vitamin C degradation was almost insignificant during the freezing process. During storage at –24°C the degradation reaction was undetected, but at –4°C it became perceivable. Although the vitamin C degradation rate was low in this case, it might be important with a longer storage time. Frozen foods should thus be stored below –18°C to reduce chemical reactions and the rate of vitamin C degradation. The lower retention values of samples frozen with air at –18°C and 4.54 m/s show that additional factors such as initial fruit quality increased the vitamin C loss as well.

Table 14-6 shows the ascorbic acid loss during strawberry thawing at different temperature and times where the initial and final experimental values in the fruit exhibit larger differences in thawing processes at higher temperatures and longer times. The ascorbic acid concentration in the exudate was also smaller for the strawberries thawing at higher temperature. Different thawing process conditions modified the temperature gradient and water movement inside the fruits which affected the substance losses. The oxidation reaction and the diffusion phenomenon are the two important problems involved in the process of vitamin C loss. Furthermore both increased at higher temperatures, yielding not just a negative additive but synergistic effect on the degradation reaction. For this reason the vitamin C value was smaller inside the strawberries thawed at high temperatures as well as in corresponding exudate (since the oxidation rate inside and outside the fruit was also higher).

Table 14-5 Ascorbic acid in strawberries after freezing and storage under different conditions.

T	$V = 0.508$ m/s $T = -24°C$			$V = 4.54$ m/s $T = -24°C$			$V = 0.508$ m/s $T = -18°C$			$V = 4.54$ m/s $T = -18°C$		
°C	C_i kg %	C_f kg %	R	C_i kg %	C_f kg %	R	C_i kg %	C_f kg %	R	C_i kg %	C_f kg %	R
-24	.063	.063	0.99	.041	.041	0.99	.060	.060	1.00	.105	.074	0.71
	.050	.050	0.99	.033	.030	0.89	.045	.052	1.15	.056	.042	0.75
	.050	.050	1.00	.039	.032	0.82	.039	.053	1.14	.070	.074	1.05
-4	.090	.058	0.64	.105	.055	0.52	.086	.073	0.85	.083	.026	0.32
	.068	.041	0.61	.106	.053	0.50	.089	.075	0.84	.047	.008	0.17
	.126	.078	0.62	.065	.054	0.83	.062	.050	0.82	.067	.019	0.28

Table 14-6 Ascorbic acid in strawberry thawing.

Sample	6°C for 180 min kg %			22°C for 70 min kg %			22°C for 180 min kg %		
N°	C_i	C_s	C_L	C_i	C_s	C_L	C_i	C_s	C_L
1	.055	.055	.000	.069	.063	.021	.064	.042	.026
2	.068	.063	-----	.056	.065	-----	.072	.053	.012
3	.050	.049	.010	.062	.061	.013	.065	.053	.029
4	.070	.062	.061	.087	.074	.021			
5	.072	.064	.055	.065	.055	.015			

CONCLUSIONS

There is considerable valid information on food properties available for heat transfer processes during phase changes so that freezing and thawing times can be predicted with reasonable accuracy. The combination of both heat and mass transfer models provides additional information even though their time estimations are not always closer to experimental times than those obtained by only considering heat transfer models. Quantitative formulation of the ice crystallization process and water equilibrium in foods is still necessary for setting up exudate calculations that might relate different process conditions to food characteristics in order to control the final quality of frozen-unfrozen products, and degradation reactions remain important during storage when temperatures are not low enough and during thawing processes at high temperature.

REFERENCES

Amer M.I. (1997). *Influencia de la Variables del Proceso de Congelación en la Transformación de la Vitamina c en Frutillas.* M.S. Thesis, Universidad Nacional del Litoral.

Delgado A.E. (1997). *Determinación del Comportamiento Térmico y las Características de los Alimentos en la Conservación por Congelación.* M.S. Thesis, Universidad Nacional del Litoral.

Fennema O., Powrie W., and Marth E. (1973). *Low Temperature Preservation of Food and Living Matter.* New York: Marcel Dekker, pp. 150–239.

Hayakawa K.I. and Succar J. (1982). Heat transfer and moisture loss of spherical fresh produce. *J. Food Sci.* 47: 596–605.

Laing B.M., Shlueter D.L., and Labuza T.P. (1978). Degradation kinetics of ascorbic acid at high temperature and water activity. *J. Food Sci.* 43(5): 1440–1443.

Rubiolo de Reinick A.C. (1985). *Mathematical Modelling of Freezing and Thawing.* Ph.D. Thesis, University of Massachusetts.

Rubiolo de Reinick A.C. (1992). Estimation of average freezing and thawing temperatures versus time with infinite slab correlation. *Comp. Ind.* 19: 297–305.

Rubiolo de Reinick A.C. and Schwartzberg H. (1986). Coefficients for air-to-solid heat transfer for uniformly spaced arrays of rectangular foods. *Food Eng. Proc.* 1(26): 273–283.

Schwartzberg H.G. (1976). Effective heat capacities for the freezing and thawing of food. *J. Food Sci.* 46: 152–156.

Schwartzberg H.G. (1981). Mathematical analysis of the freezing and thawing of food. *AICHE Summer National Meeting.* Detroit.

Tocci A.M. and Mascheroni R. (1991). Simulación de la transferencia de calor y materia en la congelación de alimentos. In: *Actas del IV Congreso Latinoamericano de Transferencia de Calor y Materia.* La Serena, Chile, pp. 352–355.

CHAPTER 15

Minimal Processing of Fruits and Vegetables

R.P. Singh and J.D. Mannapperuma

INTRODUCTION

Minimal processing of fruits and vegetables encompasses technologies that do not significantly alter their fresh-like attributes. Examples include industrial processes such as washing, sorting, cutting, trimming, slicing, and dicing. Minimal processing technologies for fruits and vegetables involve boundaries extending from raw material production to the consumer. Within this context, a product may be reduced in size and weight by cutting or trimming, it may undergo a certain amount of physiological change due to ripening or senescence, and packaging or other convenience-related alterations may increase its value.

Because a number of minimal processing technologies result in wounding of plant tissue and subsequent acceleration of deteriorative processes, controlled environmental conditions are a critical requirement in the transportation, distribution, storage, and retail display of these products. To realize a desirable shelf-life for them as well, modified atmosphere packaging (MAP) has become an integral part of minimal processes (i.e., cut lettuce is now routinely sold in MAP). This chapter presents innovative approaches in minimal processing of fruits and vegetables with a particular emphasis on the development of a predictive method that could be useful in designing MAP.

FRUIT AND VEGETABLE QUALITY

The quality characteristics of fruits and vegetables are markedly influenced by the handling conditions between production and consumption. Freshness is particularly compromised with slicing, dicing, or cutting. Indeed, a disruption in the cellular structure of a tissue may result in increased respiration, elevated ethylene production, and higher enzymatic activity (Brecht, 1995). For example, by cutting and slicing, certain enzymes and substrates that were previously separated come into contact with each other, and these enzymatic reactions accelerate deterioration such as browning in cut apples (Rolle and Chism, 1987). However, minimal processing treatments such as calcium dip and MAP help to retard browning in cut fruits (Bolin et al., 1977; Poovaiah, 1986).

The peeling of fruits and vegetables removes the protective outer layer and exposes the internal contents to deteriorative mechanisms. The loss of turgor due to cutting and resulting decrease in firmness has created the need to reduce additional moisture loss and maintain firmness for extended periods within minimally processed food packaging. Aroma is also an important quality characteristic that is influenced by minimal processing techniques, as cutting and slicing may accelerate its formation. MAP is thus designed to control carbon dioxide and oxygen content in packaging headspace that can lead to alteration of aroma development pathways (Kader et al., 1989). Flavor modifications can be accomplished with MAP as well; for example, acidified storage and MAP can minimize off-flavor development in cut carrots.

The effects of MAP on the nutritional content of fruits and vegetables are not well known, but procedures involving cutting and slicing are expected to increase nutrient loss due to reductions in cell sap. Similarly, washing and soaking processes can potentially enhance the leaching of water-soluble vitamins. Microbial proliferation in minimally processed foods is another potential problem now receiving increased attention due to outbreaks of food poisoning linked to them. Simple procedures such as cutting with knives provide microbes access to the internal nutrient-rich sites of a food product where growth of *Listeria*, *Salmonella*, and *E. coli* O157:H7 occurs (Beuchat and Brackett, 1990). While the use of carbon monoxide has been shown to inhibit microbial growth, good sanitary practices, chlorinated water, and HACCP methods during minimal processing are essential to eliminate unsafe conditions (Simons and Sanguansari, 1997).

To mitigate the adverse effects of minimal processing technologies on fruits and vegetables, various approaches have been attempted. Some of these include the selection of cultivars appropriate for minimal processing; control of temperature, humidity, and atmospheric gases; use of edible coatings; and application of chemical preservatives (Romig, 1995; Baldwin et al., 1995; Hurst, 1995; Krochta and Mulder-Johnston, 1997). Time temperature indicators have also been investigated as possible monitors of temperature management during the distribution and storage of processed foods (Rooney, 1995; Taoukis and Labuza, 1989; Wells and Singh, 1988).

CONTROL OF ENVIRONMENTAL FACTORS

In the case of fruits and vegetables, control of temperature is a key environmental factor that brings enormous benefit to the stored commodity. Storage temperature affects the rate of respiration, ethylene production, and transpiration losses—factors that cause major deterioration during storage. The rates of these deteriorative reactions are significantly lowered by the use of refrigeration not only during storage, but transportation and distribution as well. However, chill-sensitive commodities must not be allowed to go below a critical chill temperature.

Modification of environmental gases around a commodity is another way to extend shelf-life. Although the beneficial effects of controlling the atmosphere of fruits and vegetables during storage were noted in 1819 by Jacques-Étienne Berard, the commercial application of this technique did not come about until more than 100 years later after the pioneering work of Franklin Kidd and Cyril West at the Low Temperature Research Station in Cambridge, England.

Controlled atmosphere (CA) storage offers high precision in controlling temperature, humidity, and atmospheric gases during the storage of a commodity in large refrigerated warehouses (Smock, 1979; Dewey, 1983). Its application to apples is now widely practiced in many parts of the world.

MODIFIED ATMOSPHERE PACKAGING

MAP can be viewed as an extension of the CA storage concept but applied on a smaller scale. In MAP, a few fruits or vegetables are packaged in containers made of semipermeable polymeric films (Brody, 1989, Exama et al., 1993; Day, 1990, Zagory and Kader, 1988). The atmosphere in the package may be either artificially modified at the time of packaging (active modification) or the respiring commodity may be allowed to create its own atmosphere during storage (passive modification). While active modification of atmosphere is accomplished by injecting gas mixtures of known concentrations at the time of packaging, passive modification depends on the rates of respiration of the packaged commodity and gas permeability characteristics of the package walls. In addition, scavenging of certain gases such as oxygen, carbon dioxide, and ethylene may be accomplished by placing special absorbing compounds inside the package (Labuza and Breen, 1989; Rooney, 1995). MAP addresses four key processes that occur simultaneously:

1. **Product respiration** as influenced by maturity stage, product temperature, and concentrations of oxygen, carbon dioxide, and ethylene in the package
2. **Transpiration** as influenced by product surface temperature and surrounding humidity in the package
3. **Gas permeation through package walls** as influenced by packaging material, temperature, and the gas concentration gradient inside and outside the package
4. **Heat transfer** as influenced by the respiration process in which the heat of respiration results in an increase in product temperature

During storage, the above processes lead to the establishment of steady-state conditions inside a package. Although mathematical prediction of conditions during the transient phase and establishment of a steady state are complicated by variability in these processes, progress has been made in the development of predictive models for estimating steady-state atmospheric conditions and selecting appropriate packaging materials (Deily and Rizvi, 1981; Singh, et al., 1994; Hayakawa et al., 1975; Henig and Gilbert, 1975; Cameron et al., 1995; Lee et al., 1991; Rooney, 1995).

Mathematical Design

Because atmospheric air contains mostly nitrogen (78.08%), oxygen (21%), and carbon dioxide (0.03%), the respiration process of a commodity in an impermeable package will result in depletion of all oxygen and generation of high levels of carbon dioxide so that reactions such as fermentation and rapid decay will occur. To prevent such a catastrophe, the objective is to allow a certain amount of carbon dioxide and oxygen permeation through the package walls. In addition, a certain amount of water vapors and other volatile gases such as ethylene may also permeate.

Permeation of gases through a polymeric film is an activated diffusion process consisting of three steps: 1) absorption of gas molecules on one side of the film, 2) diffusion through the film, and 3) desorption of molecules on the other side of the film. These absorption and desorption processes are functions of molecules condensing into the polymer, partition coefficient, and interfacial forces. The diffusion process also depends on the size and shape of the permeating molecule, number of diffusion sites, and rubber or glassy state of the film material. Because carbon dioxide is almost five times more permeable than oxygen in low-density polyethylene (LDPE), the relative difference in permeation of carbon dioxide with respect to oxygen molecules plays an important role in the selection of polymeric film for MAP application. In general, fruits and vegetables require packaging that is highly permeable to CO_2, a characteristic common to many polymeric films such as LDPE. Due to such diffusion problems proper design of polymeric packages is needed, following are some of the variables with most influence in the design of MAP packages:

- Optimum package atmosphere for extended shelf-life
- Respiration rate
- Weight of produce
- Film permeability
- Film thickness
- Film area

Consider the schematic diagram of a MAP as shown in Figure 15-1. A commodity of weight W is packaged in a permeable container with walls of thickness b. A is the surface area of the package, R_x the respiration rate for oxygen consumption, R_c the respiration rate for carbon dioxide generation, X_i, the concentration of oxygen inside the package, C_i the concentration of carbon dioxide inside the package, and X_o and C_o the outer atmospheric concentrations of oxygen and carbon dioxide, respectively.

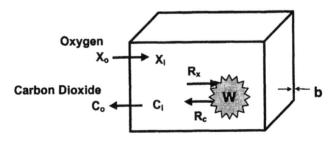

Figure 15-1 A schematic diagram of a MAP.

At steady state, conducting a mass balance on oxygen (Eq. 15-1) and carbon dioxide (Eq. 12-2; Singh et al., 1994) results in

$$WR_x = P_x A \frac{X_o - X_i}{b} \qquad (15\text{-}1)$$

$$WR_c = P_c A \frac{C_i - C_o}{b} \qquad (15\text{-}2)$$

By dividing Equation 15-1 by Equation 15-2 and rearranging terms, one obtains

$$C_i = C_o + \frac{1}{\beta}(X_o - X_i)\frac{R_c}{R_x} \qquad (15\text{-}3)$$

where

$$\beta = \frac{P_c}{P_x} \qquad (15\text{-}4)$$

Since Equation 15-3 represents a straight line, if the R_c/R_x ratio of respiration coefficients is one, Equation 15-3 has a slope of $1/\beta$.

When designing a package it is necessary to consider the following assumptions to develop information:

- The rate of respiration is constant.
- The perfect gas law is applicable.
- The package is maintained at a constant temperature.
- The permeability coefficients are constant.

As shown in Figure 15-2, Equation 15-3 can be plotted as a straight line on an x-y chart where the x-axis represents oxygen concentration and the y-axis is the concentration of carbon dioxide. The two lines shown in Figure 15-2 are for slopes 1/5 and 1/0.8. The first line is drawn for the CO_2/O_2 ratio of 5 as for LDPE, whereas the second line is for the CO_2/O_2 ratio of 0.8 for diffusion in air. The points on these lines are the concentrations of CO_2 and O_2 that can be obtained under steady-state conditions when a polymeric film with those ratios is selected. For example, if LDPE is used as a package material, then the steady-state internal package atmospheres of any combination of O_2 and CO_2 that fall on that line can be obtained (such as 10% O_2 and 2% CO_2). If a package with several large holes is used, then an internal package atmosphere of 10% O_2 and 14% CO_2 can be established. The application of this analysis depends on the selection of polymeric films that meet the necessary criterion of the CO_2/O_2 ratio. In addition, the packaging film must have the requisite permeability of oxygen and carbon dioxide that is suitable for maintaining the desired gas concentrations when a particular commodity is stored inside a package.

Post-harvest physiologists have determined recommended modified atmospheres for fruits and vegetables (Blankenship, 1985) that can also be expressed graphically to determine whether the selected packaging material will allow generation of the recommended atmosphere at steady state. For example, it is known that the recommended modified atmosphere conditions for oranges are 5 to 10% oxygen and 0 to 4% carbon dioxide; for kiwi 1 to 2% oxygen and 3 to 5% carbon dioxide; and for berries 5 to 10% oxygen and 15 to 20% carbon dioxide (Singh et al., 1994). As illustrated in Figure 15-3, this implies that LDPE is suitable for storing kiwi and oranges but not berries because the line with a slope of 1/5 crosses the boundaries of rectangles for kiwi and oranges. Further examination of recommended atmospheric conditions for MAP of fruits and vegetables reveals that LDPE can generate recommended atmospheric conditions for nectarines, peaches, plums, oranges, bananas, avocados, cauliflower, cabbage, leeks, and celery. However, published experimental data for recommended MAP conditions are available for only a few cut products such as lettuce and broccoli (Ballantyne et al., 1988a,b; McDonald et al., 1990).

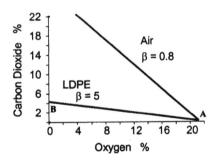

Figure 15-2 Plot of Equation 15-3 for permeability of carbon dioxide and oxygen in LDPE and air.

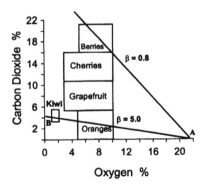

Figure 15-3 Plot of recommended conditions for MAP of selected commodities.

When designing a MAP, the type of commodity determines the respiration rates R_x and R_c and required gas concentrations for extending shelf-life X_i and C_i. The permeability ratio is thus already fixed because it identifies a potential polymeric film that will meet that specific requirement. The only three remaining items then left for a package designer to select are the film area, mass of product, and thickness of film. Equations 15-1 and 15-2 can thus be rewritten as:

$$C_i = C_o + R_c = \frac{R_c}{P_c} \phi \qquad (15\text{-}5)$$

and

$$X_i = X_o - \frac{R_x}{P_x} \phi \qquad (15\text{-}6)$$

where

$$\phi = \frac{W_b}{A} \qquad (15\text{-}7)$$

Increasing factor ϕ by augmenting the weight of the commodity in the package, enlarging the film thickness, or decreasing the surface area of the package will move a point along the straight line on Figure 15-3 from A to B. For example, if more product is put into a given size package, the internal atmosphere will become higher in carbon dioxide concentration and lower in oxygen concentration. Similar effects on gas composition will be observed if the area of the package is decreased or a thicker film is chosen.

Perforated Packages

One of the key limitations of polymeric films is that most of the more commonly used ones such as LDPE and polyvinyl chloride have a permeability ratio that is within the narrow range of 4 and 6 so that only a few products benefit from their use. In the following section, the previous mathematical analysis is extended to demonstrate how a package with perforations may be designed to overcome this limitation (Singh et al., 1994).

Because polymeric films are prone to puncture during mishandling and air is considerably more permeable than polymeric film, a small hole can significantly influence package atmosphere. However, carefully designed perforations may be introduced in a polymeric film to create atmospheres that will provide the required modified atmosphere for a given commodity. The internal atmosphere of a perforated package is influenced by permeability through both the film and perforation, and the flow of a gas is proportional to the permeance (p) of the path and concentration gradient. The resulting oxygen balance gives:

$$WR_x = p_{xf}(X_o - X_i) + p_{xp}(X_o - X_i) \qquad (15\text{-}8)$$

where p_{xf} is the permeance of oxygen in a polymeric film given by

$$p_{xf} = \frac{P_{xf}A_f}{b_f} \qquad (15\text{-}9)$$

and

$$P_x = P_{xf} + P_{xp} \qquad (15\text{-}10)$$

then

$$WR_x = P_x(X_o - X_i) \qquad (15\text{-}11)$$

Similarly, for carbon dioxide balance:

$$WR_c = P_{cf}(C_i - C_o) + P_{cp}(C_i - C_o) \qquad (15\text{-}12)$$

where p_{cf} is the permeance of carbon dioxide in a polymeric film as in:

$$P_{cf} = \frac{P_{cf}A_f}{b_f} \qquad (15\text{-}13)$$

$$P_c = P_{cf} + P_{cp} \qquad (15\text{-}14)$$

then,

$$WR_c = P_c(C_i - C_o) \qquad (15\text{-}15)$$

The characteristic equation for a perforated package similar to Equation 15-3 is:

$$C_i = C_o + \frac{1}{\beta}(X_o - X_i)\frac{R_c}{R_x} \qquad (15\text{-}16)$$

198

where β in the above equation is:

$$\beta = \frac{P_c}{P_x} = \frac{P_{cf} + P_{cp}}{P_{xf} + P_{xp}} \qquad (15\text{-}17)$$

The permeance ratio for film and perforation may then be written as:

$$\beta_f = \frac{P_{cf}}{P_{xf}} \qquad (15\text{-}18)$$

and

$$\beta_p = \frac{P_{cp}}{P_{xp}} \qquad (15\text{-}19)$$

from Equation 15-10,

$$P_x = P_{xf} + \frac{P_{cp}}{\beta_p} \qquad (15\text{-}20)$$

and combining Equations 15-14 and 15-20:,

$$P_x = P_{xf} + \frac{P_{cp} - P_{cf}}{\beta_p} \qquad (15\text{-}21)$$

or

$$P_x = P_{xf} + \frac{\beta P_x - \beta_f P_{xf}}{\beta_p} \qquad (15\text{-}22)$$

then,

$$\beta_p P_x = \beta_p P_{xf} + \beta P_x - \beta_f P_{xf} \qquad (15\text{-}23)$$

or

$$P_x = P_{xf} + \frac{\beta_f - \beta_p}{\beta - \beta_p} \qquad (15\text{-}24)$$

Combining Equations 15-9, 15-11, and 15-24 produces:

$$W = \frac{P_{xf} A_f (\beta_f - \beta_p)}{R_x b_f (\beta - \beta_p)} (X_o - X_i) \qquad (15\text{-}25)$$

199

To calculate the area for perforations, substitute Equations 15-25, 15-9, and an expression of p_{xp} (similar to Equation 15-9) in Equation 15-8 and rearranging:

$$\frac{P_{xf}A_f}{b_f} + \frac{P_{xp}A_p}{b_p} = \frac{P_{xf}A_f(\beta_f - \beta_p)}{b_f(\beta - \beta_p)} \qquad (15\text{-}26)$$

$$\frac{P_{xp}A_p}{b_p} = \frac{P_{xf}A_f}{b_f}\left(\frac{(\beta_f - \beta_p)}{(\beta - \beta_p)} - 1\right) \qquad (15\text{-}27)$$

Since $b_p = b_f$, simplifying the above gives

$$A_p = \frac{P_{xf}A_f}{P_{xp}}\left(\frac{(\beta_f - \beta)}{(\beta - \beta_p)}\right) \qquad (15\text{-}28)$$

In a packaging design preliminary study Equations 15-25 and 15-28 were used to develop a perforated package for broccoli storage where the MAP required an oxygen concentration of 2% and carbon dioxide concentration of 8%. The respiration rate for oxygen and carbon dioxide at 2.5°C is 9.5 mL/kg h. The package size selected for this study was 200 mm × 400 mm. The permeability of LDPE for oxygen is 2,120 mL-μm/m^2-hr-atm and 1,0470 mL-μm/m^2-hr-atm for carbon dioxide. The thickness of the LDPE was 100 μm. Using Equation 15-25, the weight of broccoli that will generate the required atmospheric concentrations of oxygen and carbon dioxide was determined to be 178 g, and the area of perforation was obtained from Equation 15-28 as 30,000 μm^2. These predicted values were then used in designing an actual storage study. Broccoli was stored at 2°C in a 200 mm × 400 mm package made of LDPE film with a perforation (195 μm diameter) created by puncturing the film with a hot wire (heated by discharging a capacitor). Good agreement was obtained between the predicted and experimental values of carbon dioxide and oxygen concentrations in the perforated package (Mannapperuma and Singh, 1996). Additional studies are ongoing to further develop this predictive technique for wider applications.

In summary, engineering calculations can be of great benefit in designing MAP because the estimated values can provide guidance in selecting an appropriate polymeric film. Note, however, that because these models are based on several assumptions, their reliability is also dependent on them. For wider applications of MAP for minimally processed fruits and vegetables, additional work is needed to develop polymeric films with a wider range of permeabilities and resistance to punctures, determine compounds that can be safely incorporated into a package to eliminate objectionable gases, study the effect of temperature on the permeability of gases through polymeric films, and identify varieties that are tolerant to near-zero oxygen or elevated carbon dioxide levels.

NOTATION

A = Area of film, m^2
β = permeability ratio, defined in Equation 15-4
b = film thickness, μm
C = concentration of carbon dioxide, %
ϕ = coefficient defined in Equation 15-7
p = permeance, mL/h-atm
P = permeability, $mL\text{-}\mu m/m^2\text{-}h\text{-}atm$
R = respiration rate, mL/kg-h
W = weight of commodity in a package, kg
X = concentration of oxygen, %

Subscripts:
c = carbon dioxide
f = film
i = inside the package
o = outside the package
p = perforation
x = oxygen

REFERENCES

Baldwin E.A., Nisperos-Carriedo M.O., and Baker R.A. (1995). Edible coatings for lightly processed fruits and vegetables. *HortSci.* 30(1): 35–38.

Ballantyne A., Stark R., and Selman J.D. (1988a). Modified atmosphere packaging of broccoli florets. *Int. J. Food Sci. Technol.* 23: 353–360.

Ballantyne A., Stark R., and Selman J.D. (1988b). Modified atmosphere packaging of shredded lettuce. *Int. J. Food Sci. Technol.* 23: 267–274.

Beuchat L.R. and Brackett R.E. (1990). Survival and growth of *Listeria monocytogenes* on lettuce as influenced by shredding, chlorine treatment, modified atmosphere packaging and temperature. *J. Food Sci.* 55: 755–758.

Blankenship S.M. (1985). Controlled atmospheres for storage and transport of perishable agricultural commodities. In: *Controlled Atmospheres for Storage and Transport of Perishable Agricultural Commodities: Papers Presented at the Fourth National Controlled Atmosphere Research Conference.* Raleigh, N.C: North Carolina State University, pp. 51–54.

Bolin H.R., Stafford A.E., King A.D. Jr., and Huxsoll C.C. (1977). Factors affecting the storage stability of shredded lettuce. *J. Food Sci.* 56: 60–62, 67.

Brecht J.K. (1995). Physiology of lightly processed fruits and vegetables. *HortSci.* 30(1): 18–22.

Brody A.L., ed (1989). *Controlled/Modified Atmosphere/Vacuum Packaging of Foods.* Brody A.L., Ed. Trumbull, CT: Food and Nutrition Press.

Cameron A.C., Talasilla P.C., and Joles D.W. (1995). Predicting film permeability needs for modified atmosphere packaging of lightly processed fruits and vegetables. *HortSci.* 30(1): 25–34.

Day B.P.F. (1990). A perspective of modified atmosphere packaging of fresh produce in western Europe. *Food Sci. Tech. Today* 4(4): 215–221.

Deily K.R. and Rizvi S.S.H. (1981). Optimization of parameters for packaging of fresh peaches in polymeric films. *J. Food Proc. Eng.* 5: 23–29.

Dewey D.H. (1983). Controlled atmosphere storage of fruits and vegetables, In: *Developments in Food Preservation—2.* Thorne S., Ed. London: Applied Science Publishers, pp. 1–24.

Exama A., Arul, J., Lencki R.W., Lee L.Z, and Toupin C. (1993). Suitability of plastic films for modified atmosphere packaging of fruits and vegetables. *J. Food Sci.* 58(6): 1365–1370.

Hayakawa K.E., Henig Y.S., and Gilbert S.G. (1975). Formulae for predicting gas exchange of fresh produce in polymeric film packages. *J. Food Sci.* 40: 186–93.

Henig Y.S. and Gilbert S.G. (1975). Computer analysis of the variables affecting respiration and quality of produce packaged in polymeric films. *J. Food Sci.* 40: 1033–1037.

Hurst W.C. (1995). Sanitation of lightly processed fruits and vegetables. *HortSci.* 30(1): 22–24.

Kader A.A., Zagory D., and Kerbel E.L. (1989). Modified atmosphere packaging of fruits and vegetables. *CRC Crit. Rev. Fd. Sci. Nut.* 28(1): 1–30.

Krochta J.M. and Mulder-Johnston C.D. (1997). Edible and biodegradable polymer films: Challenges and opportunities. *Food Technol.* 51(2): 61–74.

Labuza T.P. and Breen W.M. 1989. Application of "active packaging" for improvement of shelf-life and nutritional quality of fresh and extended shelf-life foods. *J. Food Proc. Pres.* 13: 1–69.

Lee D.S., Haggar P.E., Lee J., and Yam K.L. (1991). Model for fresh produce respiration in modified atmosphere based on principles of enzyme kinetics. *J. Food Sci.* 56(6): 1580–1585.

Mannapperuma J.D. and Singh R.P. (1996). Design of perforated polymeric packages for the modified atmosphere storage of fresh fruits and vegetables. Technical Report. Department of Biological and Agricultural Engineering, University of California, Davis.

McDonald R.E., Risse L.A., and Barmore C.R. (1990). Bagging chopped lettuce in selected permeability films. *HortSci.* 25: 671–673.

Poovaiah B.W. (1986). Role of calcium in prolonging storage life of fruits and vegetables. *Food Technol.* 40: 86–89.

Rolle R.S. and Chism G.W. (1987). Physiology consequences of minimally processed fruits and vegetables. *J. Food Qual.* 10: 157–177.

Romig W.R. (1995). Selection of cultivars for lightly processed fruits and vegetables. *HortSci.* 30(1): 38-40.

Rooney M.L., ed. (1995). *Active Food Packaging*. London: Blackie Academic and Professional.

Simons L.K. and Sanguansri, P. (1997). Advances in the washing of minimally processed vegetables. *Food Australia* 49(2): 75–80.

Singh R.P., Oliveira F.A., and Fernanda A.R., eds. (1994). *Minimal Processing of Foods and Process Optimization.* Eds. Boca Raton, FL: CRC Press.

Smock R.M. (1979). Controlled atmosphere storage of fruits. *Hort. Rev.* 1: 301–336.

Taoukis P.S. and Labuza T.P. (1989). Applicability of time-temperature indicators as shelf-life monitors of food products. *J. Food Sci.* 54(4): 783–788.

Wells J.H. and Singh R.P. (1988). Performance evaluation of time-temperature indicators in monitoring changes in quality attributes of perishable and semi-perishable foods. *J. Food Sci.* 53(1): 148–152, 156.

Zagory D. and Kader A.A. (1988). Modified atmosphere packaging of fresh produce. *Food Technol.* 42(9): 70–77.

CHAPTER 16

Minimal Preservation of Fruits: A CYTED Project

S.M. Alzamora, M.S. Tapia, A. Leúnda, S.N. Guerrero, A.M. Rojas, L.N. Gerschenson, and E. Parada Arias

INTRODUCTION

A multinational collaborative project titled "Development of minimal processing technologies for food preservation" was carried out from 1995 to 1997 as part of the Science and Technology for Development in Ibero-America (CYTED) Food Preservation sub-program. The main objectives of this project were:

1. To contribute with basic knowledge about:
    a. the effect of different preservation factors and their interaction on microorganisms; the organoleptic, physical, and chemical properties of certain food products; and the ultrastructural alterations of food tissues
    b. the transport phenomena involved in the different stages of preservation
2. To develop minimal processing technologies to obtain:
    a. high moisture food products with physico-chemical and sensorial properties similar to fresh foods that are stable at ambient temperature or under typical refrigeration conditions
    b. traditionally preserved food products that maintain the identity characteristics (flavor, color) of their natural state or that exhibit excellent organoleptical quality

These minimal processing technologies use a combination of different preservation factors (hurdle approach) such as: packaging in modified/controlled atmospheres, storage under refrigeration, water activity diminishment, pH control, antibrowning additives and natural antimicrobials (to totally or partially replace synthetic ones), edible films, physical nonthermal processes, slight thermal treatments, thermal treatments with assistants, irradiation, and fermentation, all appropriately selected and combined according to the product to be developed.

The six countries involved in this project were Argentina, Chile, Mexico, Portugal, Spain, and Venezuela. Table 16-1 provides information about the participating institutions and research groups.

205

Table 16-1 Summary of the overall participation in the project "Development of minimal processing technologies for food preservation."

Country	Participating Institution	Number of Researchers
Argentina	• Universidad de Buenos Aires (UBA) Facultad de Ciencias Exactas y Naturales, Departamento de Industrias	10
	• Universidad Nacional de La Plata Centro de Investigación y Desarrollo en Criotecnología de Alimentos	3
Chile	• Universidad Católica de Valparaíso y Universidad de Santiago	3
	• Pontificia Universidad Católica de Chile Departamento de Ingeniería Química y Bioprocesos	2
	• Universidad de Chile Facultad de Ciencias Agrarias Departamento de Agroindustrias y Tecnología de Alimentos	1
Spain	• Universidad Politécnica de Valencia Departamento de Ingeniería de Alimentos	17
	• Universidad de Córdoba Facultad de Veterinaria Tecnología de Alimentos	6
	• Consejo Superior de Investigaciones Científicas Instituto del Frío Departamento de Ciencia y Tecnología de Productos Vegetales	6
	• Universidad Autónoma de Barcelona Facultad de Veterinaria Unidad de Tecnología de Alimentos	8
Mexico	• Instituto Politécnico Nacional Escuela Nacional de Ciencias Biológicas Departamento de Graduados e Investigación en Alimentos	3
	• Instituto Tecnológico de Tepic Departamento de Ingeniería Química y Bioquímica	4
	• Universidad de las Américas Departamento de Ingeniería Química y de Alimentos	9
Portugal	• Universidade Católica Portuguesa Escola Superior de Biotecnología	4
Venezuela	• Universidad Central de Venezuela Instituto de Ciencia y Tecnología de Alimentos	10
	• Universidad Simón Bolívar Departamento de Procesos Bioquímicos y Biológicos	10
	• Instituto Universitario de Tecnología de Cumaná	6

The investigations performed dealt with a large variety of topics including the modeling of mass transport in solid-liquid food operations (i.e., vacuum or atmospheric dewatering and impregnation soaking); physicochemical, biochemical, and structural characterization of minimally processed plant foods; ultrastructure-mechanical behavior relationships in fruit tissues, cheese, and egg gels as affected by various processes; fruit and vegetable preservation using controlled/modified atmospheres; development and/or optimization of combined technologies for obtaining high moisture minimally processed fruits using vacuum impregnation, edible films, natural antimicrobials, or high hydrostatic pressures; response of pathogenic bacteria and spoilage flora to emerging preservation factors and their interactions with other environmental stresses; salting of fish and cheese by vacuum impregnation; and application of high hydrostatic pressures to liquid whole egg, avocado, cheese, milk, poultry meat, and yogurt.

A major part of this project was also concerned with the optimization of combined preservation technologies for obtaining high moisture fruit products (HMFP). These technologies were developed as part of a previous multinational CYTED program entitled "Bulk preservation of fruits by combined methods technology," carried out from 1991 to 1994 (Welti-Chanes and Vergara-Balderas, 1995). The major goal for the design of these combined systems was the development of simple and inexpensive conservation techniques for bulk fruit storage at ambient temperature without major capital expenditures that were energy-efficient, suitable to preserve fruits *in situ*, able to help in overcoming seasonal production constraints, and beneficial to the diversification of local fruit industries and reduction of post-harvest losses while meeting consumption trends (Alzamora et al., 1995).

Hurdles used for the stabilization of HMFP include water activity (a_w), pH, mild heat treatment, and preservatives. The preservation process combining these factors is very simple, consisting only of fruit blanching followed by an a_w depression step by osmotic dehydration with simultaneous incorporation of additives to achieve final values after equilibration of $a_w = 0.94$–0.98, pH = 3.0–4.1 (adjusted with citric or phosphoric acid), 400–1,000 ppm potassium sorbate, or sodium benzoate and 150 ppm sodium bisulphite (Fig. 16-1).

The processing of some fruits (i.e., banana purée and pomalaca) includes a slight thermal treatment after packing or a hot filling stage. The dewatering and impregnation soaking process is carried out at room temperature by placing fruit slices in concentrated aqueous solutions (moist infusion mainly of glucose, sucrose, maltodextrins, corn syrups, or their mixtures and additives) or mixing the fruit, humectant(s), and additives in required proportions (blending) as in the cases of purées and some whole fruits (i.e., strawberries).

This chapter brings together reports of progress by some of the project participants in the area of minimally processed fruits. In the optimization of these combined technologies, two aspects received consideration:

1. The equilibration stage, where vacuum techniques were used for the dewatering and impregnation soaking process and incorporation of additives.
2. The use of antimicrobials of natural origin as replacements (total or partial) for sorbates and other synthetic additives.

Advances made to accomplish fruit quality optimization in each of these aspects will be presented in the following sections.

VACUUM IMPREGNATION

The porous microstructure of food nutrients is an important property often neglected by food processors that are usually unaware that it can be successfully taken advantage of in food preservation operations and product development. Foods exhibiting this property can be impregnated (i.e., filling of pores with a suitable solution), introducing solvents and solutes of choice into their porous spaces that are occupied by a certain amount of occluded gas. The volume of this gas can be modified by penetration of the impregnation liquid as a result of capillary action or the combined effect of capillary action and pressure gradients which are imposed to the system. This penetration of liquid produced by pressure gradients which act as driving forces can be controlled by the expansion or compression of the occluded gas (Fito and Pastor, 1994). One way to accomplish this is to apply a vacuum treatment to the product for short

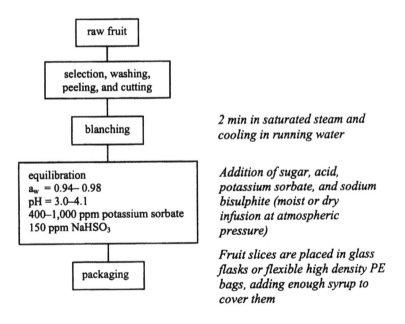

Figure 16-1 Flow diagram for the production of shelf-stable high moisture fruits (adapted from Alzamora et al., 1995).

periods (i.e., 5–10 min) while immersed in the liquid and then re-establish the atmospheric pressure, which causes the gas to be expelled from the pores being replaced by the entering liquid.

Fruits in general are good examples of foods having a microporous structure that can be vacuum impregnated. Vacuum impregnation produces faster water loss kinetics for short fruit treatments as compared with time-consuming atmospheric pseudo-diffusional processes due to the occurrence of a specific mass transfer phenomenon known as the hydrodynamic mechanism (HDM) and the increase produced in the solid-liquid interfacial area (Fito and Chiralt, 1995). This process could be appropriate in the development of new minimally processed fruit products and/or improved pretreatments for traditional preservation methods such as canning and freezing.

A proper formulation of the impregnation liquid allows expeditious compositional modifications of the solid matrix that may result in quality and stability enhancement of final products without submitting the food structure to eventual stress due to long exposure to gradient solute concentration. The use of vacuum impregnation techniques by apple processors for firming the tissue and improving the quality of canned and frozen apple slices dates back to the 1950s. Penetrated aqueous solutions may contain sugar, calcium salts, organic acids, colors, flavors, sulphurous salts, and/or a combination thereof.

During the vacuum impregnation of porous fruits, important modifications in structure and composition occur as a consequence of external pressure changes. The final products may exhibit structural, physical, and chemical properties very different from those of atmospherically infused fruits. For a better understanding of the complexity involved in vacuum solid-liquid operations, analysis is needed of the sequence in which the typical phenomena take place when a porous fruit is immersed in a liquid and subatmospheric pressures are applied to the system (Fito et al., 1996; Sousa et al., 1998):

- During the first moments, there is an expansion of gas trapped in the pores (partially flowing out) and the solid matrix may be deformed. Nearly all fruits lose native liquid during this stage.
- When the pressure equilibrium is reached, the flow of gas from the pores stops, the HDM occurs, and external liquid begins to enter the pores as an effect of capillary pressure. At equilibrium, the gas volume has diminished and the same volume of liquid has been penetrated by the HDM.
- When atmospheric pressure is restored in the system, forces due to differences between external and internal pressures make the liquid partially fill the intercellular spaces by the HDM and may also produce solid matrix deformation. This impregnation depends on the quantity of gas removed from the porous material since it is reversible and controlled by the compression or expansion of the occluded gas.

Gas expansion during the vacuum step may cause some deformation of the solid matrix. When atmospheric pressure is applied back to the system, the residual gas compression permits relaxation of the deformed solid depending on its viscoelastic properties. Hence, three phenomena are coupled in vacuum impregnation: 1) the gas flowing out, 2) the deformation-relaxation of the solid matrix and the liquid flowing in, and 3) the characteristics of the impregnated product depending on which of them is the limiting step along the process. The major factors determining the kinetics and final equilibrium of the system include (Fito et al., 1996):

- Tissue structure (pore size, distribution and connections between pores)
- Relaxation time of the solid matrix (viscoelastic properties of the fruit)
- Mass transport rate of the HDM (viscosity of the solution, fruit structure, size and shape of pores, pressure gradient)
- Size and shape of the sample

Assuming the relevant role of fruit structure and its mechanical properties, it is obvious that each fruit will exhibit a different response to the application of subpressures and process variables. Accordingly, different values of the effective porosity (ε_e), volumetric fraction of liquid that penetrates the tissue by HDM (X), and final solid matrix deformation at various low pressures have been reported for many fruits (among others, banana, apple, mango, papaya, strawberry, kiwifruit, and apricot) (Fito and Chiralt, 1995; Sousa et al., 1998; Shi, 1994).

The mathematical approach to describe the phenomenon of de-watering and the impregnation soaking process (DIS) produced by the application of an initial vacuum pulse has been carried out by Fito and co-workers as part of their attempts to perform the physicochemical and structural characterization of raw and vacuum processed fruits and identify the mechanisms involved (Fito and Pastor, 1994; Fito and Chiralt, 1995; Fito et al., 1996). The latest advances on this topic are addressed in Chapter 12, but here some recent results are introduced on the color and textural responses of fruits to vacuum impregnation as well as microbiological aspects of concern in relation to organism penetration by HDM in plant tissues.

Textural Aspects

Texture is one of the major quality attributes for fruit acceptability, as it is well known that microstructure relationships govern the mechanical behavior of plant tissues (Jackman and Stanley, 1995; Ilker and Szczesniak, 1990). Since biological tissues are highly organized, contributions from the different structures (i.e., polymers, cell wall, cell, tissue, and organ) and their interactions determine the textural characteristics of fruit tissue and/or organs (Waldron et al., 1997). The mechanical properties of cell walls, turgor pressure generated within cells by osmosis, and cell to cell adhesion (through the middle-lamellar pectic polysaccharides and/or plasmodesmata) are key structural determinants of parenchymatous edible fruit tissue texture. Recent cell wall models are based on the existence of three independent but interacting networks: the cellulosexyloglucan, pectin, and protein, all contributing to the overall mechanical properties of the cell wall (Carpita and Gibeaut, 1993). And since the cell is polymeric in nature, it is also viscoelastic (i.e., it exhibits both viscous and retarded elastic deformations in response to stress) (Brett and Waldrom, 1996).

The viscoelastic behavior of fruit tissues is normally subject to drastic changes as a result of an a_w lowering and/or blanching treatment. Previous research aimed to understand the structural–mechanical property relationships of various minimally processed fruits (strawberry, kiwifruit, pineapple) focused on the effect of a_w depression by atmospheric impregnation, type of humectant, level of a_w reached, addition of calcium, and existence of a previous blanching step. Although cellular membranes, cell walls, and cell separation were affected in a specific way by the different treatments, a close correlation between structural cell modifications and instrumental textural changes was obtained (Vidales et al., 1998; Muntada et al., 1998; Alzamora and Gerschenson, 1997; Alzamora et al., 1997).

When the DIS process was performed under vacuum, fruits showed different textural characteristics than those processed at atmospheric pressure. Vacuum-treated fruits (e.g., mango, papaya, apple, melon, and kiwifruit) were perceived as notoriously more juicy even than fresh fruit, and in general exhibited better textural quality.

Research to provide some insight into the phenomena involved in the textural changes occurring during both impregnation treatments has recently been performed on kiwifruit and melon infused with glucose to a_w values ranging between 0.97–0.98 (Muntada et al., 1998; Rojas et al., 1998). Both infusion treatments (atmospheric and vacuum) resulted in a loss of turgor pressure either due to plasmolysis or disruption of the tonoplast and plasmalemma, as well as a sharp decrease in the rheological elasticity of the fruit tissues as shown by relaxation assays. This loss in elasticity is also attributed to the air-liquid exchange during the vacuum operation.

For kiwifruit (*Actinidia deliciosa* cv. Hayward), the atmospheric process (immersion of kiwifruit halves in a 0.97 a_w glucose aqueous solution for 5 d, with a final a_w of 0.97) significantly decreased the compression force (F). Glucose infusion under vacuum (immersion of kiwifruit halves in a 59.0% w/w glucose aqueous solution for 10 min at 60 mm Hg, followed by 10 min at atmospheric pressure, with a final a_w of 0.98) produced failure forces similar to that of the fresh fruit and > 300% greater than those observed after atmospheric infusion ($F_{vacuum} \cong 188$ N versus $F_{atm} \cong 38$ N). The final moisture attained in both cases was equal, but a very different failure pattern was determined for both treatments since the pattern for vacuum conditions was more similar to that exhibited by raw fruit. Atmospheric impregnated kiwifruits showed a low F and a force drop followed by an oscillating force stage while liquid extraction took place. Vacuum impregnated kiwifruits showed a steady force that increased until fracture, followed by a force drop. The force then increased to high values at further deformation, ending at the predetermined deformation of the assay.

When kiwifruit samples with different treatments were examined using light microscopy (LM) and transmission electron microscopy (TEM), histological analyses revealed thin-walled parenchymatous small and large cells, some nearly spherical and others with an angular polyhedral shape and limited intercellular spaces. A large vacuole limited the cytoplasm to a thin peripheral layer pressed against the cell wall between the tonoplast and plasmalemma. Many cell walls exhibited various small localized areas of intense staining. In TEM observations, these regions contained plasmodesmatal connections between cells. Cell walls of raw fruit were stained intensely near the wall surface, but this was reduced in the region of the middle lamella. There was a slight thinning of the wall containing the plasmodesmata. In contrast, atmospheric solute impregnation caused extensive plasmolysis of cellular membranes, degradation of the cell walls and middle lamella, and a decrease in cell to cell contact. The cell walls of the treated tissues were difficult to identify and had very low optical density. In TEM, glucose-treated tissues showed severe reduction in overall staining since they were lower in the plasmodesmatal region. Alternatively, microscopy of vacuum-treated tissues showed a moderate optical density of cell walls since the staining was of similar density in the wall and plasmodesmatal regions. A conspicuous middle lamella was also observed in some areas. As seen in LM, original cell shapes and arrangements were retained, although some cell rupture occurred. The slight histological and ultrastructural changes within the kiwifruit due to vacuum treatment were consistent with the greater firmness observed, and indicated that glucose impregnation was less detrimental to kiwifruit texture when performed under vacuum.

Rojas et al. (1998) analyzed the effect of glucose and/or calcium infusion (in vacuum or atmospheric) on the texture and ultrastructure of melon (*Cucumis melo* L., honeydew variety) and found that the effect of the DIS process with glucose solutions to depress a_w to 0.98 depended on the pressure treatment. When impregnation was performed in vacuum (immersion in a 55% w/w glucose aqueous solution 10 min at 213 mbar followed by 10 min at atmospheric pressure, with a final a_w of 0.98), the a_w reduction produced a similar resistance to puncture in the melon and raw fruit. However, atmospheric treatment (immersion in a 0.98 a_w glucose aqueous solution for 4 d, with a final a_w of 0.98) determined the need of a smaller force to penetrate the fruit. Resistance to puncture was significantly greater when Ca^{2+} was present during atmospheric or vacuum impregnation, and there was good correlation between the instrumental measurement of texture and ultrastuctural flesh tissue changes (Fig. 16-2). Fresh tissues showed darkly stained walls with greater intensity toward the margin and central zone, marginal arrangement of cytoplasm, numerous invaginations of plasmalemma (arrows), and the formation of vesicles, although the plasmalemma and tonoplast were left intact (Fig. 16-2a). Slight plasmolysis of the cellular membranes was caused by vacuum treatment, although the cell walls showed good optical density with the middle lamella clearly visible (Fig. 16-2b). The addition of calcium (final concentration of calcium lactate in the fruit was $\cong 0.08$ w/w) during vacuum treatment resulted in darkly stained walls with longitudinally packed miofibrils (Fig. 16-2c). Atmospheric infusion produced folding and rupture (solid arrow) of cell walls and disruption of the tonoplast and plasmalemma with formation of coarse granules assembled along the cell walls (open arrows). Lack of staining indicated loss of cell wall material (Fig. 16-2d, e). The cell walls of the atmospheric-treated tissues with the addition of calcium (final concentration of calcium lactate in the fruit $\cong 0.1$ w/w) showed good electron density with a clear reticulate pattern, although the membranes appeared broken with vesicle formation (arrow) (Fig. 16-2f). The greater integrity of the kiwifruit and melon tissue cell walls when the DIS process was performed under vacuum was clearly due to the better textural quality of the fruits.

Color Aspects

The major visual changes in minimally processed fruits of high a_w during processing and/or storage are caused by both enzymatic and/or non-enzymatic development of brown substances and decolorations due to pigment degradation and/or leaching. Color stability is reportedly dependent upon the type of fruit, preservation system used, storage temperature, and chemical and/or physical depletion of the antibrowning agent in the fruit tissue (Alzamora et al., 1995). The DIS process performed at low operating pressures also greatly affects tissue color. During vacuum treatment air leaves the pores and thereby diminishes the oxygen concentration and hence the oxidative reaction rates, which also results in a final product whose natural color has been greatly intensified.

Combined factors preservation technology involving blanching and vacuum solute (sucrose, potassium sorbate, ascorbic and citric acids, zinc chloride) impregnation can minimize color changes in minimally processed kiwifruit slices for at least one month of storage. Atmospheric impregnation was also studied in order to compare both impregnation techniques. The effect of blanching time, zinc content, and storage temperature was evaluated using Response Surface Methodology. A Box-Behnken design was adopted for fitting a second-order model that related those processing variables to a color function (Brown Index).

Kiwifruits were sliced $\cong 10$ mm thick and half slices were blanched in saturated vapor for 0, 30, or 60 seconds. For vacuum impregnation the samples were immersed in an aqueous soaking solution with the following composition: 65% w/w sucrose ($a_w = 0.84$); 0, 250, or 500 ppm zinc chloride; 15% w/w potassium sorbate; and 640 ppm ascorbic acid. The pH was adjusted to $\cong 3.2 \pm 0.2$ by adding citric acid. A pulse of 10 min was applied at a vacuum pressure of 160 mm Hg, and the kiwifruit slices were impregnated at about 4% (final $a_w = 0.97$).

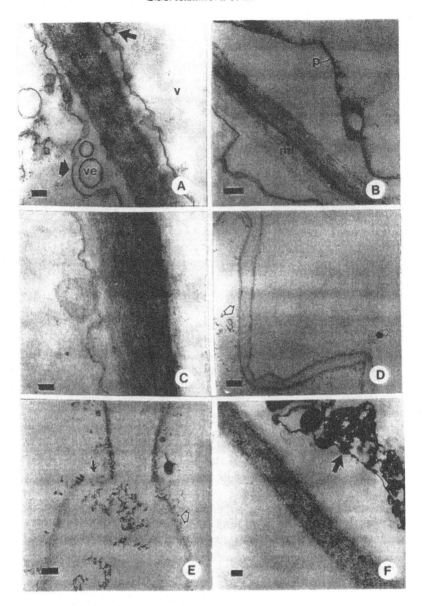

Figure 16-2 Ultrastructure (TEM) of melon flesh tissue as a function of infusion treatment (A: fresh control; B: vacuum infused; C: vacuum infused with Ca^{2+} addition; D,E: atmospheric infused; F: atmospheric infused with Ca^{2+} addition). Scale: A–C, E: 200 nm; D: 1 μ; F: 500 nm.

The atmospheric process involved immersing the fruit slices in an aqueous solution for 3 d at room temperature ($\cong 22°C$) until the a_w of the fruit was in equilibrium with the a_w of the aqueous solution ($a_w = 0.97$). The solution composition in this case was: 30% w/w sucrose ($a_w = 0.97$); 0, 100, or 200 ppm zinc chloride; 0.1% w/w potassium sorbate; 300 ppm ascorbic acid; and pH $\cong$ 3.2 ± 0.2 (adjusted with citric acid). After the impregnation treatment, the samples were withdrawn from the solution, covered with a solution of similar composition to that employed for the atmospheric treatment, and stored in the dark at 5, 15, or 25°C for 33 d.

A different dependence of the response variable on the process variables was obtained for both impregnation procedures. Figures 16-3 and 16-4 illustrate the response surfaces generated after storage periods of 0, 15, and 33 d for vacuum and atmospheric impregnation treatments, respectively, when stored at 15°C. As can be seen, after the impregnation process and before storage (time zero), large differences came into view between both processes (Figs. 16-3a and 16-4a), indicating the impregnation procedure had a high effect on the color changes from the beginning of the storage. In the response surfaces corresponding to the atmospheric impregnated kiwifruit slices (Fig. 16-4), higher zinc chloride and blanching time levels gave greater Brown Index values. For the vacuum process the opposite was found in that many combinations of these variables improved the color quality, which was observed during the entire storage period. Although similar in shape, the response surfaces for the vacuum infused fruit were vertically displaced to greater Brown Index values as the storage time increased, while the surfaces corresponding to the atmospheric impregnated slices changed slightly. In general, as the storage time and temperature increased, the color changes in the kiwifruit slices were more deleterious. After storage, the total chlorophyll had been degraded between 70 and 90% depending on the pretreatments, but no consistent relation between the changes in the total chlorophyll content and color were found. However, chlorophyll content values were much bigger for vacuum processed slices than atmospheric impregnated ones, showing a protective effect of the vacuum treatment towards the retention of this pigment. The lack of correlation between the chlorophyll concentration and Brown Index can be explained on the basis of 1) the low levels of initial total chlorophyll and very high conversion of chlorophylls to pheophytins in the fresh fruit, and 2) the color deterioration was produced not only by chlorophyll breakdown but other complex mechanisms such as degradation of some xantophylls and non-enzymic browning reactions involving ascorbic acid (Cano and Marín, 1992; Wong and Stanton, 1989; Cano et al., 1993).

The color evolution of apricot processed in vacuum conditions (70 mbar) at 30°C in a 69°Brix sucrose aqueous solution ($a_w = 0.88$) for 150 min was studied by García-Redón et al. (1994). The product (25.85°Brix, 69.49% moisture content, a_w 0.97) was stored at 5°C in polyethylenepolyamide (80:20) vacuum closed bags. A small decrease of L, a, and b coordinates with time was observed, with this variation occurring faster during the first 10–20 d of storage. The

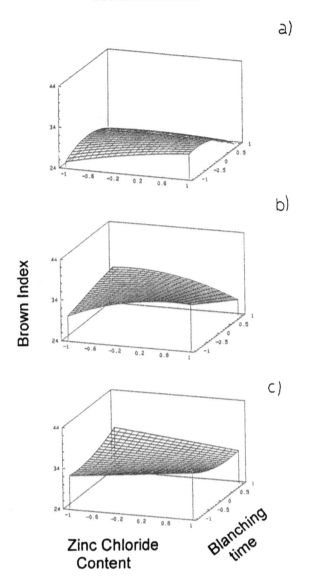

Figure 16-3 Response surfaces for the effect of zinc chloride content and blanching time on vacuum infused kiwifruit stored at 15°C for: a) 0 d, b) 15 d, and c) 30 d.

maximum ΔE value of 10 referred to just-processed fruit, reached after four months of storage. When 600 ppm of sodium methabisulphite was added to the infusion solution, no differences were detected between the untreated samples and samples containing the antibrowning additive. These results indicated good color preservation in apricots using the vacuum process without the addition of sulphites whose use is being increasingly restricted by legislation because of the allergic reactions sometimes exhibited by consumers.

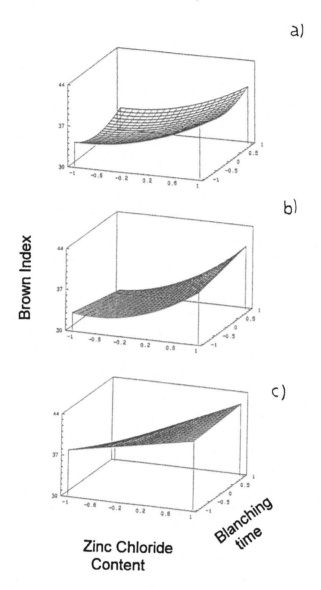

Figure 16-4 Response surfaces for the effect of zinc chloride content and blanching time on atmospheric infused kiwifruit when stored at 15°C for: a) 0 d, b) 15 d, and c) 33 d.

MICROORGANISM PENETRATION AND THE HYDRODYNAMIC MECHANISM

Primary plant cell walls have a hydrophilic character with continuous channels of water extending through them. Apoplastic transport involves movement of molecules through the aqueous part of the cell wall matrix provided they are not immobilized by electrostatic or other forms of binding to the wall polymers (Brett and Waldron, 1996). However, the diameter of the pores in the cell wall (generally in the 35–55 angstrom range) imposes a restriction on the size of the molecules that can penetrate without difficulty (Carpita et al., 1979). The mean separation of the wall polymers is such that the pores easily allow the passage of salts, sugars, amino acids, and some movement of small proteins and polysaccharides, though large proteins are generally immobile. In the same way, intact plant cell walls are extremely effective physical barriers against attack by microorganisms because their pores are far too small to permit bacteria (size range ≅0.1 to 5 μm), yeasts, molds, and even viruses (size range ≅0.02 to 0.3 μm) to penetrate through to the protoplast. Microorganism penetration in cell walls therefore requires physicochemical or enzymic degradation and/or alteration of wall structures.

It has also been reported that despite the differences in composition and structure of the cell walls of various plant tissues, the limiting pore diameters may be similar (Carpita et al., 1979). But in porous fruits it is the intercellular air spaces rather than the wall pores that may play the major role in microorganism penetration. For instance, mature cells of apple parenchyma tissues can be 50–500 μm in diameter with interconnecting air spaces (≅20–30% of tissue volume) ranging from 210–350 μm across (Lapsley et al., 1992). These spaces are large enough for microorganisms to pass through. Senescing apples also exhibit important changes in cell wall composition with disintegration of the middle lamella and enlargement of intercellular spaces, which permits the passage of large molecules and microorganisms (Ben-Arie et al., 1979).

In order to analyze if HDM would facilitate the incorporation of microorganisms inside tissues, some impregnation tests were performed using Granny Smith apple cylinders for the solid phase and an isotonic sucrose solution (to avoid other mass transport mechanisms) as the impregnation liquid phase. The sugar solution was inoculated with *Saccharomyces cerevisiae*, *Phoma glomerata*, or *Lactobacillus acidophilus*. The system (apple plus inoculated isotonic solution) was placed under vacuum for 10 min and then returned to atmospheric pressure to investigate different subatmospheric pressures (75, 125, 225, 325, or 425 mm Hg). Systems prepared in the same way but maintained for 10 min at atmospheric pressure were used as controls. After treatment, the fruit samples were removed and their microbial concentration monitored by conventional plating.

For the three microorganisms assayed, impregnation made at the lower pressures (75 mm Hg and 125 mm Hg) increased the colony forming units count (CFUs) by approximately 0.5 logarithms as compared to the controls (Fig. 16-5). As working pressure increased, there was no appreciable difference in counts between the vacuum treated samples and controls. We also experimentally determined the effective porosity (ε_e) of the apple cylinders (i.e., the fraction of the total fruit volume which is occupied by gas) at each pressure, as well as the corresponding volumetric fraction of liquid transferred by the HDM (X). The ε_e values decreased with the increase in pressure, which was 17.8, 16.8, 14.1, 12.0,

218

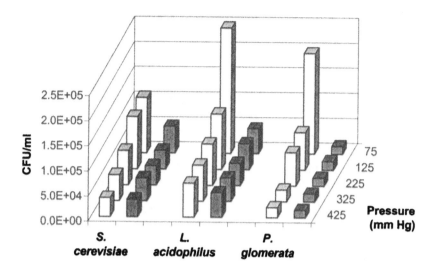

Figure 16-5 Microorganism counts (average of twenty plates) from apple slices after vacuum infusion (10 min at room temperature) at different pressures. ☐ vacuum treated; ■ control.

and 8.0 for 75, 125, 225, 325, and 425 mm Hg, respectively. The low ε_e and consequently low X explained the similarity between the microorganism concentration of the controls and that of the samples subjected to the higher subatmospheric pressures. Moreover, from the ε_e and X values and application of the mathematical model for HDM through a mass balance, we calculated the microorganism concentration in the vacuum infused apples tissues and found that the predicted data was in good agreement with the experimental results. Microorganism penetration was also visualized using scanning electronic microscopy.

In summary, the HDM and coupled deformation-relaxation of the solid matrix may result in facilitated microorganism penetration in fruit tissues. In this respect, microbiological stability may be carefully evaluated to assure the safety of fruits preserved using vacuum techniques.

USE OF NATURAL ANTIMICROBIALS

Governmental restrictions and consumer demand for more natural foods have focused attention on naturally occurring antimicrobial systems derived from animals, plants, or microorganisms. In addition, with the greater emphasis on the development and marketing of refrigerated foods by the food industry, use of antimicrobial agents to provide the desired safety is growing in importance due to the potential for microorganism growth during temperature abuse situations. A growing number of such natural systems are being explored as potential substitutes (partial or total) for common synthetic preservatives to help combat food poisoning (Gould, 1996; Davidson and Parish, 1989; Davidson, 1997, Schillinger et al., 1996). However, despite the increasing interest in the use of natural antimicrobials, they have not yet been exploited commercially. Some of the possible reasons are outlined below as problems one might encounter in their industrial application:

1. Industry reluctance to conduct substantial and expensive programs of toxicological testing for the introduction of a new antimicrobial
2. *In vitro* tests may not duplicate the variability of real foods since many antimicrobials are less effective inhibitors in food matrices than *in vitro*. Some examples of the interaction between natural antimicrobials and food matrices are:
    a. binding to food components such as proteins and fats chemical degradation
    b. inactivation and/or biological stabilization by other ingredients or components
    c. physical losses by mass transport from a food to the environment
    d. poor solubility and/or uneven distribution in foods
3. Unknown concentration of the antimicrobial compound tested and/or ignorance of the principal component exhibiting antimicrobial activity in the natural antimicrobial system
4. Lack of standardization and/or uniform procedures for isolation, purification, stabilization, and incorporation of these natural systems into foods
5. Poor understanding of the effect of the additive, synergistic, or antagonistic combination of the natural antimicrobial with other preservation factors or techniques.
6. Adverse effects on sensory characteristics

Based on these considerations, the potential use of natural preservatives as one of the hurdles for fruit preservation (instead of potassium sorbate and sulphites) were studied. The antimicrobial activities of extracts from several types of plant and plant parts used as flavoring agents in foods have been recognized for many years (Wilkins and Board, 1989). Among these, vanillin (4-hydroxy-3-methylbenzaldehyde), a major constituent of vanilla beans, was selected because of its several major advantages over other current spices. Since it is an effective GRAS (generally recognized as safe) flavoring crystalline agent widely used for ice-creams, drinks, confectionary, and cookies, its sensorial characteristics are well accepted. Furthermore, it has been found to be compatible with the organoleptic characteristics of various fruits (apple, plum, pear, mango, papaya, pineapple, strawberry, banana) in concentrations up to 3,000 ppm (Cerrutti and Alzamora, 1996). Vanillin cannot serve as a substrate for microbial growth and toxin production since it occurs as crystals; therefore, its antimicrobial activity depends on vanillin source and obtainment. Applying vanillin to minimally processed fruits involves study in a laboratory environment treating model fruit systems as well as the actual fruits (Cerrutti and Alzamora, 1996; López-Malo et al., 1997a, b; Cerrutti et al., 1997).

Growth of *Saccharomyces cerevisiae, Zygosaccharomyces rouxii, Debaryomyces hansenii* and *Z. bailii* was inhibited in culture media (pH 4.0, a_w 0.99) containing 2,000 ppm vanillin for 40 d of storage at 27°C (Cerrutti and Alzamora, 1996). The effect of lowering a_w to 0.95 combined with the action of 1,000 ppm vanillin resulted in the inhibition of *S. cerevisiae* and *D. hansenii*. Tests with 2,000 ppm of vanillin were conducted in apple and banana purées to corroborate the *in vitro* results. This vanillin concentration was effective not only in preventing growth in the apple purée (pH 3.5), but was also germicidal for the four yeasts assayed. However, 2,000 ppm of vanillin in banana purée (pH adjusted to 4.0 by cytric acid addition) was not effective in inhibiting yeast growth. The addition of 3,000 ppm of vanillin resulted in the inhibition of *S. cerevisiae, Z. rouxii,* and *D. hanseni,* but not *Z. bailii.* Vanillin inhibitory concentrations for *Aspergillus flavus, A. niger, A. ochraceus,* and *A. parasiticus* in laboratory media and five fruit-based (apple, banana, mango, papaya, and pineapple) agars with pH 3.5 and a_w 0.98 were generally lower than the 2,000 ppm vanillin since the smaller inhibiting effect was on the mango and banana agars (López-Malo et al., 1995). The lower effect found on the mango and papaya systems compared with other fruits was partially attributed to the higher fat and/or protein content of these fruits. These substances are known to bind and/or solubilize phenolic compounds, reducing their availability for antimicrobial activity (McNeil and Schmidt, 1993).

The combined effects of pH (3.0–4.0) and vanillin (500–1,000 ppm) on the growth of *A. flavus, A. niger, A. ochraceus,* and *A. parasiticus* were evaluated in a potato-dextrose agar adjusted to a_w 0.98 by López-Malo et al. (1997b). A combination of vanillin with pH reduction had an additive or synergistic effect on mold growth depending on the *Aspergillus* specie. The pH reduction alone was not sufficient to delay germination after more than 46 h but the addition of vanillin increased the lag times from 48% to near 1,000%.

The combined effects of pH (3.0–4.0), vanillin concentration (350–1200 ppm), and incubation temperature (10–30°C) on the growth of these molds were also analyzed in a potato-dextrose agar with a_w 0.98 by López-Malo et al. (1997a). The germination time and radial growth rates were significantly affected by the three studied variables ($P < 0.001$), but the inhibitory conditions (no growth after 30 d) depended on the type of mold. *Aspergillus niger,* the most resistant species, could be inhibited with 1,000 ppm of vanillin at pH $\leq$ 3.0 and an incubation temperature $\leq$ 15°C or 500 ppm of vanillin at pH $\leq$ 4.0 and $\leq$ 10°C. For the most sensitive *A. ochraceus,* the inhibitory conditions in systems containing 500 ppm vanillin were pH 3.0 and $\leq$ 25°C or pH 4.0 with $\leq$ 15°C. If the pH was raised to 4.0, the vanillin concentration could be $\leq$ 1,000 ppm with $<$ 15°C or 500 ppm of vanillin if the incubation temperature was reduced to 10°C.

The effect of pH on the antimicrobial activity of natural phenolic compounds is not clearly understood. Sykes and Hooper (1954) attributed this greater effect at acid pH values to the increased solubility and stability of these compounds at low pH. Juven et al. (1994) postulated that at low pH values, the phenol molecule is mostly undissociated, binding better to the hydrophobic regions of the membrane proteins and dissolving better in the lipid phase of the membrane.

The potential utilization of vanillin as an antimicrobial instead of potassium sorbate and sodium bisulphite in the formulation of a previously combined technology developed for obtaining a shelf-stable strawberry purée was analyzed by Cerrutti and Alzamora (1996). The addition of ascorbic acid (500 ppm), a controlled pH of 3.0, the reduction of a_w to 0.95, and a mild heat treatment (blanching) were the other hurdles that combined to accomplish the desired microbiological stability of the fruit. The response to the combined preservation system of key microorganisms was addressed by studies on the evolution of native flora and challenge testing with microorganisms of concern in the preserved strawberry purée (*Saccharomyces cerevisiae, Zygosaccharomyces rouxii, Z. bailii, Schizosaccharomyces pombe, Pichia membranaefaciens, Botrytis* spp., *Byssochlamys fulva, Bacillus coagulans* and *Lactobacillus delbrueckii*). Combinations of these hurdles prevented the growth of both native and inoculated flora for at least 60 d of storage at room temperature.

These results showed that the addition of vanillin in combination with a slight reduction of a_w and pH may be a promising form of natural strawberry preservation. Vanillin could thus be a natural alternative to sulphite and potassium sorbate. In the same way, the use of vanillin for mold control in the refrigerated storage of food with lowered pH and a_w could represent a safety factor for possible temperature fluctuations during handling, transportation, distribution, and retail sale.

REFERENCES

Alzamora S.M. and Gerschenson L.N. (1997). Effect of water activity depression on textural characteristics of minimally processed fruits. In: *New Frontiers in Food Engineering. Proceedings of the Fifth Conference of Food Engineering.* Barbosa-Cánovas G.V., Lombardo S., Narsimhan G., and Okos M.R., Eds. Los Angeles: AIChE.

Alzamora S.M., Cerrutti P., Guerrero S., and López-Malo A. (1995). Minimally processed fruits by combined methods. In: *Food Preservation by Moisture Control. Fundamentals and Applications.* Barbosa-Cánovas G.V. and Welti-Chanes J., Eds. ISOPOW Practicum II. Lancaster, PA. Technomic Publishing, pp. 463–492.

Alzamora S.M., Gerschenson L.N., Vidales S.L., and Nieto A.B. (1997). Structural changes in the minimal processing of fruits: Some effects of blanching and sugar impregnation. In: *Food Engineering 2000*. Fito P., Ortega-Rodríguez E., Barbosa-Cánovas G.V., Eds. New York: Chapman & Hall, pp. 117–139.

Ben-Arie R., Kislev N., and Frenkel C. (1979). Ultrastructural changes in the cell walls of ripening apple and pear fruit. *Plant Physiol.* 64: 197–202.

Brett C.T. and Waldron K.W. (1996). *Physiology and Biochemistry of Plant Cell Walls*. Cambridge, UK: Chapman and Hall.

Cano M.P. and Marín M.A. (1992). Pigment composition and color of frozen and canned kiwifruit slices. *J. Agric. Food Chem.* 40: 2141–2146.

Cano M.P., Marín M.A., and De Ancos B. (1993). Pigment and color stability of frozen kiwifruit slices during prolonged storage. *Z. Lebensm. Unters. Forsch.* 197: 346–352.

Carpita N.C. and Gibeaut D.M. (1993). Structural models of primary cell walls in flowering plants: consistency of molecular structure with the physical properties of the walls during growth. *Plant J.* 3: 1–30.

Carpita N., Sabularse D., Montezinos D., and Delmer D.P. (1979). Determination of the pore size of cell walls of living plant cells. *Sci.* 205: 1144–1147.

Cerrutti P. and Alzamora S.M. (1996). Inhibitory effects of vanillin on some food spoilage yeasts in laboratory media and fruit purées. *Int. J. Food Microbiol.* 29: 379–386.

Cerrutti P., Alzamora S.M., and Vidales S.L. (1997). Vanillin as antimicrobial for producing shelf-stable strawberry purée. *J. Food Sci.* 62: 608–610.

Davidson P.M. (1997). Chemical preservatives and natural antimicrobial compounds. In: *Food Microbiology: Fundamentals and Frontiers*. Doyle M.P., Beuchat L.R., and Montville T.J., Eds. Washington, DC: ASM Press, pp. 520–556.

Davidson P.M. and Parish M.E. (1989). Methods for testing the efficacy of food antimicrobials. *Food Technol.* 43(1): 148–155.

Fito P. and Pastor R. (1994). On some non diffusional mechanisms occurring during vacuum osmotic dehydration. *J. Food Eng.* 21: 513–519.

Fito P. and Chiralt A. (1995). An update on vacuum osmotic dehydration. In: *Food Preservation by Moisture Control: Fundamentals and Applications*. Barbosa-Cánovas G.V. and Welti-Chanes J., Eds. ISOPOW Practicum II. Lancaster, PA: Technomic Publishing, pp. 351–374.

Fito P., Andrés A., Chiralt A., and Pardo P. (1996). Coupling of hydrodynamic mechanism and deformation-relaxation phenomena during vacuum treatments in solid porous food-liquid systems. *J Food Eng.* 21: 229–240.

García-Redón E., Fito P., Salazar D., and Chiralt A. (1994). Vacuum osmotic dehydration of apricot (*Prunus armeniaca* cv. Canino). In: *Proceedings of the International Symposium on the Properties of Water. Practicum II.* Universidad de las Américas, Puebla, Mexico.

Gould G.W. (1996). Industry perspectives on the use of natural antimicrobials and inhibitors for food applications. *J. Food Prot.* Supplement, pp. 82–86.

Ilker R. and Szczeniak A.S. (1990). Structural and chemical bases for texture of plants foodstuffs. *J. Tex. Stud.* 21: 1–36.

Jackman R.L. and Stanley D.G. (1995). Perspectives in the textural evaluation of plant foods. *Trends Food Sci. Technol.* 6: 187–194.

Juven B.J., Kanner J., Schved F., and Weisslowicz H. (1994). Factors that interact with the antibacterial action of thyme essential oil and its active constituents. *J. Appl. Bact.* 76: 626-631.

Lapsley K.G., Escher F.E., and Hoehn E. (1992). The cellular structure of selected apple varieties. *Food Struct.* 11: 339-349.

López-Malo A., Alzamora S.M., and Argaiz A. (1995). Effect of natural vanillin on germination time and radial growth rate of molds in fruit based systems. *Food Microbiol.* 12: 213-219.

López-Malo, A., Alzamora S.M., and Argaiz A. (1997a). Effect of vanillin concentration, pH and incubation temperature on *Aspergillus flavus, Aspergillus niger, Aspergillus ochraceous* and *Aspergillus parasiticus* growth. *Food Microbiol.* 14: 117-124.

López-Malo, A., S.M. Alzamora, and A. Argaiz (1997b). Vanillin and pH synergistic effects on mold growth. *J. Food Sci.* 63: 143-146.

McNeil V.I. and Schmidt K.A. (1993). Vanillin interaction with milk protein isolates in sweetened drinks. *J. Food Sci.* 58: 1142–1147.

Muntada V., L.N. Gerschenson, S.M. Alzamora, and M.A. Castro (1998). Solute infusion effects on texture of minimally processed kiwifruit. *J. Food Sci.* 63(4): 616-626.

Rojas A.M., Castro M.A., Gerschenson L.N., and Alzamora S.M. (1998). Firmness and structural characteristics of glucose impregnated melon. In: *Proceedings of ISOPOW 7.* Ross Y.H., Ed. Helsinki, Finland, pp. 204–207.

Schillinger U., Geisen R., and Holzapfel W.H. (1996). Potential of antagonistic microorganisms and bacteriocins for the biological preservation of foods. *Trends Food Sci. Technol.* 7: 158-164.

Shi X.Q. (1994). *Vacuum Osmotic Dehydration of Food: Some Applications in Fruit Preservation.* Ph.D. Thesis, Universidad Politécnica de Valencia, España.

Sousa R., Salvatori D., Andrés A., and Fito P. (1998). Vacuum impregnation of banana (*Musa acuminata* cv. Giant Cavendish). *Food Sci. Technol Int.* 4(2): 127–131.

Sykes G. and Hooper M.C. (1954). Phenol as the preservative in insulin injections. *J. Pharmacol. Pharmacy* 6: 552-558.

Vidales S.L., Castro M.A., and Alzamora S.M. (1998). The structure-texture relationship of blanched glucose impregnated strawberries. *Food Sci. Technol. Int.* 4: 169–178.

Waldron K.W., Smith A.C., Parr A.J., Ng A., and Parker M.L (1997). New approaches to understanding and controlling cell separation in relation to fruit and vegetable texture. *Trends Food Sci. Technol.* 8: 213–221.

Welti-Chanes J. and Vergara-Balderas F. (1995). Fruit preservation by combined methods: An Ibero-American research project. In: *Food Preservation by Moisture Control: Fundamentals and Applications*. ISOPOW Practicum II. Barbosa-Cánovas G.V. and Welti-Chanes J., Eds. Lancaster, PA: Technomic Publishing, pp. 449–462.

Wilkins K.M. and Board R.G. (1989). Natural antimicrobial systems. In: *Mechanisms of Action of Food Preservation Procedures*. Gould G.W., Ed. New York: Elsevier Applied Science, pp. 285–360.

Wong M. and Stanton D.W. (1989). Nonenzymic browning in kiwifruit juice concentrate systems during storage. *J. Food Sci.* 54: 669–673.

CHAPTER 17

A Review of Nonthermal Technologies

*S. Bolado-Rodriguez, M.M. Góngora-Nieto U. Pothakamury,
G.V. Barbosa-Cánovas, and B.G. Swanson*

INTRODUCTION

The thermal methods used today to preserve foods are largely chosen due to traditional reasons. Heating does inactivate spoilage microorganisms, but also causes a loss of nutrients and flavors. In contrast, nonthermal processes can be used for the inactivation of food spoilage and pathogen microorganisms with little or no effect on taste, flavor, vitamin, or protein contents. Furthermore, they can extend product shelf-lives and reduce processing costs.

The HHP technique is the uniform application of hydrostatic pressures on the order of 4,000–9,000 atm throughout a food product. Since late 1800s, high pressure technology has been gaining importance in the food industry because of the advantages of inactivating microorganisms and enzymes to produce high quality foods (Mertens and Knorr, 1992).

Light pulses expose food surfaces or packaging materials to short-duration pulses (1 μs–0.1 s) of intense non-ionizing flashing lights with energy densities of about 0.1 to 50 J/cm^2, and wavelengths between 170 and 2,600 nm. The benefits of the process were described in 1988 (Anonymous, 1988). PurePulse Technologies, Inc., who has named the process PureBright™ pulsed-light, leads industry implementation.

The nonthermal inactivation of microorganisms using strong electric fields known as the PEF process was demonstrated by Doevenspeck in 1961. Studies in this area report the use of charging voltages up to 40 kV and pulse rates up to 1000 Hz. Today it is possible to inactivate enzymes as well as pathogenic and spoilage microorganisms without causing perceptible alterations in food color, flavor, texture, and nutrients. This nonthermal technology also allows an extension of shelf-lives while maintaining the fresh quality of food products.

Oscillating magnetic fields (OMF) is a nonthermal electrotechnology to preserve food that relies on the application of one or more pulses of an OMF to a packaged or non-packaged food. The intensity of these fields is between 5–50 Tesla, with a frequency in the 5–500 kHz range. The process was described in the world patent application by Hofmann (1985). The scientific literature is scarce in this area, and no sufficient evidence yet exists to define its effectiveness against microorganisms without affecting food quality (Mertens and Knorr, 1992).

HIGH PRESSURE TECHNOLOGY

The attempts to use high hydrostatic pressures to treat foods started at the turn of this century. Since the early 1980's important efforts have taken place to investigate the relationships between this process and the physical and sensorial characteristics of foods, as well as the mechanisms by which the inactivation of microorganisms and enzymes take place. Today HHP processing is a reality in the food industry where many products are commercially available and under development. In general, an HHP process starts when a food is subjected to a high pressure (1,000 to 9,000 atm) in either batch or semicontinuous mode. This pressure is maintained for a specified time period that depends on the type of food and process temperature. At the end of the processing time the chamber is decompressed to remove the treated batch. A new batch of food is then placed in the pressure vessel and the cycle begins again (Zimmerman and Bergman, 1993).

HHP is unique among other preservation techniques in that the applied pressure is isostatic, so the pressurization process is instantaneous and uniform throughout the product. All the food is treated equally and at the same time, with no part escaping preservation. Furthermore, there is no variation in pressure throughout the food or package, so rupture or breakage will not occur. Unlike thermal treatments, HHP is not time/mass dependent, and thus reduces processing times (Zimmerman and Bergman, 1993).

In the food processing industry HHP has displaced other thermal methods. In Japan high pressure processing is done commercially on a meaningful scale with fruit jams, jellies, salad dressing, yogurts, fruit mixes, fruit juices, and concentrates. In the United States most of the activity has been confined to research labs and pilot plants, where good results have been obtained for the processing of spaghetti in sauce, Spanish rice, yogurt with fruit, and fruit mixes (Demetrakakes, 1996). This technology has opened new frontiers for the industrialization and commercialization of such delicate foods as guacamole, made from avocado fruit.

Engineering Aspects

The HHP technology was adapted from the metal and ceramic industries. The necessary changes have included the use of higher pressure, holding time, and potable water as a pressure medium (Fig. 17-1). The equipment used for the pressurization of foods should be able to tolerate pressures of about 4,000–9,000 atm, with holding times ranging between 5 and 20 minutes and cycling rates of about 100,000 cycles/year. Current pressure vessel design efforts are aimed at useful lives of 1,000,000 cycles (Farkas and Ting, 1997). Because of the high pressure, vessels have a finite fatigue life which can be improved by using an appropriate design and higher-strength materials, as well as pre-stressing the vessel. At pressures as high as required for food processing, there is a limit to the size the vessels that can be forged, and consequently, to their effective working volumes. Efforts performed in the design and development of HHP equipment are now focused on increasing actual working volumes (Table 17-1) and keeping a HHP. Other crucial requirements for the commercial application of HHP technology in the food industry are the design of adequate packages, reduction of the batch cycle time, and use of safe, easy-to-clean equipment with accurate process controls.

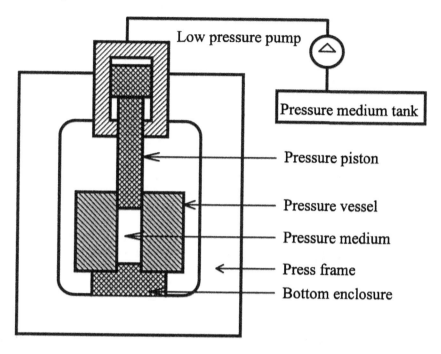

Figure 17-1 Schematic representation of a high pressure system.

Table 17-1 Pressure and working volumes of different commercial HHP vessels (adapted from Barbosa-Cánovas et al., 1997).

Company	Volume (Liter)	Maximum Operating Pressure (atm)
Mitsubishi	0.6	7,000
	6–210	4,000
Kobe Steel Ltd.	9,400	1,960
ABB Autoclave Sys.—Quintus	100–500	9,000
Engineering Pressure Systems	9,000	1,000
	37	6,900
	3.5	13,800

The pressurization process can be carried out in any type of hydraulic fluid, but in the field of food processing the use of potable water is preferred for its compatibility with food, ease of operation, and low compressibility. Furthermore, the energy efficiency of this method is high because the water is relatively incompressible (just 15% at 6,000 atm) (Hayashi, 1989), and high pressure vessels using water do not present the same operating hazards as vessels using compressed gases (Farr, 1990).

Biological Effects

In Microorganisms

The biological changes induced by high pressure in microorganisms include filament formation, cessation of motility, and inactivation. Filament formation is a very marked morphological change in microorganisms such as *Escherichia coli*, *Vibrio spp.*, and *Serratia marinorubra* subjected to high pressure. Most of the motile bacteria are immobilized by prolonged pressurization at 200–400 atm. At 100 atm, *E. coli*, *Vibrio*, and *Pseudomonas* have flagella, whereas at 400 atm, these microorganisms lose their flagella (Zobell, 1970). Cessation of motility is reversible in some bacteria (Kitching, 1957).

In general, moderately high pressures decrease the rate of growth and reproduction, while very high pressures cause inactivation of microorganisms. Inactivation is caused by an increase in the permeability of the cell membrane and the resulting loss of intracellular contents, inhibition of energy-producing reactions, and denaturation of essential enzymes for cell growth and reproduction. The baro-sensitivity of microorganisms increases on the order of gram-positive bacteria, yeasts, and gram-negative bacteria. Shigehisa et al. (1991) conducted experiments with inoculated pork slurries and showed that at a pressure of 3,000 atm, the inactivation of *Streptococcus fecalis* and *Staphylococcus aureus* was less than 1 log cycle, while the inactivation of *Saccharomyces cerevisiae* and *Candida utilis* was between 1 and 2 log cycles. Under the same conditions the inactivation of the gram-negative *Salmonella typhimurium* and *Pseudomonas aeruginosa* reached 6 log cycles. Other studies conducted by Mussa et al. *(*1997) with *Listeria monocytogenes* inoculated in pork chops and pressurized at 3,000 atm resulted in a log reduction of almost 3 cycles. Bacterial inactivation by HHP is therefore affected by parameters such as the composition and pH of the growth media, osmotic pressure, and water activity. Of particular importance is the amount of free water present, because at least 40% is needed to kill vegetative microbes (Earnshaw, 1996). In addition, the presence of preservatives and use of mild heat treatments have been proven highly effective against *L. monocytogenes*, *S. aureus*, *S. typhimurium* and *E. coli* O157:H7. The inactivation of these pathogens when individually inoculated in peptone solution was increased from 1–2 log cycles (30 min. 30 kPsi and 30°C) to over 8 log cycles within 5–6 min when pediocin AcH (3,000 AU/ml) in combination with moderate temperatures (up to 50°C) and HHP (up to 50 kPsi) were used (Kalchayanand et al. 1997) (Fig. 17-2).

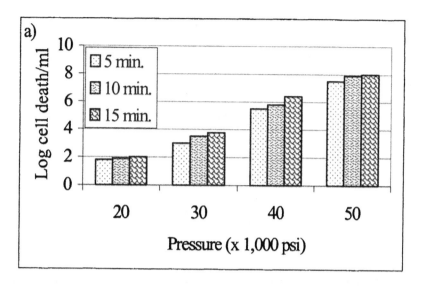

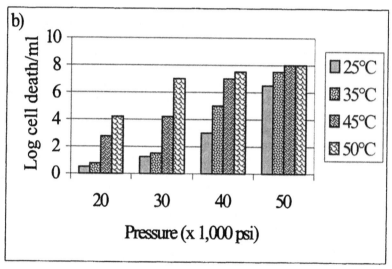

Figure 17-2 Pooled viability loss of pathogens (*S. aureus, L. monocytogenes, E. coli 0157:h7, S. typhimurium*) as a function of: a) pressure and time and b) pressure and temperature (Adapted from Kalchayanand et al., 1997).

Bacterial spores have shown a high resistance to HHP because of their small size, low water content, and water impermeable coat. At low temperatures the inactivation of spores increases with any increase in pressure, and low pressures cause germination and heat sensitization, but no appreciable inactivation of germinated spores. Medium pressures in the range of 1,000–

3,000 atm cause considerable germination and inactivation of a large proportion of germinated spores. High pressure causes less germination, but inactivation in only a small proportion of germinated spores. Above 65°C, heat rather than pressure cause the inactivation of spores. In the case of *B. stearothermophilus*, pulse and oscillatory pressurization are more effective in spore inactivation than continuous pressurization. Hayakawa et al. (1994) observed by electron microscopy that damage in spore coats is attributable to the adiabatic expansion of the water or aqueous solutions in cells. They concluded that pulse pressurization is useful to decrease spore inactivation time, and that the depressurization rate can be another important parameter of the process. Peck (1997) gave a list of eight procedures to ensure the safety of HHP to extend shelf-life of low-acid foods, where the presence of *Clostridium botulinum* is a major constraint. The most important rules are to limit the shelf-life to less than five days if the product is held at 5–10°C, adjust the pH of the product to less than 5.0, and reduce the water activity to below 0.97.

In Enzymes

The exposure of enzymes to high pressure can either generate a total inactivation, stimulation, or have no effect. Enzymatic reactions catalyzed by thermolysin or cellulase are stimulated up to 15 times by high pressure (Hoover, 1993; Kunugi, 1992). The proteolytic activity of enzymes in meat which gives an index of lysosomal membrane disruption and influences tenderization is enhanced by the application of high pressures. The total activities of cathepsin B, D, L, and acid phosphatase in muscle increase when subjected to pressures ranging between 1,000 and 5,000 atm (for 5 min at 2°C). While cathepsin H and aminopeptidase B are not affected by pressures up to 1000 atm, at higher treatment levels the activity of both enzymes decreases gradually with an increase in pressure (Homma et al., 1994) (Fig. 17-3).

In comparison, the activity of succinate, formate, malate dehydrogenases, and aspartase in *E. coli* are completely inactivated when subjected to 1,000 atm (Morita, 1957). In the case of *Pectinesterase* in juices such as Satsuma mandarin, inactivation takes place when the pressurization reaches 3,000–4,000 atm.

Enzyme inactivation takes place as a result of the alteration of intramolecular structures or conformational changes at the active site. The inactivation of some enzymes is reversible when they are pressurized to 1,000–3,000 atm. Reactivation after decompression depends on the degree of molecule distortion. The chances of reactivation decrease with increases in pressure beyond 3,000 atm (Jaenicke, 1981; Suzuki and Suzuki, 1963).

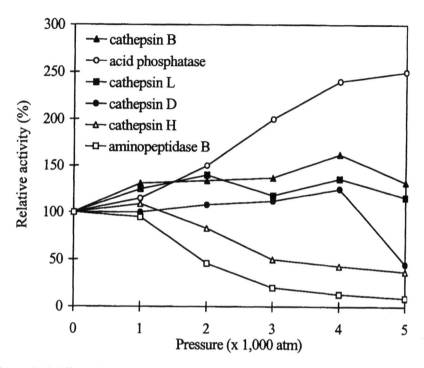

Figure 17-3 Effect of pressure on proteolytic enzyme activity in meat (adapted from Homma et al., 1994).

The inactivation of enzymes is influenced by the pH, substrate concentration, subunit enzyme structure, and temperature of pressurization (Hoover et al., 1989). While the activity of polyphenol oxidase increases five times when slices of Bartlett pears are pressurized at 4,000 atm and 25°C for 10 min, when working with the homogenates of apples, bananas, and sweet potatoes as substrates, there is no activation of this enzyme (Asaka and Hayashi, 1991).

In Biochemical Reactions

The biochemical reactions that are most affected by high pressure are the ones with a change in volume (Hedén, 1964) because pressure causes a decrease in the available molecular space or an increase in interchain reactions. Under HHP, reactions implying a volume reduction are accelerated, and those implying a volume increase are prevented (Deplace and Deplace, 1996).

High pressure also causes denaturation of protein molecules by disruption of hydrophobic ion pair bonds. In addition, HHP has been shown to disrupt the tertiary and quaternary structure of most globular proteins, with relatively less influence on their secondary structures. Large hydration changes and decreases in protein volumes follow the structural and conformational changes in proteins.

The denaturation reaction depends on the protein structure, pressure range, temperature, pH, and solvent composition. Oligomeric proteins are dissociated by relatively low pressures (2,000 atm), whereas single chain protein denaturation occurs beyond 3,000 atm. Studies conducted by Galazka et al. (1997) on the effect of HHP on the functionality of bovine serum albumen with the addition of dextran sulfate (DS) suggest a protecting effect of the protein against aggregation in the presence of DS. He concludes that this is a means to control protein functionality during pressurization.

It is important to note that fermentation products formed by HHP are quite different from those obtained at atmospheric pressure [i.e., fermentation of glucose results in increased amounts of low molecular weight acids such as formic (Johnston et al., 1992)]. Fermentation reactions are also retarded by high pressure. The high acidity in yogurt caused by continuous fermentation during storage can be prevented by a treatment at 2,000–3,000 atm for 2 min at 10°C. The lactic acid bacteria are thus maintained at the initial population and further growth is prevented due to the injury caused by the process.

In Microbial Cells

Nucleic acids are much more baro-resistant than protein molecules. The structure of DNA is largely a result of hydrogen bonding. Since high pressure favors hydrogen bonding, the DNA molecules are more stabilized at high pressures, while high temperatures cause them to denature. DNA transcription and replication is disrupted by high pressure due to the involvement of enzymes. The reversibility of this disruption decreases with the amount of pressure applied (Landau, 1967). However, HHP damaged cells may be able to repair themselves when held under favorable conditions of nutrient media composition and temperature. The injured cells may have a holding time for repair up to 60 days at refrigerated temperatures (Nandathur et al., 1997).

Cell membranes are assumed to contain protein units embedded in lipid bilayers, which play an important role in cellular transport. It is generally felt that for microorganisms, the primary effect of pressure damage is on cell membranes (Kristel et al., 1996). High pressure reduces (though not uniformly) the intermolecular distance between acyl chains within the bilayer, causing isothermal phase transitions which change the molecular structure, and thus alter the permeability. HHP also decreases the efflux potassium and sodium, which increases the permeability of the cell membrane and causes the contents of the cell to leak out and disrupt cell functions. In addition, HHP denatures proteins and inhibits the uptake of essential amino acids for cell growth. If the applied pressure is low, the cell regains its original permeability, but destruction of the cell wall is irreversible when the applied pressure is high, and finally results in cell inactivation (Farr, 1990).

Applications

High pressure can be used for increasing the shelf-life of foods by inactivating microorganisms, spores, and undesirable enzymes. Interest in this technology as a preservative method results from the minor effect it has on the composition, as well as sensory and nutritional characteristics of foods. When Donsi et al. (1996) processed orange juice at 3,500 atm for 1 min at 30°C, they obtained a shelf-life of no less than two months. The overall quality of the HHP juice was excellent in comparison to traditional thermal processing in that the composition of vitamins, sugars, and organic acids had no substantial modification, and the pH, aroma, and color compounds essentially remained the same as in the fresh samples. Another potential application is with milk treatment prior to cheese making, where an increase in cheese yields was verified with no detrimental effects on flavor. The microbiological quality was shown to be comparable to pasteurized milk (Drake et al., 1997).

Although there is no effect on food composition and nutritional characteristics, some foods do change in structure with high pressure exposure. However, these changes would be desirable in specific applications like the production of surimi gels because of deformation without fracture more easily than those thermally obtained (González et al., 1997). The tenderization of meat, freezing without ice crystals, faster thawing, improvement of starch digestibility, gelation of eggs, rapid tempering of chocolate, and simplified processing of fruit jams, sauces, and marmalades are other possibilities.

HHP can also be used to improve the extraction yields of other metabolites such as flavor or pigments at ambient temperatures. With some plant-derived foods, it is possible to selectively extract undesirable substances which are not normally consumed (i.e., the cyanide from fruit stones and seeds).

In commercial and industrial implementation, the biggest obstacle to HHP processing is cost. Those factors influencing the cost of HHP include processing conditions which fatigue the life of the equipment, loading efficiency, maintenance, and consumable parts. In general, the greater the pressure used and time needed, the higher the cost. Because the process is relatively expensive, products manufactured in this way will need to have a retail value above \$4/gal [i.e., inherent added values or increased profitability (Ting, 1997)].

New products that may be processed in the coming years include those with high levels of acid, in need of chilling, heat-sensitive, and concentrates. The ideal products for nonthermal high pressure processing will be those that absolutely cannot be heat-treated, and if left untreated will not have a long enough shelf-life to travel [pineapple is a good example (Demetrakakes, 1996)].

PULSED LIGHT TREATMENT

This technology which is also called short pulses of intense light and/or high intensity light pulses uses short duration flashes of broad spectrum white light to inactivate a wide range of microorganisms, including bacterial and fungal spores (Dunn, 1996). The spectrum of light used for sterilization purposes includes wavelengths in the ultraviolet (UV) to near infrared regions. The surface material to be sterilized is exposed to at least one pulse of light with an energy density in the range of about 0.01–50 J/cm^2.

The PureBright™ process developed by PurePulse Technologies, Inc. applies non-ionizing flashes of broad spectrum light. Such flashes have intensities of about 20,000 times that of sunlight at the surface of the earth. The duration of the pulses ranges from 1 μs–0.1 s (Anonymous, 1994), and the flashes are typically applied at a rate of 1–20 per second.

In August of 1996, the FDA approved the use of high intensity pulsed light as an irradiation processing method (Part 179) after amending the food additive regulations to provide the safe use of a source of high intensity pulsed light to control microorganisms on the surface of foods. This action was in response to a food additive petition filed in February 1994 by Foodco Corp., now known as PurePulse, Inc. The regulations in 21 CFR Section 179.41 list the conditions under which pulsed light may be safely used for the treatment of foods. The irradiation sources should consist of xenon flash lamps with wavelengths covering the range of 200 to 1,100 nm, and pulse durations no longer than 2 ms. The treatment should be used for surface microorganism control, with the total cumulative treatment not exceeding 12.0 J/cm^2.

With the data provided by PurePulse, the FDA agency found that treated foods will not undergo significant nutrient reductions. From a microbiological standpoint, the agency determined that the types and amounts of photo-products that might be produced and subsequently consumed are not of any toxicological significance. The proposed treatment was thus deemed effective in reducing the number of microorganisms on the surface of treated foods, which are at least as safe as the untreated foods currently marketed.

Engineering Aspects

A system to generate light pulses consists of two main parts: the power and lamp units. The power unit generates high voltage, high current pulses to energize the lamps in the system, and operates by converting line voltage AC power into high voltage DC power. The high voltage DC is used to charge the system's capacitor. Once the capacitor is charged to a preset point, a high voltage switch discharges the light energy from the capacitor to the lamps. In general, gas-filled flash lamps or spark-gap discharge apparatus can be used. The lamp firing sequence may be controlled by an internal controller or interfacing with the packaging/processing machine controller. Each pulse of light can be easily monitored through fiber optics and feedback circuits to assure that proper treatment is provided and the process can be essentially fail-safe.

The process can be applied to sterilize pumpable foods such as water, saline and dextrose solutions, and fruit juices. The treatment equipment consists of a reflective cylindrical enclosure which defines the treatment chamber. The chamber may be designed to include a reflector assembly as an outer wall, or external reflector to reflect back the illumination traversing the food product. In those fluids with significant absorption where significant reduction in flux density occurs, it is important to maintain a minimum flux density throughout the treatment zone, and mixing must occur to ensure that the entire fluid is subjected to the appropriate flux density (Dunn et al., 1991). Furthermore, it is preferable if the fluid has a minimum transparency to UV light such that at least half of the incident light at 260 nm is transmitted through 0.25 cm into the fluid (Barbosa-Cánovas et al., 1997).

Full or filtered spectrum light may be used depending on the selected application and degree of sterilization expected. Filtered spectrum light is preferred because it eliminates wavelengths known to cause undesirable reactions in foods. Glass or liquid filters are used to obtain a filtered spectrum.

The treatment effectiveness of light pulses is based on the penetration of light through the surface of materials according to the equation given by Dunn et al. (1991):

$$I = (1 - R)\, I_0\, e^{-x} \qquad (17\text{-}1)$$

where I is the energy intensity of light transmitted to a distance below the surface, R the surface coefficient of reflection, I_0 the intensity incident upon the surface, and x the extinction coefficient (a determination of the material's opacity).

Treatment of packaged products by pulsed light minimizes the risk of further re-contamination. Many plastics can be used to efficiently transmit light to the product. Some examples are polyethylene, polypropylene, nylon, EVA, and EVAOH. Polyaromatic hydrocarbon-rich plastics (such as PET, polycarbonate, polystyrene, or PVC as normally formulated) do not generally transmit pulsed light well enough to allow the treatment of products (Dunn et al., 1997a).

When the surface of the material to be treated is opaque, or the light absorption coefficient is low, absorption-enhancing agents have to be utilized. FDA approved colors such as carotene, red dye #3, lime green, black cherry, and mixtures thereof may be used as absorption enhancing agents. Natural or cooking oils are also permitted. Mixtures of two or more components having different maximum absorptions may be used to increase the optical absorption over the desired spectrum (Dunn et al., 1991).

Effects on Microorganisms

PurePulse, Inc. has developed most of the research in the pulsed light area. They have patented this sterilization method and are the authors of most of the scientific literature about its biological effects. The effectiveness of light pulses is due to their rich broad-spectrum content, short duration, and high power. They also contain non-ionizing wavelengths with approximately 25% of their light from the UV region (Dunn et al., 1997a). The antimicrobial effects of UV wavelengths are primarily mediated through absorption by highly conjugated carbon-to-carbon double-bond systems in proteins and nucleic acids. When UV-rich light pulses with at least 30% of the wavelengths shorter than 300 nm are used, microorganism and virus inactivations are achieved through combined photothermal and photochemical mechanisms. If the quality and flavor of the food product is adversely affected by the UV bandwidth and the UV portion is filtered out, the inactivation of microorganisms and/or viruses is primarily photothermal (Mertens and Knorr, 1992). Comparison of the antimicrobial effects obtained using pulsed light with those using non-pulsed or continuous wave conventional UV lamp sources shows a significantly higher inactivation for pulsed light. With relatively resistant *Aspergillus niger* spores dried on a packing surface, conventional ultraviolet light from a high intensity mercury vapor lamp kills from 3.5 to 4.5 logs of the spores in a 6–10 second period. Pulsed lights can kill 7 logs of these spores in a fraction of a second (Clark et al., 1997).

The PureBright™ system can reduce a vegetative microorganism population by about 9 log cycles, while a spore population can be reduced by 7 log cycles on a smooth non-porous surface. On porous and complex surfaces such as meat, an approximately 2–3 log cycle reduction is obtained. It has also been observed that mold spores rather than bacterial spores are more resistant to UV light (Anonymous, 1994). A variety of microorganisms including *E. coli, S. aureus, B. subtilis,* and *S. cerevisiae* were inactivated by using between 1–35 pulses of light with an intensity ranging from about 1–12 J/ cm^2 (Dunn et al., 1991) (Fig. 17-4).

On the surface of different packaging materials a single light pulse inactivates *S. aureus* with an intensity as small as 0.75 J/cm^2. Two pulsed light flashes can achieve more than 6 log CFU/cm^2 reductions of this microorganism, while a single light pulse of 1.25 J/cm^2 completely inactivates it. *Bacillus cereus* and *Aspergillus* spores were inactivated with light intensities greater than 2 J/cm^2. More than a 7 log reduction of *A. niger* spores was achieved with just a few pulsed light flashes on the packaging surface (Dunn et al., 1991).

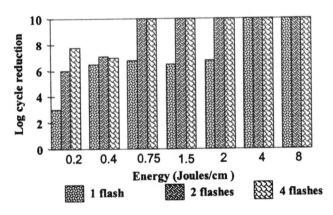

Figure 17-4 Inactivation of *Staphylococcus aureus* using full spectrum light [Ref. 2 US Patent 5,034,235] (adapted from Dunn et al., 1991).

Exposure to light pulses has also been shown to reduce the populations of *Listeria* and *Salmonella* inoculated in meat with minimal changes in nutrient contents (Rice, 1994). With the PureBright™ process, *Salmonella* on chicken wings (inoculated with 10^5 or 10^2 CFU/wing) was reduced by 2 log cycles. *Listeria* was reduced by 2 log cycles on hot dogs (inoculated with 10^3 or 10^5 CFU/wiener) after pulsed light treatment.

Dunn et al. (1991) inoculated curds of commercial dry cottage cheese with *Pseudomonas* and subjected them to light with an energy density of 16 J/cm^2 and pulse duration of 0.5 ms. After two flashes the viability of the microbial population was reduced by 1.5 log cycles and the temperature at the surface of the curd closest to the light source increased by only 5°C. Sensory evaluations using experienced taste testers showed no effect on the taste of the cheese.

Commercial eggs off the shelf from a grocery store and raw unwashed eggs from a farm were inoculated with *S. enteritidis* by immersion for 10 min, and then treated with 8 flashes at 0.5 J/cm^2 per flash (Dunn, 1996). Pulsed light eliminated the microbial contamination from the surface of the shelled eggs by as much as 8 log cycles.

A variety of biochallenged indicator organisms known to be highly resistant to the range of traditional sterilization methods for aseptically filled seal products have also been inactivated by pulsed light. Dunn et al. (1997b) worked with *Bacillus pumilus*, a suggested indicator spore for gamma ray or ionizing radiation sterilization; *B. subtilis* strain niger variety globigii spores, a suggested indicator for ethylene oxide, hydrogen peroxide, and dry heat sterilization; and *B. stearothermophilus*, an indicator for autoclave steam sterilization. These microorganisms and spores were inoculated and sealed in containers of 0.5 to 15 ml, and after the containers were treated with 10–20 flashes at 1 J/cm^2, all the inoculated spore samples were found to be sterile.

Applications

The potential applications of pulsed light technology include the surface sterilization of fresh meat, poultry, and fish with a significant reduction of *E. coli, Salmonella, Listeria,* and spoilage microorganisms such as *Pseudomonas*. In the case of meat products, thin slices can permit light penetration. It is even possible that prepared and processed meat products such as sausages and ground meat patties could be treated to increase their shelf-life under refrigeration without the necessity for freezing.

In the pulsed light treatment of fresh fruits and vegetables such as potatoes, tomatoes, apples, and bananas, an important reduction in the microbial and enzymatic activity (up to 60 to 70% with one flash of light at 1 J/cm^2) takes place which substantially increases their marketable life. For example, fresh tomatoes treated at PurePulse with pulsed light and stored under refrigeration remained acceptable for 36 d. The pulsed light treatment of baked goods such as pastas and rice entrees can also significantly increase the shelf-life of these products. Dunn et al. (1997a) found that white bread slices treated by light pulses through their packaging material maintained a fresh appearance for more than two weeks, while untreated white bread slices became moldy. The shelf-life of preservative-free breadsticks was increased from 6–20 days with a treatment of 4 J/cm^2. Similar results were obtained with chocolate cupcakes and pizza.

High levels of kill for the bacterial spores *Cryptosporidium* oocysts, *Klebsiella terragena*, and various viruses have been demonstrated in water (Dunn, 1996). The applications being investigated include municipal water, waste water, and packaged drinking water before, during, and after packaging (Dunn et al., 1995). A pulsed light dose response test conducted using multiple strains of potentially pathogenic waterborne organisms such as *Cryptosporidium, E. coli, Salmonella sps., Klebsiella, Poliovirus,* and *Rotavirus* led to the design and fabrication of a point-of-entry PureBright™ water treatment unit that will soon be commercialized.

Pulsed light can also be used to sterilize the packaging material surfaces for the food used in the medical and pharmaceutical industries, medical/dental devices and instruments, hospital air, and other environments. Since the pulsed light technique does not use heat, chemicals, or ionizing radiation, it is considered a clean process in which no chemical residues are left in the product and there is no need for evacuation steps. Another advantage is its capability to be implemented in a continuous process so that it can sterilize products on-line at high throughput rates since one to a few flashes yields high levels of microbial kill. The limitation of this technology is that it is not ionizing and does not penetrate opaque material, which in these cases confines its application to surface treatments.

Because the implementation of any new methodology is an economic strain, it is important to consider what type of budgeting may be involved with the incorporation of the PureBright™ system. Conservative estimates from PurePulse are about 0.1 to 0.2 cents U.S./ft^2 of treated area using 4 J/cm^2, which includes equipment amortization, lamps, electricity, and maintenance (Dunn et al., 1996; Clark et al., 1997).

PULSED ELECTRIC FIELDS TECHNOLOGY

The PEF processing of food involves the application of short pulses (duration of micro- to milliseconds) of high electric field intensity on the order of 20–40 kV/cm. In a continuous process the residence time of the food in the treatment zone is adjusted to obtain the required dosage (i.e., appropriate number of pulses). Food may be processed at ambient or refrigerated temperatures. In order to obtain a higher shelf-life, it is recommended that the product be aseptically packed and stored under refrigerated conditions. Furthermore, PEF should be applied to preheated liquid foods to increase both the total log colony-forming-unit reduction and shelf-life stability (Vega-Mercado et al., 1997). The PEF process is safe because no dangerous chemical reactions have been detected; moreover, it is reliable because the same results can be obtained repeatedly.

Based on the dielectric rupture theory, an external electric field can induce an electric potential difference across cell membranes known as the transmembrane potential (Barbosa-Cánovas et al., 1997). When the

240

transmembrane potential reaches a critical or threshold value higher than a cell's natural potential of ≈1 V (Castro et al., 1993), electroporation or pore formation in the membrane occurs, and cell membrane permeability thus increases (Glaser et al., 1988). It is important to know, however, that the threshold transmembrane potential depends on the specific microorganism as well as the medium in which the microorganisms are present (Barbosa-Cánovas et al., 1997). The increase in membrane permeability is reversible if the external electric field strength is equal to or slightly exceeds the critical value. Gásková et al. (1996) report that the pulse-induced increase in membrane permeability for small species of inorganic ions is sufficient to cause cell death. Cell membrane breakdown or irreversible electroporation occurs under the application of high intensity electric field pulses (Ho and Mittal, 1996). Observations by Barsamian and Barsamian (1997) on fruit tissue samples and Lubicki and Jayaram (1997) on the bacterium *Yersinia enterocolitica* revealed irreversible membrane disruption through microscopy. The formation of pearl chains and cell fusion have been obtained by subjecting the coelomic cells of *Lapito mauritii* to electric field strengths as low as 125–300 V/cm. In other experiments where cells of *E. coli* suspended in SMUF were subjected to 64 pulses at 60 kV/cm (Pothakamury et al., 1995), the scanning electron microscopic (SEM) technique did not indicate significant differences between treated and untreated cells (Fig. 17-5). However, transmission electron microscopy (TEM) showed some shrinkage of the cytoplasmic membrane away from the outer membrane which indicated a loss of the cell membrane's semipermeability (Fig. 17-6).

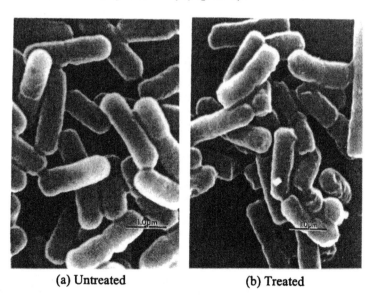

(a) Untreated	(b) Treated

Figure 17-5 Untreated (a) and PEF treated (b) cells of *Escherichia coli* in SMUF as seen with scanning electron microscope (Pothakamury, 1995).

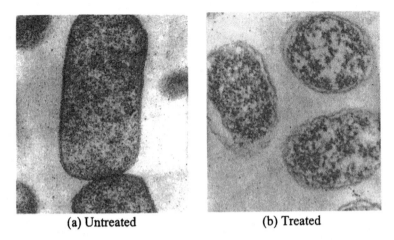

(a) Untreated   (b) Treated

Figure 17-6 Untreated (a) and PEF treated (b) cells of *Escherichia coli* in SMUF as seen with transmission electron microscope (Pothakamury, 1995).

PEF technology has the potential to economically and efficiently improve energy usage, as well as provide consumers with microbiologically safe, minimally processed, nutritious, fresh-like foods (Ott, 1997). Potential applications of PEF include cold sterilization of liquid foods and increasing juice yields by the disruption of plant cell membranes (Angersbach et al., 1997).

Engineering Aspects

A high intensity PEF processing system is a simple electric system consisting of a high voltage source, capacitor bank, switch, treatment chamber, voltage, current and temperature probes, and aseptic packaging equipment (Fig. 17-7). The hazard analysis control points (HACCP) and hazard and operability studies (HAZOP) are key elements for the design and construction of a PEF facility. The work of Vega-Mercado et al. (1996a) analyzes these areas. They found the following critical control points (CCP) in need of special attention: 1) the raw material receiving area, 2) the treatment chamber, and 3) the packaging machine. The main concern of the individuals working in a PEF facility is the voltage intensity, which indicates that proper safeguards must be in place to prevent possible electro-shock.

A power source is used to charge the capacitor bank, and a switch to discharge energy from the capacitor bank across the food held in a treatment chamber. The high intensity electric fields can have the form of exponential, square-wave, oscillatory, bipolar, or instant reverse charge pulses.

The treatment chamber is one of the most important and complicated components of the processing system. It consists of two electrodes held in position by insulating material to form an enclosure containing the food. Parallel plates and wires, concentric cylinders, and a rod-plate are some possible

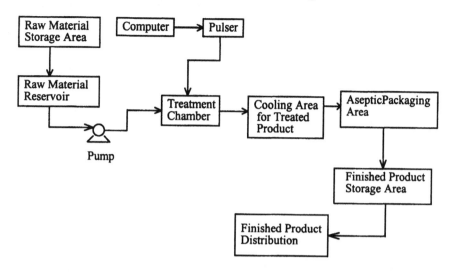

Figure 17-7 Schematic representation of PEF food processing.

electrode configurations. The type of electrode arrangement also influences microbial inactivation (Ohshima et al., 1997). One of the major technical problems with the PEF technology is the dielectric breakdown of foods, which is characterized by a spark. Dielectric breakdown is attributed to local electric field enhancement due to differences in dielectric properties. This phenomenon can be minimized during pulsing by using smooth electrode surfaces, round electrode edges, and designing the chamber to provide a uniform electric field. Several chamber designs are suggested: generally, a static chamber is suitable for preliminary laboratory scale studies, while for pilot plant or industrial scale operations a continuous chamber is preferred. Table 17-2 presents some chamber designs and characteristics of currently developed equipment. Lubicki and Jayaram (1997) presented new designs without direct exposition of the treated suspension to the electrodes that are still under research.

When electrical energy is applied in the form of short pulses, it destroys the cell membrane by mechanical effects with no significant heating of the food. Although the pulse repetition rate generates heat, the processing system generally includes a means to provide cooling of the treatment chamber.

The voltage, current, and electric field strength can be measured using an oscilloscope. Fiber-optic temperature instruments are being employed in order to directly measure treatment chamber temperatures (Qin et al., 1998). The treated food should be filled into packages or storage containers using aseptic packaging equipment. An integrated pilot scale system with a fluid handling system and aseptic packaging machine has been designed and set up by researchers at The Ohio State University (Zhang, 1997a).

Table 17-2 Specifications for static and continuous PEF chambers.

Static Chambers

Electrode	Spacer	Electrode Area (cm²)	Maximum E. Field Strength (kV/cm)	Reference
Carbon	Polyethylene		25	Sale and Hamilton, 1967
Stainless steel	Nylon		25	Dunn and Pearlman, 1987
Carbon		50	30	Grahl et al., 1992
Stainless steel	Polysulfone	27	70	Zhang et al., 1995
Stainless steel	Plexiglass	7 (3cm D.)	28	Gásková et al., 1996
Stainless steel	Delrin	165 (14.5cm D.)	87	Ho et al., 1997

D electrode diameter

Continuous Chambers

Electrode	Spacer	Reference
Parallel metallic plates	Dielectric insulator	Dunn and Pearlman, 1987
Electrode reservoir zones	Dielectric insulator	Dunn and Pearlman, 1987
Parallel stainless steel plates	Polysulfone with zigzag channels	Zhang et al., 1995
Coaxial stainless steel	Plexiglas	Qin et al., 1995b
Carbon with gold, platinum, and metal oxides	—	Bushnell et al., 1996
Co-field chamber: tubular shaped food grade stainless steel	Polycarbonate, ceramic, glass, or plastic	Yin et al., 1997

Biological Effects

In Microorganisms

Inactivation of microorganisms is affected by such factors as electric field strength, treatment time and temperature, pulse waveshape, type, concentration and growth stage of microorganisms, and characteristics of the medium. Microbial inactivation occurs when the applied electric field exceeds the critical transmembrane potential. Once this happens, microbial inactivation increases with an increase in the applied electric field strength. Hülsheger et al. (1981) related the microbial survival rate with the electric field strength according to the following equation:

$$ln\ s = -\ b_e\ (E - E_c) \tag{17-2}$$

where b_e is the rate constant, E the applied electric field, and E_c the extrapolated value of E for 100% survival, known as the critical or threshold electric field.

With the pulse duration held constant, microbial inactivation increases with an increase in the number of pulses (Lubicki and Jayaram, 1997), but each additional pulse has a less killing effect than the previous one (Liu et al., 1997). Martín-Belloso et al. (1997), working with *E. coli* suspended in liquid egg, showed that the inactivation increases with an increase in the pulse duration for a constant number of pulses. However, a large increase in the pulse duration may result in a great increase in the temperature of the food. Therefore, a compromise should be made for the increase in pulse duration. Hülsheger et al. (1981) related the microbial survival rate to the treatment time (defined as the product of the number and width of pulses) with the following equation:

$$ln\ s = -\ b_t\ ln\ \left(\frac{t}{t_c}\right) \tag{17-3}$$

where b_t is the kinetic rate constant, t the treatment time, and t_c the extrapolated value of t for 100% survival.

The work of Gásková et al. (1996) indicates that a substantial killing effect on *S. cerevisiae* can be obtained if rectangular electric pulses of 4–28 kV/cm with a minimum duration (10 μs) are used. Zhang et al. (1994) found that *S. cerevisiae* inactivation reaches a saturation with ten pulses of 25 kV/cm.

Both moderate temperature and electric field treatment exhibit synergistic effects on the inactivation of microorganisms. The critical potential for membrane breakdown decreases when temperature increases (Liu et al., 1997). The minimum survival ratio of *S. cerevisiae* was found to increase with the temperature from about 10^{-6} when 300 J/ml was applied at 50°C to 10^{-1} at 10°C. Heat sterilization without pulse treatment was not observed below 45°C (Ohshima et al., 1997). Although the inactivation has been shown to increase with an increase in temperature, it should be noted that the temperatures must be held far below those used in pasteurization.

Electric field pulses may also be applied in different forms. Oscillatory pulses are the least efficient for microbial inactivation, and square wave pulses are more energy efficient and lethal than exponentially decaying pulses. Bipolar pulses are more lethal than monopolar pulses because they cause a stress in the cell membrane and enhance its electric breakdown, and they also offer the advantages of minimum energy utilization, reduced deposition of solids on electrode surfaces, and reduced food electrolysis (Qin et al., 1994).

The sensitivity of microbial cells to electric field treatment increases with an increase in the size of the cell. Bacteria are more resistant to electric fields than yeasts, and among bacteria, in general those that are Gram positive are more resistant than those that are Gram negative. Spores are the most resistant to electric field treatment. Grahl and Märkl (1996) found a negligible inactivation of *Clostridium tyrobutyricum* (endospores), *Bacillus cereus* (endospores), and *Byssochlamys nivea* when an electric field strength of 22.4 kV/cm and 30 pulses was used. Gould (1995) considered that the inactivation of spores may also result from an indirect bactericidal effect of electrolysis products formed during the electrical treatment. Furthermore, it was found that spores of *B. subtilis* and *B. cereus* could be inactivated with very long treatment times, high temperatures, or long time gaps between pulses (Marquez et al., 1997).

The initial number of microorganisms in a food may or may not have an effect on its electric field induced inactivation. For example, inactivation was not affected when the concentration of *E. coli* in a simulated milk ultrafiltrate (SMUF) medium was varied from 10^3 to 10^8 cfu/ml. On the other hand, increasing the concentration of *S. cerevisiae* in apple juice resulted in a slightly lower inactivation. The effect of microbial concentration on inactivation may be related to the cluster formation of yeast cells and/or the possible hiding of microorganisms in low electric field regions (Qin et al., 1996).

The bacterial growth stage is an important factor for PEF inactivation since during proliferation, the area between the mother and daughter cells and sensitive parts of the cell envelope are susceptible to applied electric fields. Logarithmic phase cells are thus more sensitive than lag and stationary phase cells (Pothakamury et al., 1996) (Fig. 17-8).

Electric field induced microbial inactivation increases with a decrease in the ionic strength of the medium (Fig. 17-9). Foods with large electrical conductivities are difficult to work with because this means a smaller peak electric field will be generated across the treatment chamber. Using the same field intensity and number of pulses, Martin et al. (1997) found that the inactivation of *E. coli* in skim milk was lower than when it was inoculated in buffer solutions due to the complexity of the milk's composition, its lower resistivity, and the presence of proteins and fat, all factors which can be said to have a protective effect on the bacteria against electrical field pulses.

Benz et al. (1979) found no effect of a medium's pH on the breakdown of lipid bilayer membranes prepared from oxidized cholesterol. However, Vega-Mercado et al. (1996) reported a slightly greater inactivation of *E. coli* in SMUF at a low pH (5.7) than at a high pH (6.8). Furthermore, inactivation is not influenced by a bacteria being cultured in the presence or absence of oxygen, nor does Na and K in a treatment medium produce a greater killing effect. Ca and Mg, on the other hand, induce a protective effect against electric field treatment (Hülsheger et al., 1981). Particles and gas bubbles in a food are also a problem, as the former prevents a non-uniform distribution of the applied electric field, and the latter causes dielectric breakdown.

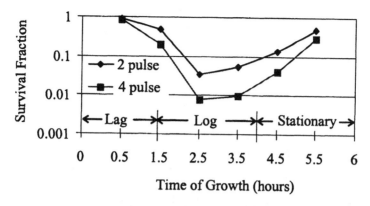

Figure 17-8 Cells of *E. coli* harvested at different growth stages suspended in SMUF and subjected to an electric field of 36 kV/cm at 7°C (adapted from Pothakamury et al., 1996).

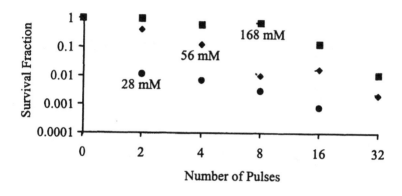

Figure 17-9 The effect of ionic strength on the inactivation of *E. coli* suspended in SMUF at 40 kV/cm and 10°C (two samples per each experimental condition) (adapted from Vega-Mercado et al., 1996b).

In Enzymes

The effect of PEF on enzymatic activity is not yet clear. The literature shows different results depending on the treatment conditions and enzyme studied. Ho et al. (1997) tested the influence of high electric field pulses on eight enzymes, applying 30 pulses of 13–87 kV/cm field intensity, 0.5 Hz pulse frequency, 2 μs pulse width, at a process temperature of 20°C. For some enzymes, activities were reduced after the pulse treatments: lipase, glucose oxidase, and heat-stable α-amylase exhibited a vast reduction of 70–85%; peroxidase and polyphenol oxidase showed a moderate 30–40% reduction; whereas alkaline phosphatase only displayed a slight 5% reduction. On the other hand, the enzyme activities of lysozyme and pepsin were increased under a certain range of voltages. In order to explain these results, the authors propose that the electric field pulses change the three-dimensional structure of the globular protein. In general, comparing the results with microbial depletion, the enzyme inactivation under PEF demands a higher voltage for significant reduction. Knowing that some enzymes are useful to the food industry provides opportunities for achieving efficient microbial and enzymatic control.

The treatment medium seems to be another important factor in enzymatic inactivation. Cells of *E. coli* inactivated by electric fields lose the ability to synthesize β-galactosidase, but the activity of this enzyme is not affected by electric fields in *E. coli* cells cultivated with benzene (Hamilton and Sale, 1967). Vega-Mercado et al. (1996c) studied the inactivation of a protease obtained from *Pseudomonas fluorescens* M3/6. The protease was inactivated up to 80% after 20 pulses at a rate of 0.25 Hz, with an electric field intensity of 18 kV/cm in a model system consisting of tryptic soy broth and yeast extract. However, in skim milk, a 60% inactivation was obtained after 98 pulses at a rate of 2 Hz with an electric field intensity of 14 kV/cm. The reduction in proteolytic activity was also a function of the electric field, number of pulses, and pulsing rate (Fig. 17-10).

In some cases, enzyme inactivation by PEF treatment can be even more effective than thermal pasteurization. Proteases and lipases are heat stable enzymes which cause spoilage in ultra-high temperature processed milk during storage at 3 to 6°C. One of these enzymes is plasmin (fibrinolysin E.C.3.4.21.7), otherwise known as milk alkaline protease. Although pasteurization decreases the initial plasmin activity in milk, proteolytic activity increases during the storage of processed milk. The plasmin activity in SMUF was found to decrease by 90% with electric fields of 30 kV/cm after 50 pulses at 10°C. Electric field treated plasmin solutions showed no significant changes in activity after 24 h of storage at 4°C (Vega-Mercado et al., 1995).

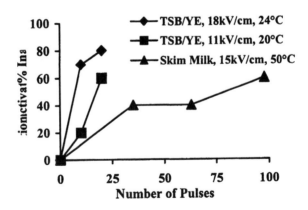

Figure 17-10 Inactivation of a protease from *P. fluorescens* M3/6 in triptych soy broth enriched with a yeast extract (TSB/YE, pulsing rate of 0.25 Hz) and skim milk (pulsing rate 2 Hz) using 2 μs pulses (adapted from Vega-Mercado et al., 1996c).

Applications

The PEF technology has promising potential to be used as a method of nonthermal food preservation. Model foods such as chloride, sulfate or phosphate solutions of sodium, SMUF, phosphate buffer solutions, semisolid potato dextrose agar, and real foods including apple juice, beaten eggs, milk, orange juice, yogurt, and green pea soup have been subjected to PEF treatment. These experiments demonstrated not only the inactivation of microorganisms and enzymes, but significant shelf-life extensions with minimum changes in the physical and chemical properties of the foods. The sensory properties of foods are also not degraded by PEF. A sensory evaluation found no significant differences between thermally processed and electric field treated apple juice and milk, and electric field processed eggs and pea soup were even preferred over at least one commercially available product (Barbosa-Cánovas et al., 1997).

Electric field treatment is associated with minimum energy utilization and high energy efficiency. Industrial scale processes are being assayed using exponential decay pulses. Although the peak power of each pulse is very high, the duration is very short, so the total energy in each pulse is relatively low and the average total power requirement is modest. For example, in the treatment of apple juice, the energy utilized with PEF is 90% less than the amount of energy used in the high temperature short time (HTST) processing method (Qin et al., 1995a).

Some of the many groups working on the industrial application of this nonthermal preservation process are PurePulse in the US, Thompson in France, and several universities in Europe, Canada, and US (Dunn and Pearlman, 1987; Bushnell et al., 1993, Barbosa-Cánovas, 1999.). As a result of the effort to develop this technology different patents claim PEF processes to preserve fluid foods such as dairy products, fruit juices, liquid eggs; emulsions such as salad dressings; and food ingredients (Dunn, 1996) by treatment with high voltage pulses. A 300 l/h continuous flow CoolPure Pilot System is now in use at their laboratory in San Diego, and they have designed, built, and delivered similar production scale and pilot units to food processors (Clark et al., 1997).

Furthermore, Washington State University (WSU) has a comprehensive program to pasteurize foods by high intensity PEF. A group there designed and constructed 12 and 25 ml volume temperature controlled static chambers, as well as parallel-plate and coaxial continuous treatment chambers (Zhang et al., 1995; Qin et al. 1995b). Model food such as SMUF and potato dextrose agar in the form of semisolid and real foods such as apple juice, peach juice, milk, orange juice, beaten eggs, and green pea soup have been successfully treated (Barbosa-Cánovas et al., 1997 and 1999).

Other research groups are making great efforts in the design and development of PEF treatment chambers, system setup, and scale-up. Yin et al. (1997) developed a co-field treatment chamber that consists of a series of interconnected tubular sections aligned in such a way that the product will first flow through an electrode, then through an insulator, and exit the system through a second electrode. The insert insulator section serves the purpose of holding the electrodes and enhancing the electric field in the treatment region. Due to the configuration of this chamber, the electric field is not uniform, but the homogeneity of the treatment can be significantly improved by the connection of several co-field chambers in series. One of the advantages of this single tubular flow chamber is that it will minimize electrode damage and deposition, therefore reducing maintenance and cleaning costs. Co-field chambers have been successfully implemented in a pilot plant PEF system that includes a final aseptic filling of the treated product (Zhang et al., 1997b). As a result of all these efforts, several patent applications on the use of PEF have been filed (Qin et al., 1997; Zhang et al., 1996; Yin et al., 1997).

Besides food preservation, there are many other possible applications of high intensity PEF. Some of these are an increase in juice yields by the disruption of plant cell membranes and tissues (Angersbach et al., 1997), the treatment of microorganism-charged residual water (Mittal and Choudhry, 1997), biofouling prevention, and medical applications such as tumor or cancer treatment (Schoenbach et al., 1997).

OSCILLATING MAGNETIC FIELDS

Microorganism spoilage is often the most important cause of food deterioration. Therefore, any process that inhibits microbial growth has the potential to preserve foods. Previous works have shown the capability of high intensity OMF to pasteurize food with an improvement in the quality and shelf-life compared with conventional pasteurization processes. Despite these encouraging first results, there has not been much research on the biological effect of high magnetic fields compared to low magnetic fields (Ueno, 1996). The lack of appropriate equipment capable of producing the needed high magnetic fields may be one of the reasons for the slow progress in this area. Furthermore, the literature reveals numerous examples of apparently conflicting observations on the effects of magnetic fields on biological systems which need clarification.

Generation

Electromagnets generate magnetic fields by supplying current to electric coils that are coaxially placed. The field strength can be adjusted by controlling the amount of current delivered to the coils in the magnet, and easily reversed by changing the direction of the current. One disadvantage of large electromagnets is their large power consumption, which requires high power supplies and cooling systems to reduce the heat generated by the magnet coils. Improved magnets have been developed with new designs, considerably increasing the magnetic field achieved by the traditional resistive magnets [i.e., the 30 T water cooled Bitter magnet working at the US National High Magnetic Field Laboratory facility (NHMFL)] (Crow et al., 1996).

Stable magnetic fields between 5 and 20 T are best produced in superconducting solenoid liquid that is helium cooled, but the maximum field achievable is limited by the properties of the material because superconductivity ceases in strong magnetic fields (Kovacs et al., 1997). These magnets are not appropriate for generating time varying fields, but they do produce very stable magnetic fields with low energy consumption on the order of watts compared to mega watts of resistive magnets. Intensive research is focused on the development of new superconductor materials capable of generating high intensity magnetic fields (Okada et al., 1996) that work with temperatures higher than liquid helium (Seuntjens and Snitchler, 1997). Hybrid magnets combining a superconducting magnet on the outside and a water-cooled normal magnet mounted inside can generate fields above 40 T, producing higher and flexible fields with minimal energy. The NHMFL has hybrid magnets capable of continuous operation at 45 T, and is developing one to operate at 60 T.

Much higher fields can be readily generated in pulsed form with a short duration compared to those produced with steady fields. The pulsed current can be obtained from a capacitor bank, but is limited by the Maxwell stress caused by the electromagnetic force between the current and field (Miura, 1991). Fields of 20 to 50 T can be produced by small solenoids in pulsed operations for the duration of a few milliseconds (Kovacs et al., 1997). The improvement of magnet design with an optimized glass or carbon fiber reinforcement can generate 40 T–81 T magnetic fields with 50 to 10 ms pulse widths (Heralch et al., 1996; Crow et al., 1996). Some commercial laboratories present these types of instruments with a charging unit, capacitor bank to store the energy, and a cooled magnet (Fig. 17-11). Oxford Instruments offers the PulseLab with peak fields in excess of 45 T and pulse widths of about 10 msec (Anonymous, 1997). Maxwell Laboratories developed the Magneform 7000 series[TM] coil, which gives peak fields up to 12 T and has been used to conduct research in the area of microbial inactivation (Barbosa-Cánovas et al., 1997).

Figure 17-11 Washington State University OMF laboratory scale unit coil and three types of field shapers.

251

Biological Effects

In Microorganisms

In general, magnetic fields influence the direction of migration and alter the growth and reproduction of microorganisms. Even magnetic fields of extremely low frequency, such as the Earth's magnetic field, can exert some influence over microorganisms. In some aquatic bacteria like the magnetotactic, the geomagnetic field lines force it to orientate and migrate along the field (Frankel et al., 1997). Furthermore, the influence of magnetic fields can also affect the locomotion of several protozoa (Brown, 1962).

Magnetic fields increase DNA synthesis, change the orientation of biomolecules and biomembranes, and alter ionic drifts across the plasma membrane, resulting in a modified rate of cell reproduction. Literature reports do not provide a clear understanding of the conditions under which magnetic fields have stimulatory, inhibitory, or no effects on the growth and reproduction of microorganisms. There is also little known on the possible influence of such process parameters as temperature, osmotic pressure, time, type of magnetic field, microorganism, stage of growth, or growth media. The recent review published by Kovacs et al. (1997) presents a comprehensive and important collection of studies about the effect of magnetic fields on microorganisms.

Van Nostran et al. (1967) observed that the rate of *S. cerevisiae* reproduction decreased considerably when exposed to a homogeneous field of 0.46 T, but the combined effect with salt and temperature increased the number of cells. A magnetic field with a heterogeneity of 2,300 Oe/cm did not affect the growth rate of *S. aureus* during an incubation at 37°C. However, with a heterogeneity gradient of 5,200 Oe/cm, the cell population increased after 3 to 6 h compared to the control. After 6 h, growth was inhibited, and after 7 h, the cell population was lower than that of the control culture (Gerencser et al., 1962). These different responses may be caused by an additional force or Maxwell stress induced by the field heterogeneity (Kovacs et al., 1997). The decreased growth rate was attributed to the increase in the inactivation of the cells rather than the decrease in the bacterial fission rate. In another study Yoshimura (1989) observed no changes in the growth of yeast cells treated with a static magnetic field of 0.57 T, while the growth of the cells was inactivated when the field applied was oscillating. A recent report (Ruzic et al., 1997) considered the effect of different factors over the growth of fungi under extremely low frequency sinusoidal magnetic fields using different media. A culture of *Pisolithus tinctorius* was exposed to fields of 0.01 mT at 46 Hz, and 0.025 and 0.1 mT at 50 Hz. Since the experiments with 46 Hz did not demonstrate any influence on the fungi growth (Adey, 1980; Broers et al., 1992), the authors suggested that the effects of the magnetic fields could be revealed only at a suitable frequency or power (called frequency or power windows). The same study demonstrated that using a liquid medium of pH 3, the transport of nutrients was more stimulated than when a solid medium with a pH of 6 was used. This result was attributed to changes in the membrane permeability-fluidity.

Moore (1979) studied the effect of pulsed magnetic fields with strengths ranging from 0.015 to 0.06 T and a frequency of 0.3 Hz on the growth of four bacterial and one yeast culture. The Gram-negative bacteria (*Pseudomonas aeruginosa* and *Halobacterium halobium*) showed a greater stimulatory growth response to magnetic fields than the Gram-positive bacteria (*B. subtilis* and *S. epidermidis*) or the yeast (*Candida albicans*). Pretreatment of the culture medium did not produce any observable difference in growth.

Studies conducted by Nakamura et al. (1997) and Tsuchiya et al. (1996) during the aerobic growth of *B. subtilis* and *E. coli* under homogeneous (7 T) and inhomogeneous (5.2–6.1 T) magnetic fields indicated the potential use of this technology to control microbial growth and metabolism. The effect of the inhomogeneous magnetic fields was much stronger than that of the homogeneous. Under the log phase, the high magnetic fields adversely affected the growth of the bacterial cells. During the stationary phase, the magnetic fields reduced the rate of vegetative cell compared with that under geomagnetic fields. Some possible explanations were given based on the constantly changing chemical conditions and extracellular nutrients which are susceptible to magnetic fields (i.e., water, ion solutions, macromolecules, lipid bilayers of the membrane).

Cells of the bioluminescent bacteria *Photobacterium phosphoreum* were exposed to a 20 Hz, 6 mT magnetic field for 20 min. Working with a temperature of 20°C, no change in the intensity of luminescence was observed. When the temperature was raised to 37°C, the luminescence increased at first and then decreased. After recovering from heat stress, exposed cells showed irregular oscillations for more than 20,000 seconds, which may indicate compensatory regulation. From these results it was determined that magnetic fields may act on bacteria as a co-stressing factor which activates a process or reaction already initiated by other stresses (Mittenzwey et al., 1996).

In Enzymes

Very few tests on enzymes exposed to high magnetic fields have been performed, and most of these were obtained in vitro using static magnetic fields. These published works on enzyme activity have resulted in a variety of effects, so no clear generalization yet exists. Some of these studies were conducted by Rabinovitch et al. (1967a) who subjected ribonuclease to magnetic fields up to 15 T for 5 to 6 min, and by Maling et al. (1965) treating peroxidase and aldolase-fructose 1.6-diphosphate exposed 10 min under 17 T. The results of these studies are in agreement with numerous reports on in vitro inactivation (Grissom, 1995; Kovacs et al., 1997; Hwang and Grissom, 1994) that show no appreciable static homogeneous magnetic field effects on enzyme activity reactions. Experiments in vivo with thymidine kinase under 0.2 to 1.4 T for 30 min also showed no effect (Feinendegen and Muhlensiepen, 1987). The literature also contains reports of some work with lactic dehydrogenase and ribonuclease subjected to 1.4 T and 0.4 T/cm heterogeneous magnetic fields for 40 min with no change in the reaction rate (Muller et al., 1971).

The activity of ethanolamine ammonia lyase was found to decrease when it was exposed to a 0.25 T magnetic field (Harkins and Grissom, 1994). On the other hand, the activity of carboxidismutase increased between 14–20% when it was exposed to a magnetic field of 2 T, but decreased as soon as the magnet was turned off. The activation of this enzyme was attributed to its stabilization by an increase in hydrogen bonding.

Different authors have obtained different magnetic field responses when inactivating the trypsin enzyme. Cooke and Smith (1964) working for 1 to 3 h with a 0.8 T and 0.022 T/cm heterogeneous magnetic field observed an increase in enzyme activity. In contrast, when Rabinovitch et al. (1967b) exposed the enzyme-substrate trypsin-BAPA that had been pretreated for 65 to 220 min at 20.8 T to a magnetic field of 22 T for 9 min, no effect was observed. Experiments conducted by Nazarova et al. (1982) and Vadja (1980) revealed the same independence from magnetic fields. Karavaec et al. (1974) reported a decrease in trypsin activity when it was exposed 1.5–2 h under a 0.8–09 T magnetic field. These results indicate that enzymatic activity is dependent on many factors, including field intensity, duration of magnetic exposure, pH, temperature, concentration, stirring speed, buffer composition, and ionic strength.

The results obtained by Nakamura et al. (1997) on alkaline phosphatase activity under inhomogeneous high magnetic fields suggest that in vitro and in vivo responses are not fully comparable. The activity of this commercially available enzyme was not affected by the magnetic exposure, while studies on *B. subtilis* and *E. coli* showed their intracellular alkaline phosphatase activity was reduced under high magnetic fields. When Nossol et al. (1993) studied the effect of magnetic fields on cytochrome c oxidase, they found the existence of "windows" or distinct ranges of field strength where the rate of enzymatic activity increased.

In Tissues and Membranes

The most important field effects are orientation, rotation and movement of cells, pearl chain formation, and deformation and fusion or destruction of cells (Kaiser, 1996). Biological membranes exhibit strong orientation in magnetic fields because of the intrinsically anisotropic structure of the membranes. Whether the cell membrane orients parallel or perpendicular to the applied magnetic fields depends on the overall anisotropy of the biomolecules (such as proteins that are associated with the membrane). Higashi et al. (1995) observed that normal erythrocytes in suspension are affected by fields of 1 T, and almost all of them orient themselves when exposed to a field of 4 T. This is attributed to diamagnetism of the erythrocyte membrane components alone.

Dihel et al. (1985) reported that the plasma membrane is a very important site for alterations when microorganisms are exposed to magnetic fields. Measurements of membrane fluidity demonstrated that magnetized membranes exhibited less fluidity than controls. These changes in membrane fluidity resulted in an alteration in the membrane-associated enzymatic activity, and

Ca^{2+} binding to the phospholipid head groups. Alterations in ion flux across the plasma membrane were also determined to alter cell division rates.

DNA synthesis in stationary phase cells of human fibroblasts was found to increase when subjected to a magnetic field of 2.3 to 56 μT and frequency in the range of 1.5 to 4 kHz (Liboff et al., 1984). When mouse fibroblasts were treated for 1 h with sinusoidal 50 Hz and a 2 mT magnetic field, the mean DNA content of the exposed cell population decreased, and the cell proliferation was reduced (Schimmelpfeng and Dertinger, 1993).

Many authors have reported magnetic field induced changes in specific gene transcriptions, which suggests an important role for magnetic field exposure in altering cellular processes. Exposure of several cell types to electromagnetic fields that differ in waveform, amplitude, and frequency have been found to induce general changes in gene transcription. Lin et al. (1996) studied the effects of continuous and single limited exposures of HL60 cells to 60 Hz electromagnetic fields, and the results showed an increase in transcript levels. Tuinstra et al. (1997) also obtained an altered cellular transcription. These results suggest that alternative sites and/or interactive mechanisms to the membrane may be capable of inducing a bio-response when cells are exposed to magnetic fields.

Applications

Hofmann (1985) showed the potential of OMF for food preservation when he subjected food sealed in a plastic bag to 1 to 100 pulses of an OMF with an intensity between 2 and 100 T and a frequency between 5 and 500 kHz. A single pulse generally decreases the microorganism population by about two orders of magnitude; however, additional pulses may be used to effect a greater degree of sterilization without any detectable change in food quality (Table 17-3). The Hofmann patent (1985) suggests that OMF couples energy into the magneto-active parts of large microorganisms' molecules like DNA or proteins. These critical biomolecules are broken by the treatment, which effectively destroys the microorganism or at least renders it reproductively inactive.

Table 17-3 Magnetic field inactivation of food spoilage microorganisms (adapted from Hofmann 1985).

Food Product	Temp (°C)	Field Intensity (Tesla)	N° of Pulses	Frequency (kHz)	Initial CFU/ml	Final CFU/ml
Milk	23.0	12.0	1	6.0	25,000	970
Yogurt	4.0	40.0	10	416.0	3,500	25
Orange Juice	20.0	40.0	1	416.0	25,000	6
Roll dough		7.5	1	8.5	3,000	1

The most important requirement for the application of this technology is that the food material has high electrical resistivity (greater than 10 to 25 ohms-cm). The applied magnetic field intensity depends on the electrical resistivity and thickness of the food being magnetized, with larger magnetic field intensities used for smaller electrical resistivities and greater thicknesses. The temperature of the food increases by 2 to 5°C, but the organoleptic properties remain unaltered. Important advantages of the OMF technology are that no special preparation is required before treatment of the food, which can be packaged prior to processing to reduce the possibility of cross-contamination. However, metal packaging cannot be used under a magnetic field.

Food preservation by the application of magnetic fields is safe to perform. High intensity magnetic fields exist only within the coil and immediately around it. Within a very short distance from the coil, the intensity drops drastically. For example, if the intensity inside the coil is about 7 T, within about 2 m beyond the coil the intensity drops off to about $7 \ 10^{-5}$ T, which is comparable to geomagnetic field intensity. Thus, the operator positioned a reasonable distance from the coil is out of danger and the process may be operated without any shielding.

The potential application of magnetic fields to decontaminate and reuse water on space flights was tested by Chizhov et al. (1975). A one-day-old culture of *E. coli* was inoculated in water, on which magnetic fields were shown to have a considerable bactericidal effect. However, the inactivation of microorganisms was lower when water contained higher concentrations of microorganisms.

Besides the mentioned works, no other technical literature on this nonthermal preservation method is presently available. The conditions necessary for magnetic fields to exhibit inhibitory, stimulatory, or no effects on microorganisms are not clearly understood. Though several mechanisms are proposed to explain the inhibitory effects of magnetic fields on microorganisms, there is little explanation for the stimulatory effects.

Additional research is required to optimize the OMF which inactivate microorganisms without denaturing the nutritional components of food. Both the effects of magnetic fields on the quality of food and the mechanism of microorganism inactivation must be studied in more detail before a commercialization process can be considered.

REFERENCES

Adey W.R. (1980). Frequency and power windowing in tissue interactions with weak electromagnetic fields. *Proc. IEEE* 68(1): 119–125.

Angersbach A., Esthiaghi N.M., and Knorr D. (1997). Impact of high intensity electric field pulses on plant cells and tissues. In: *New Frontiers in Food Engineering. Proceedings of the Fifth Conference of Food Engineering.* Barbosa-Cánovas G.V., Lombardo S., Narishman G., and Okos M., Eds. New York: AIChE, pp. 247–254.

Anonymous (1988). *Methods and Apparatus for Preservation of Foodstuffs.* World Patent 8,803,369. Applied by Maxwell Laboratories Inc., San Diego, CA.

Anonymous (1994). *The PureBright™ Process.* PurePulse Proprietary Information.

Anonymous (1997). Oxford Instruments Pulse Lab, Technical data.

Asaka M. and Hayashi R. (1991). Activation of polyphenol oxidase in pear fruits by high pressure treatment. *Agric. Biol. Chem.* 55(9): 2439–2440.

Barbosa-Cánovas G.V., Góngora-Nieto M.M., Pothakamury U.R., and Swanson, B.G. (1999) *Preservation of Foods with Pulsed Electric Fields.* San Diego CA: Academic Press.

Barbosa-Cánovas G.V., Pothakamury U.R., Palou E., and Swanson B.G. (1997). *Nonthermal Preservation of Foods.* New York: Marcel Dekker, Inc.

Barsamian S.T. and Barsamian T.K. (1997). Dielectric phenomenon of living matter. *IEEE Trans. Dielec. Elec. Insul.* 4(5): 629–643.

Benz R., Beckers F., and Zimmermann U. (1979). Reversible electrical breakdown of lipid bilayer membranes: A charge-pulse relaxation study. *J. Membrane Biol.* 48: 181–204.

Broers D., Kraepelin G., Lamprecht I., and Schulz O. (1992). Mycotypha africana in low–level athermic ELF magnetic fields. *Bioelectrochem. Bioenerg.* 27: 281–292.

Brown F.A. Jr. (1962). Responses of the planarian, Dugesia, and the protozoan, Parameci very weak horizontal magnetic fields. *Biol. Bull.* 123: 264.

Bushnell A.H., Dunn J.E., Clark R.W., and Pearlman J.S. (1993). *High Pulsed Voltage Systems for Extending the Shelf-Life of Pumpable Food Products.* U.S. Patent 5,235,905.

Bushnell A.H., Clark R.W., and Dunn J.E. (1996). *Process for Reducing Levels of Microorganisms in Pumpable Food Products Using a High Pulsed Voltage System.* U.S. Patent 5,514,391.

Castro A.J., Barbosa-Cánovas G.V., and Swanson B.G. (1993). Microbial inactivation of foods by pulsed electric fields. *J. Food Proc. Pres.* 17: 47–73.

Chizhov S.V., Sinyak Y.Y., Shikina M.I., Ukhanova S.I., and Krasnoshchekov V.V. (1975). Effect of magnetic fields on *E. coli* (in Russian). *Moscow Kosmicheskaya Biologiya I Aviakosmicheskaya Meditsina* 9(5): 26–31.

Clark W., Bushnell A., Dunn J., and Ott T. (1997). Pulsed light and pulsed electric fields for food preservation. In: *New Frontiers in Food Engineering. Proceedings of the Fifth Conference of Food Engineering.* Barbosa-Cánovas G.V., Lombardo S., Narishman G., and Martin Okos, Eds. New York: AIChE, pp. 212–215.

Cooke E.S. and Smith M.J. (1964). Increase of trypsin activity. In: *Biological Effects of Magnetic Fields, Vol. 1.* Barnothy M.F., Ed. New York: Plenum Press, pp. 246–254.

Crow J.E., Parkin D.M., Schneider-Muntau H.J., and Sullivan N.S. (1996). The United States National hihg magnetic field laboratory facilities, science and technology. *Physica B* 216:146-152.

Demetrakakes P. (1996). The pressure is on. *Food Proc.* 2: 79–80.

Deplace C. and Deplace G. (1996). High hydrostatic pressure with and without thermal control. In: *Tecnologías Avanzadas en Esterilización y Seguridad de Alimentos y Otros Productos.* Rodrigo M., Martinez A., Fisman S.M., Rodrigo C., and Matew A., Eds. Valencia, Spain: IATA.

Dihel L.E., Smith-Sonnenborn J., and Middaugh R.C. (1985). Effects of extremely low frequency electromagnetic fields on cell division rate and plasma membrane of Paramecium tetraurelia. *Bioelectromag.* 6: 61–71.

Doevenspeck H. (1961). Influencing cells and cell walls by electrostatic impulses. *Fleischwirtschaaft* 13: 986–987.

Donsi G., Ferrari G., and Di Matteo M. (1996). High pressure stabilization of orange juice: Evaluation of the effects of process conditions. *Ital. J. Food Sci.* 2: 99–106.

Drake M.A., Harrison S.L., Asplund M., Barbosa-Cánovas G.V., and Swanson B.G. (1997). High pressure treatment of milk and effects on microbiological and sensory quality of cheddar cheese. *J. Food Sci.* 62(4): 843–845.

Dunn J. (1996). Pulsed light and pulsed electric field for foods and eggs. *Poultry Sci.* 75: 1133–1136.

Dunn J., Bushnell A, Ott T., and Clark W. (1997a). Pulsed white light food processing. *Cereal Foods World* 42(7): 510–115.

Dunn J., Bushnell A., and Clark W. (1997b). Pulsed light processing and sterilization. In: *Biotechnology International.* Universal Medical Press, Inc., pp. 233–238.

Dunn J., Clark R.W., Asmus J.F., Pearlman J.S., Boyer K., Pairchaud F., and Hofmann G. (1991). *Methods for Preservation of Foodstuffs.* U.S. Patent 5,034,235.

Dunn J., Ott T., and Clark W. (1995). Pulsed light treatment of food and packaging. *Food Technol.* 49(9): 95–98.

Dunn J.E. and Pearlman J.S. (1987). *Methods and Apparatus for Extending the Shelf Life of Fluid Food Products.* U.S. Patent 4,695,472.

Earnshaw R. (1996). High pressure food processing. *Nutr. Food Sci.* 2: 8–11.

Farkas D.F. and Ting E.Y. (1997). Manufacturing issues in the application of high pressure to food processing. In: *New Frontiers in Food Engineering. Proceedings of the Fifth Conference of Food Engineering.* Barbosa-Cánovas G.V., Lombardo S., Narishman G., and Okos M., Eds. New York: AIChE, pp.199–204.

Farr D. (1990). High pressure technology in the food industry. *Trends Food Sci. Tech.* 1: 14–16.

Feinendegen L.E. and Muhlensiepen H. (1987). In vivo enzyme control through a stationary strong magnetic field: The case of Thymidine kinase in mouse bone marrow cells. *Int. J. Radiat. Biol. Relat. Stud. Phys. Chem. Med.* 52(3): 469–480.

Frankel R.B., Bazylinski D.A., Johnson M.S., and Taylor B.L. (1997). Magneto-aerotaxis in *Marine coccoid* bacteria. *Biophys. J.* 73: 994–1000.

Galazka V.B., Dickson E., Summer I.G., and Ledward D.A. (1997). Effect of high pressure on the functionality of β-Lactoglobulin and Bovine serum albumin and their mixtures with dextran sulfate. In: *New Frontiers in Food Engineering. Proceedings of the Fifth Conference of Food Engineering.* Barbosa-Cánovas G.V., Lombardo S., Narishman G., and Okos M., Eds. New York: AICHE, pp. 123–128.

Gásková D., Sigler K., Janderová B., and Plásek J. (1996). Effect of high voltage electric pulses on yeast cells: Factors influencing the killing efficiency. *Bioelectrochem. Bioenerg.* 39: 195–202.

Gerencser V.F., Barnothy M.F., and Barnothy J.M. (1962). Inhibition of bacterial growth by magnetic fields. *Nature* 196: 539–541.

Glaser R.W., Leikin S.L., Chernomordik L.V., Pastushenko, V.F., and Sokirko A.I. (1988). Reversible electrical breakdown of lipid bilayers: Formation and evolution of pores. *Biochim.Biophys. Acta* 940: 275–287.

González C., Ibarz A., Ma L., Barbosa-Canovas G.V., and Swanson B.G. (1997). Textural properties of high hydrostatic pressure and heat formed surimi gels. In: *New Frontiers in Food Engineering. Proceedings of the Fifth Conference of Food Engineering.* Barbosa-Cánovas G.V., Lombardo S., Narishman G., and Okos M., Eds. New York: AIChE, pp. 132–137.

Gould G.W. (1995). *New Methods of Food Preservation.* London: Blackie Academic and Professional.

Grahl T. and Märkl H. (1996). Killing of microorganisms by pulsed electric fields. *Appl. Microbio. Biotechnol.* 45: 148–157.

Grahl T., Sitzman W., and Märkl H. (1992). Killing of microorganisms in fluid media by high voltage pulses. In: *10th Dechema Annual Meeting of Biotechnologists. Biotechnology Conference Series, 5B.* Karlsruhe, Germany, pp. 675–678.

Grissom C.B. (1995). Magnetic field effects in biology: A survey of possible mechanisms with emphasis on radical-pair recombination. *Chem. Rev.* 95: 3–25.

Hamilton W.A. and Sale A.J.H. (1967). Effects of high electric fields on microorganisms II. Mechanism of action of the lethal effect. *Biochim. Biophys. Acta* 148: 789–800.

Harkins T.T. and Grissom C.B. (1994). Magnetic field effects on bìsubï12 ethanolamine ammonia lyase: Evidence for a radical mechanism. *Sci.* 263(5149): 958.

Hayakawa I., Kanno T., Tomita M., and Fujio Y. (1994). Application of high pressure for spore inactivation and protein denaturation. *J. Food Sci.* 59(1): 159–163.

Hayashi R. (1989). Application of high pressure to food processing and preservation: Philosophy and development. In: *Engineering and Food.* Vol. 2. Spiess W.E.L. and Schubert H., Eds. London: Elsevier Applied Science, p. 815.

Hedén C.G. (1964). Effects of high hydrostatic pressure on microbial systems. *Bacteriol. Rev.* 28: 14–29.

Heralch F., Agosta Ch. C., Bogaerts R., Boon W., Deckers I., De Keyser A., Harrison N., Lagutin A., Li L., Trappeniers L., Vanacken J., Bockstal L.V., and Esch A.V. (1996). Experimental techniques for pulsed magnetic fields. *Physica B*. 216: 161–165.

Higashi T., Yamagishi A., Takeuchi T., and Date M. (1995). Effects of static magnetic fields on erythrocyte rheology. *Bioelectrochem. Bioenerg.* 36: 101–108.

Ho S.Y. and Mittal G.S. (1996). Electroporation of cell membranes: A review. *Crit. Rev. Biotechnol.* 16(4): 349–362.

Ho S.Y., Mittal G.S. and Cross J.D. (1997). Effect of high field electric pulses on the activity of selected enzymes. *J. Food Eng.* 31: 69–84.

Hofmann G.A. (1985). *Deactivation of Microorganisms by an Oscillating Magnetic Field*. U.S. Patent 4,524,079.

Homma N., Ikeuchi Y. and Suzuki A. (1994). Effects of high pressure treatment on the proteolytic enzymes in meat. *Meat Sci.* 38: 219–228.

Hoover D.G. (1993). Pressure effects on biological systems. *Food Tech.* 47(6): 150–155.

Hoover D.G., Metrick C., Papineau A.M., Farkas D.F., and Knorr D. (1989). Biological effects of high hydrostatic pressure on food microorganisms. *Food Tech.* 43(3): 99–107.

Hülsheger H., Potel J., Niemann E.G. (1981). Killing of bacteria with electric pulses at high electric strength. *Radiat. Environ. Biophys.* 20:53–65.

Hwang C.C. and Grissom C.B. (1994). Unusually large deuterium isotope effect in Soybean Lipoxygenase is not caused by a magnetic isotope effect. *J. American Chem. Soc.* 116(2): 795.

Jaenicke R. (1981). Enzymes under extreme conditions. *Ann. Rev. Biophys. Bioeng.* 10: 1.

Johnston D.E., Austin B.A., and Murphy R.J. (1992). Effects of high hydrostatic pressure on milk. *Milchwissenchaft* 47(12): 760–763.

Kaiser F. (1996). External signals and internal oscillation dynamics: Biophysical aspects and modelling approaches for interactions of weak electromagnetic fields at the cellular level. *Bioelectrochem. Bioenerg.* 41: 3–18.

Kalchayanand N., Sikes A., Dunne C.P., and Ray B. (1997). Effectiveness of hydrostatic pressure in combination with pressurization time and temperature and bacteriocin on viability loss kinetics of foodborne pathogens. In: *Institute of Food Technologists Annual Meeting*. Orlando, FL. Paper 59E–22.

Karavaec V.G., Misuno A.I., Rubenchik A.Y., and Shimanovich L.E. (1974). Toward the problem concerning the effect of a constant magnetic field on proteolytic enzymes. In: *Teplomasso Perenos Teplofiz Svoistva Veshestv*. Shashkov A.G., Ed. pp. 184–187.

Kitching J.A. (1957). Effects of high hydrostatic pressure on the activity of flagellates and ciliates. *J. Exptl. Biol.* 34: 494–510.

Kovacs P.E., Valentine R.L., and Alvarez P.J. (1997). The effect of static magnetic fields on biological systems: Implications for enhanced biodegradation. *Crit. Rev. Environ. Sci. Technol.* 27(4): 319–382.

Kristel J.A., Hauben E. Y., Wuytack C., Soontjens F., and Michels C.W. (1996). High-pressure transient sensitization of *Escherichia coli* to Lysozyme and Nisin by disruption of outer-membrane permeability. *J. Food Prot.* 59(4): 350–355.

Kunugi S. (1992). Effect of pressure on activity and specificity of some hydrolytic enzymes. In: *High Pressure and Biotechnology. Vol. 224. Proceedings of the First European Seminar on High Pressure and Biotechnology.* Hayashi R., Heremans K., and Masson P., Eds. Colloques INSERM John Libbey Eurotext Ltd. La Grande Motte, France 13-17, September, p. 129.

Landau J.V. (1967). Induction, transcription and translation in *Escherichia coli*: A hydrostatic pressure study. *Biochim. Biophys. Acta* 149: 506–512.

Liboff A.R., Williams T., Strong D.M. and Wistair R. (1984). Time varying magnetic fields: Effect on DNA synthesis. *Sci.* 223: 818–820.

Lin H., Blank M., Jin M., and Goodman R. (1996). Electromagnetic field stimulation of biosynthesis: Changes in c-myc transcript levels during continuous and intermittent exposure. *Bioelectrochem. Bioenerg.* 39: 215–220.

Liu X., Yousef A.E., and Chism G.W. (1997). Inactivation of *Escherichia coli* O157:H7 by the combination of organic acids and pulsed electric fields. *J. Food Safety* 16: 287–299.

Lubicki P. and Jayaram S. (1997) High voltage pulse for the destruction of Gram-negative bacterium *Yersinia enterocolitica*. *Bioelectrochem. Bioenerg.* 43: 135–141.

Ma L., Chang F.J., and Barbosa-Cánovas G.V. (1997). Inactivation of *E. coli* in liquid whole eggs using pulsed electric fields technology In: *New Frontiers in Food Engineering. Proceedings of the Fifth Conference of Food Engineering.* Barbosa-Cánovas G.V., Lombardo S., Narishman G., and Okos M., Eds. New York: AICHE, pp. 216–221.

Maling J.E., Weissbluth M., and Jacobs E.E. (1965). Enzyme substrate reactions in high magnetic fields. *Biophys. J.* 5: 767–776.

Marquez V.O., Mittal G.S., and Griffiths M.W. (1997). Destruction and inhibition of bacterial spores by high voltage pulsed electric fields. *J. Food Sci.* 62(2): 399–401.

Martin O., Qin B.L., Chang F.J., Barbosa-Cánovas G.V., and Swanson B.G. (1997). Inactivation of *Escherichia coli* in skim milk by high intensity pulsed electric fields. *J. Food Proc. Eng.* 20: 317–336.

Martín-Belloso O., Vega-Mercado H., Qin B.L., Chang F.J., Barbosa-Cánovas G.V., and Swanson B.G. (1997). Inactivation of *Escherichia coli* suspended in liquid egg using pulsed electric fields. *J. Food Proc. Pres.* 21: 193–208.

Mertens B. and Knorr D. (1992). Development of non thermal processes for food preservation. *Food Technol.* 46(5): 124–133.

Mittal G.S. and Choudhry M. (1997). Pulsed electric field sterilization of waste brine solution. In: *Proceedings of ICEF7.* Seventh International Congress on Engineering and Food, The Brighton Center, U.K., pp. O13–O16.

Mittenzwey R., Sumuth R., and Mei W. (1996). Effects of extremely low frequency electromagnetic fields on bacteria—the question of a co-stressing factor. *Biochem. Bioenerg.* 40: 21–27.

Miura N. (1991). Generation of megagauss magnetic fields and their application to solid state physics, physics and engineering applications of magnetism. In: *Springer Series in Solid State Sciences, 92.* Ishikawa Y. and Miura N., Eds. Berlin: Springer-Verlag, pp. 19–47.

Moore R.L. (1979). Biological effects of magnetic fields: Studies with microorganisms. *Can. J. Microbiol.* 25(10): 1145–1151.

Morita R.Y. (1957). Effect of hydrostatic pressure on succinic, malic and formic dehydrogenases in *Escherichia coli. J. Bacteriol.* 74: 251–255.

Muller K., Haberditzl W., and Pritze B. (1971). Examination of the influence of magnetic fields on chemical reactions. *Z. Phys. Chem. Leipzig.* 248: 185–192.

Mussa D.M., Ramaswamy H.S., and Smith J.P. (1997). Assesment of safety of ultra high pressure processed pork through destruction kinetics of *Listeria monocytogenes.* In: *New Frontiers in Food Engineering. Proceedings of Fifth Conference of Food Engineering.* Barbosa-Cánovas G.V., Lombardo S., Narishman G., and Okos M., Eds. New York: AIChE, pp. 205–211.

Nakamura K., Okuno K., Ano T., and Shoda M. (1997). Effect of high magnetic fields on the growth of *Bacillus subtilis* measured in a newly developed superconducting magnet biosystem. *Bioelectrochem. Bioenerg.* 43: 123–128.

Nandathur S.R., Aleman G.D., Farkas D.F., and Raghubeer E.V. (1997). Inactivation and long term regeneration of vegetative microbes in high pressure treated ranch style salad dressing. In: *Institute of Food Technologists Annual Meeting.* Orlando, FL. Paper 59E–22.

Nazarova N.M., Livshitz V.A., Anzin V.B., Veselago V.G., and Kuznetsov A.N. (1982). Hydrolysis of globular proteins with trypsin in a strong magnetic field. *Biofizika.* 27(4): 720–721.

Nossol B., Buse G., and Silny J. (1993). Influence of weak static and 50 Hz magnetic fields on the redox activity of cytochrome c oxidase. *Biolectromag.* 14 (4): 361–372.

Ohshima T., Sato K., Terauchi H. and Sato M. (1997). Physical and chemical modifications of high-voltage pulse sterilization. *J. Electrostat.* 42: 159–166.

Okada M., Tanaka K., Fukushima K., Sato J., Kitaguchi H., Kumakura H., Kiyoshi T., Inoue K., and Togano K. (1996). Bi-2212/Ag Superconducting insert magnet for high magnetic field generation over 22 T. *Japan J. Appl. Phys.* 35(2): L623–L626.

Ott T. (1997). The pulse is with you: New nonthermal preservation processes. *Cereal Foods World* 42(6): 460.

Peck M.W. (1997). *Clostridium botulinum* and the safety of refrigerated processed foods of extended durability. *Trends Food Sci. Technol.* 8(6): 186–192.

Pothakamury U.R. (1995). *Preservation of Foods by Nonthermal Processes.* Ph.D. Thesis, Washington State University.

Pothakamury U.R., Vega H., Zhang Q., Barbosa-Cánovas G.V., and Swanson B.G. (1996). Effect of growth stage and temperature on inactivation of *E. coli* by pulsed electric fields. *J. Food Prot.* 59: 1167–1171.

Qin B.L., Zhang Q., Barbosa-Cánovas G.V., Swanson B.G., and Pedrow P.D. (1994). Inactivation of microorganisms by pulsed electric fields with different voltage waveforms. *IEEE Trans. Dielec. Elec. Insul.* 1(6): 1047–1057.

Qin B.L., Chang F.J., Barbosa-Cánovas V.G., and Swanson B.G. (1995a). Nonthermal inactivation of *Saccharomyces cerevisiae* in apple juice using pulsed electric fields. *Lebensm.-Wiss. u.-Technol.* 28: 564–568.

Qin B.L., Zhang Q., Barbosa-Cánovas G.V., Swanson B.G., and Pedrow P.D. (1995b). Pulsed electric field treatment chamber design for liquid food pasteurization using finite element method. *Trans. ASAE*, 38 (2): 557–565.

Qin B.L., Pothakamury U.R., Barbosa-Cánovas G.V., and Swanson B.G. (1996). Non thermal pasteurization of liquid foods using high-intensity pulsed electric fields. *Crit. Rev. Food Sci. Nutrit.* 36(6): 603–627.

Qin B.L., Barbosa-Cánovas G.V., Pedrow P.D., Olsen R.G., Swanson B.G., and Zhang Q. (1997). *Continuous Flow Electrical Treatment of Flowable Food Products.* U.S. Patent 5,662,03.

Qin B.L. Barbosa-Cánovas G.V., Swanson B.G., Pedrow P.D., and Olsen R. (1998). Inactivating microorganisms using a pulsed electric field continuous treatment system. *IEEE Trans. Industry Applic.* 34(1/2): 43–50.

Rabinovitch B., Maling J.E., and Weissbluth M. (1967a). Enzyme-substrate reactions in very high magnetic fields. I. *Biophys. J.* 7: 187–204.

Rabinovitch, B., Maling, J.E., and Weissbluth, M. (1967b). Enzyme-substrate reactions in very high magnetic fields. II. *Biophys. J.* 7: 319–327.

Rice J. (1994). Sterilizing with light and electrical impulses *Food Proc.* 7: 66.

Ruzic R., Gogala N., and Jerman I. (1997). Sinusoidal magnetic fields: Effects on the growth and ergosterol content in Mycorrhizal fungi. *Electro- and Magnetobiol.* 16(2): 129–142.

Sale A.J.H. and Hamilton W.A. (1967). Effect of high electric field on microorganisms I. Killing of bacteria and yeast. *Biochim. Biophys. Acta.* 148:781–788.

Schimmelpfeng J. and Dertinger H. (1993). The action of 50 Hz magnetic and electric fields upon cell proliferation and cyclic AMP content of cultured mammalian cells. *Bioelectrochem. Bioenerg.* 30: 143–150.

Schoenbach K.H., Peterkin F.E, Alden R.W., and Beebe S.J. (1997). The effect of pulsed electric fields on biological cells: Experiments and applications. *IEEE Trans. Plasma Sci.* 25(2): 284–292.

Seuntjens J.M. and Snitchler G. (1997). Practical high temperature superconductor composites for high energy physics applications. *IEEE Trans. Applied Supercond.* 7(2): 1817–1820.

Shigehisa T., Ohmori T., Saito A., Taji S., and Hayashi R. (1991). Effects of high hydrostatic pressure on characteristics of pork slurries and inactivation of microorganisms associated with meat and meat products. *Int. J. Food Microbiol.* 12: 207–216.

Sitzmann W. (1990). Keimabtotung mit hilfe elecktrischer hochspannungsimpulse in pumpfahigen nahrungsmitteln. *Vortrag anlablich des Seminars Mittelstansforderung in der Biotechnologie.* Ergebnisse des indirektspezifischen Programma des BMFT 1986–1989. KFA Julich.

Sitzmann W. (1995). High voltage pulsed techniques for food preservation. In: *New Methods in Food Preservation.* Gould G.W., Ed. New York: Blackie Academic & Professional, pp. 236–252.

Suzuki C. and Suzuki K. (1963). The gelation of ovalbumin solutions by high pressure. *Arch. Biochem. Biophys.* 102(3): 367.

Thorne S. (1986). *The History of Preservation of Food.* Cumbria: Parthenon Publishing Group, Ltd.

Ting E. (1997). Ultrahigh pressure food pasteurization: A mechanical engineer's view. *Cereal Foods World* 2(6): 461.

Tsuchiya K., Nakamura K., Okuno K., Ano T., and Shoda M. (1996). Effect of homogeneous and inhomogeneous high magnetic fields on the growth of *Escherichia coli. J. Ferment. Bioeng.* 81: 344–347.

Tuinstra R., Greenebaum B., and Goodman E.M. (1997). Effects of magnetic fields on cell-free transcription in *E. coli* and HeLa extracts. *Bioelectrochem. Bioenerg.* 43: 7–12.

Ueno S. (1996). *Biological Effects of Magnetic and Electromagnetic Fields.* New York: Plenum Press.

Vadja T. (1980). Magnetic field effects on trypsin (EC 3.4.21.4) activity. *Radiat. Envron. Biophys.* 18(4): 275–280.

Van Nostran F.E., Reynolds R.J., and Hedrick H.G. (1967). Effects of high magnetic fields at different osmotic pressures and temperatures on multiplication of *Saccharomyces cerevisiae. Appl. Microbiol.* 15: 561–563.

Vega-Mercado H., Powers J.R., Barbosa-Cánovas G.V. and Swanson B.G. (1995). Plasmin inactivation with pulsed electric fields. *J. Food Sci.* 60(5): 1132–1136.

Vega-Mercado H., Luedecke L.O., Hyde G.M., Barbosa-Cánovas G.V., and Swanson B.G. (1996a). HACCP and HAZOP for a pulsed electric field processing operation. *Dairy, Food Environ. San.* 16(9): 554–560.

Vega-Mercado H., Pothakamury U.R., Chang F.J., Barbosa-Cánovas G.V., and Swanson B.G. (1996b). Inactivation of *E. coli* by combining pH, ionic strength and pulsed electric field hurdles. *Food Res. Int.* 29(2): 117–121.

Vega-Mercado H. (1996c). Inactivation of a Protease from *Pseudomonas fluorescens M3/6* Using Pulsed Electric Fields. Food Preservation by Pulsed Electric Fields. Ph.D. Thesis, Washington State University.

Vega-Mercado H., Martin-Belloso O., Qin B.L., Chang F.J., Góngora-Nieto M., Barbosa Cánovas G.V., and Swanson B.G. (1997). Nonthermal food preservation: Pulsed electric fields. *Trends Food Sci. Technol.* 8: 151–157.

Yin Y., Zhang Q.H., and Sastry S.H. (1997). *High Voltage Pulsed Electric Field Treatment Chambers for the Preservation of Liquid Food Products*. U.S. Patent 5,690,978.

Yoshimura N. (1989). Application of magnetic action for sterilization of food. *Shokuhin Kaihatsu* 24(3): 46–48.

Zhang Q., Monsalve-Gonzalez, A., Qin B.L., Barbosa-Cánovas G.V., and Swanson B.G. (1994). Inactivation of *Saccharomyces cerevisiae* by square wave and exponential-decay pulsed electric fields. *J. Food Proc. Eng.* 17: 469–478.

Zhang Q., Barbosa-Cánovas G.V., and Swanson B.G. (1995). Engineering aspects of pulsed electric field pasteurization. *J. Food Eng.* 25(2): 268–281.

Zhang Q., Qin B., Barbosa-Cánovas G.V., Swanson B.G. and Pedrow P.D. (1996*). Batch Mode Food Treatment Using Pulsed Electric Fields*. U.S. Patent 5,549,041.

Zhang Q.H. (1997a). Integrated pasteurization and aseptic packaging using high voltage pulsed electric field. In: *Proceedings of ICEF7*. Seventh International Congress on Engineering and Food, The Brighton Center, UK. pp. K13–K15.

Zhang Q.H., Qiu X., and Sharma S.K. (1997b). Recent developments in pulsed electric processing. In: *New Technologies Yearbook*. Chandrana D.I., Ed. Washington DC: National Food Processors Association, pp. 31–42.

Zimmerman F. and Bergman C. (1993). Isostatic pressure equipment for food preservation. *Food Technol.* 47(6): 162–163.

Zobell C.E. (1970). Pressure effects on morphology and life processes of bacteria. In: *High Pressure Effects on Cellular Processes*. Zimmerman A.M., Ed. New York: Academic Press.

CHAPTER 18

Minimally Processed Foods with High Hydrostatic Pressure

A. López-Malo, E. Palou, G.V. Barbosa-Cánovas, B.G. Swanson, and J. Welti-Chanes

INTRODUCTION

Consumer trends are changing, as high quality foods with fresh-like attributes are increasingly in demand (Gould, 1995). Consequently, less extreme treatments and/or additives are required. Gould (1992) identified fresh appearance and less heat and chill damage, acid, salt, sugar, and fat as just some food characteristics that must be attained in response to consumer demands. To satisfy these requirements, changes in the traditionally used preservation techniques must be achieved. Since these changes have important and significant microbiological implications, the safety, quality, and marketability of foods will be focused on the use of alternative emerging technologies.

Minimally processed foods (MPF) and partially processed foods (PPF) (fresh-like foods) are closely related to the recent changes in consumption patterns (Tapia de Daza et al., 1996) as well as certain needs in the catering industry (Ahvenainen, 1996). In many countries where there are no refrigerated storage and transport facilities, MPF or PPF may act as a mechanism to regulate fruit and vegetable production and their supply to the final transformation industries (Alzamora et al., 1993, 1995; Argaiz et al., 1995). Minimally processed foods and PPF include a series of products and processes that may be grouped in diverse food categories such as minimally, invisibly , carefully, or partially processed, and high moisture shelf-stable. Although not representative of the same types of products, all these terms can be grouped as MP based on the combination of preservation factors known as hurdle technology (Tapia de Daza et al., 1996), giving a wider concept than those definitions used by Rolle and Chism (1987), Shewfelt (1987), Huxsoll and Bollin (1989), Wiley (1994), and Ohlsson (1994). The original concept of minimal processing considered only food products that maintain their freshness by keeping biological tissues alive (Rolle and Chism, 1987; Shewfelt, 1987), but now it includes those that maintain the characteristics of fresh foods by inactivating the cellular metabolism in biological tissues (Huxsoll and Bollin, 1989; Wiley, 1994; Ohlsson, 1994). Therefore, high moisture fruit products (HMFP) preserved by hurdle technology can be classified as MP.

HIGH MOISTURE FRUIT PRODUCTS

The definition of HMFP is related to that given by Wiley (1994) for MP refrigerated fruits (MPRF). According to this author, the differences between MPRF and fresh fruits or those that have been preserved by cold, irradiation, dehydration, or thermal treatments are based on product quality, preservation method, storage form and conditions, and packaging procedure. Minimally processed refrigerated fruits keep their freshness characteristics to a greater extent than any of the other fruit products mentioned by Wiley, having an expected average shelf-life of at least 21 days compared to the 4–7 days of a MPF. The only other difference between MPF and MPRF is the use of refrigeration as a preservation factor, as both types of products should comply with the key requirements mentioned by Ahvenainen (1996) to obtain high quality MP products. In addition to the preparation and manipulation steps of MP fruits, HMF require the use of preservation factors such as blanching, reduction of water activity (a_w) and pH, and incorporation of antimicrobial agents and other additives to substitute the need for refrigeration. High moisture fruit products represent an improved version of many intermediate moisture (IMF) products and an alternative to MPRF in which refrigeration is not needed to increase the stability and shelf-life of the product. When Tapia de Daza et al. (1996) compared HMFP, MPRF, and IMF products developed from fruits, they found that the a_w levels of MPRF and HMFP are similar and generally higher than those of IMF. Incorporating a_w depressing solutes in HMFP therefore should not affect the freshness of the fruit, and not needing refrigeration as a preservation factor is set as an economic and technological objective. The use of additives and blanching also distinguishes HMFP from MPRF.

Stability

When generating a new type of product as in the case of HMFP, the quality changes and stability critical points should be clearly recognized. Ahvenainen (1996) stated that such modifications to MP vegetable products may be attributed to physiological, biochemical, microbiological and nutritional changes, and safety aspects. However, due to the use of the proposed preservation factors (a_w, pH, blanching, and preservative incorporation) for HMFP, it is adequate to just address the sensory changes and possible textural and structural modifications.

As presented in Figure 18-1, the development of a simple technology has been achieved to obtain HMFP (Alzamora et al., 1995). The sequence and number of operations are clear, and the preservation factors (hurdles) used are those mentioned before (a_w, pH, incorporation of additives, and a mild blanching). The flow chart shows the case of fruit pieces, but when the fruit is transformed into a puree, a pulping stage is included before equilibration. However, the type and level of hurdles are fixed within the ranges shown in Figure 18-1 (Alzamora et al., 1993, 1995). Although a_w reduction is carried out mainly by osmotic dehydration at atmospheric pressure in sugar syrups, Fito and

Chiralt (1997) have reported the application of vacuum osmotic dehydration to reduce process time and improve product quality.

Microbiological Aspects

In HMFP, pH plays an important role in final stability compared to pasteurized fruits and vegetables (Tapia de Daza et al., 1995) or those products named in a generic way by Leistner (1992) as pH-SSP (pH-shelf stable products). The a_w values of HMFP are in general higher than 0.95, so the pH value, blanching treatment, and addition of antimicrobial agents are the set of hurdles that help to avoid microbiological problems. When the evolution of microbial flora during the elaboration stages of HMFP with pineapple (Alzamora et al., 1989), papaya (López-Malo et al., 1994), and mango (Díaz de Tablante et al., 1993) were evaluated, these fruits were found microbiologically stable for 3–8 months at selected temperatures between 20 and 35°C.

The deteriorative microorganisms that may cause problems in HMFP are the osmophilic preservative-resistant yeasts *Zygosaccharomyces rouxii, Z. bailii,* and *Schizosaccharomyces pombe* (Tapia de Daza et al., 1996). Problems related to pathogenic microorganisms are not probable in HMFP mainly due to low pH values (3.0–4.0). Native flora evolution studies have demonstrated that HMFP are stable (Tapia de Daza et al., 1995, 1996); however, possible contamination of the products with osmo-resistant/osmophilic yeasts that may develop preservative-resistance could be a significant problem. This actually did occur (Fig. 18-2) when mango (0.97 a_w, pH 3.35, 150 ppm sodium bisulfite, and 1,000 ppm potassium sorbate) and papaya (0.98 a_w, pH 3.5, 150 ppm sodium bisulfite, and 1,000 ppm potassium sorbate) were challenged with *Z. bailii*, as the microorganism was able to survive and grow.

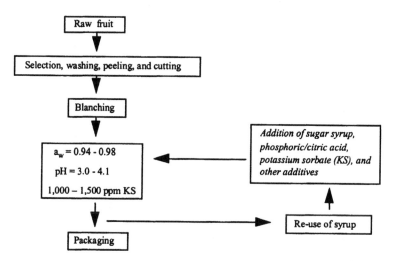

Figure 18-1 Flow diagram for the production of shelf-stable HMFP (adapted from Alzamora et al., 1995).

Color Changes

In most MP fruit and vegetable products the main deterioration problem is related to enzymes which maintain their normal activity or in many cases can be increased due to manipulation and cutting. Enzymatic activity mainly affects color and texture due to a food's physiological activity. Since the blanching stage included in the manufacture procedure of HMFP has both superficial disinfecting effects on the product and reduction of enzyme activity, these types of treatments should be very carefully applied.

The most important enzyme in MP fruits and vegetables is polyphenoloxidase (PPO), which causes browning. The other two enzymes of interest in these products are the pectolitics that cause fruit softness, and the lipoxidases that lead to oxidation problems with the development of off-flavors. Control of enzymatic activity can be achieved not only by thermal inactivation, but also by making the substrate on which the enzyme acts totally or partially unavailable or simply avoiding the presence of oxygen or other co-factors needed for the enzymatic reaction. To reduce enzyme activity, incorporating some chemical compounds or limiting the quantity of oxygen within the package system of the final product has successfully been used.

Color changes linked to the browning of products are normally due to the joined action of enzymatic and non-enzymatic reactions that in many cases are difficult to distinguish, so the global effect of the studied variables on the browning process is normally evaluated. Sulfite levels are of particular interest since the use of such compounds has been restricted in many regulatory agencies around the world. Therefore, sulfite substitutes to inhibit enzymatic and nonenzymatic browning should be more deeply investigated. A fitting example is the study performed by Monsalve-González et al. (1993) to control

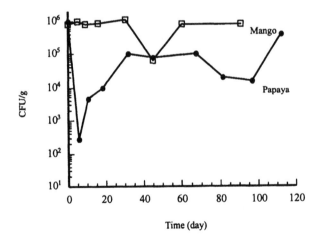

Figure 18-2 Evolution of *Zygosaccharomyces bailii* in HMFP from mango and papaya preserved by combined methods (adapted from Tapia de Daza et al., 1995).

browning in non-blanched apples (a_w 0.96-0.97, pH 3.9–4.2, and 1,000 ppm of ascorbic acid), where it was observed that the color of the fruit remained without change during storage at 25°C for 75 days when they were immersed in 100 ppm of sodium bisulfite solutions or solutions containing 200 ppm of 4-hexylresorcinol.

New Hurdles

The problems that have been identified in HMFP are the use of sulfites as antibrowning and antimicrobial agents, and the possibility of preservative- and heat-resistant yeast and mold growth. Thus, additional inactivation or stress factors should be used to control enzyme and microbial activity. High hydrostatic pressure treatments can be considered as a new hurdle that can be used in combination with other traditional microbial stress factors such as pH, a_w, and preservatives (Gould and Jones, 1989; Leistner, 1995). As such it is a promising choice to circumvent the problems created by browning and specific microorganisms; however, the effect of HHP needs to be investigated. The results of the research reflected in the remainder of this chapter thus focus on the use of HHP as an additional hurdle for the manufacture of HMFP.

HIGH HYDROSTATIC PRESSURE

The tremendous amount of information generated in this decade raises evidence about the effects of high pressure on microbial inactivation, chemical and enzymatic reactions, as well as structural and functionality changes in biopolymers (Hoover et al., 1989; Farr, 1990; Ledward, 1995). The basis of HHP is the Le Chatelier principle, which states that any reaction, conformational change, or phase transition that is accompanied by a decrease in volume will be favored at high pressures, while reactions involving an increase in volume will be inhibited (Ledward, 1995; Cheftel, 1995; Palou et al., 1999). However, due to the inherent complexity of foods and possibility of changes and reactions that can occur under pressure, predictions and generalizations of the effects of HHP treatments are difficult.

Pressure is an important thermodynamic variable that primarily affects the volume of a system (Heremans, 1995), and thus can affect a wide range of biological structures, reactions, and processes (Earnshaw, 1996). The influence of pressure on the reaction rate may be described by the transition state theory, which states that the rate constant of a reaction in a liquid phase is proportional to the quasi equilibrium constant for the formation of active reactants (van Eldik et al., 1989; Tauscher, 1995). Based on this assumption, Heremans (1995), van Eldik et al. (1989), and Tauscher (1995) reported that at constant temperature, the pressure dependence of the reaction velocity constant (k) is due entirely to the activation volume of the reaction ΔV^*:

$$\left(\frac{\delta \ln k}{\delta P}\right)_T = -\frac{\Delta V^*}{RT} \tag{18-1}$$

where P is the pressure, R the gas constant ($8.314 \text{ cm}^3\text{MPaK}^{-1}\text{mol}^{-1}$), and T temperature (K).

Microorganisms; chemical, biochemical, and enzymatic reactions; as well as some functional properties of biomolecules are affected by HHP. In an aqueous system, water molecules surrounding an ionized group align themselves according to the influence of the electrostatic charge, giving a more compact arrangement. Ionization of the acidic or basic groups found in many biomolecules such as proteins involves a volume decrease and therefore will be enhanced by increased pressure (Hoover et al., 1989; Palou et al., 1999).

A high pressure system consists of a high pressure vessel and its closure, pressure generation system, temperature control device, and material handling system (Mertens, 1995). Once loaded and closed, the vessel is filled with a pressure transmitting medium. Air is removed from the vessel by means of a low pressure fast fill and drain pump in combination with an automatic de-aeration valve to generate HHP. High pressures can also be generated by direct or indirect compression or heating the pressure medium (Deplace and Mertens, 1992; Palou et al., 1999).

Effects on Microorganisms

Sensitivity to pressure depends on the type of microorganisms, the most important of which are the vegetative and spore forms. In general, the vegetative forms are inactivated by pressures between 400 and 600 MPa, while the spores of some species may resist pressures higher than 1,000 MPa at ambient temperatures. The relative pressure sensitivity of vegetative microorganisms has made them the obvious first targets for the preservation of foods by high pressure, particularly those with low pH and intrinsic preservation systems already operating that ensure the pressure resistant food poisoning or spoilage sporeformers which may survive are unable to grow (Gould, 1995). However, the extent of microbial inactivation achieved at a particular pressure treatment depends on a number of interacting factors including type and number of microorganisms, magnitude and duration of HHP treatment, temperature, and composition of the suspension media or food (Patterson et al., 1995; Palou et al., 1997a, b).

The pressure sensitivity of microorganisms varies with the stage of the growth cycle at which the organisms are subjected to HHP treatment. In general, cells in the exponential phase are more sensitive to pressure treatments than those in the lag or stationary phases of growth (Earnshaw, 1995; Patterson et al., 1995). To illustrate, Table 18-1 presents the response of *Z. bailii* cells from different phase growths during HHP treatments (Palou et al., 1998a). A significant reduction in the pressure resistance of *Z. bailii* with cells from the exponential growth phase can be observed, showing that a pressure around 300 MPa reduces 50% of the initial inocula for cells in the stationary phase, while treatments at < 172 MPa produce the same effect on cells from the exponential growth phase.

Table 18-1 Pressure effects on *Zygosaccharomyces bailii* cells from stationary or exponential growth phase.

Pressure[1] (MPa)	Survivors (%)	
	Stationary Phase	**Exponential Phase**
138	100.0	--
172	98.0	29.3
207	--	15.8
242	--	6.7
245	87.6	--
276	65.0	3.3
310	43.3	1.5
345	15.4	--
431	1.0	--

[1]Treatments that account only for the pressure come-up time

For vegetative cells it has been reported (Cheftel, 1995; Palou et al., 1999) that the increase in holding time can be noticed only when pressures above 200–300 MPa are used. Therefore, a pressure threshold exists for each microorganism and depends on food composition and the levels of other variables such as temperature. Palou et al (1997b) noted for *Z. bailii* that even for 10 min holding time, pressure treatments at 172 MPa have no effect on yeast counts. However, the effect of pressurization time was noticeable for pressures of 345 MPa (Fig. 18-3). Food characteristics such as pH, a_w, preservatives, and nutrients may influence the pressure threshold which can increase or decrease depending on the microorganism and levels of other processing factors. For *Z. bailii*, the presence of 1,000 ppm potassium sorbate was found to have significant effects on yeast counts when combined with HHP treatments (Fig. 18-3).

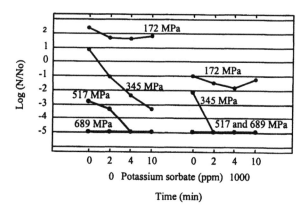

Figure 18-3 Combined effects of high pressure (MPa), time of exposure and potassium sorbate concentration on the log of the *Zygosaccharomyces bailii* survival fraction (N/N_0) at a_w 0.98 .

Palou et al. (1997a) reported the effect of reduced a_w (or increasing soluble solids concentration) on *Z. bailii* inhibition suspended in laboratory model systems adjusted to pH 3.5 (Table 18-2). Since this yeast is osmotolerant and its growth occurs in media containing up to 60% (w/w) glucose with an a_w of 0.85, the soluble solids concentration in the model systems considerably lowered the decimal reductions obtained after high pressure treatment as the soluble solids concentration increased. Smaller decimal reductions were observed for the experiments with sugar concentrations > 40%, although for the tests run with less than 20% soluble solids, complete inhibition (4.5 decimal reductions) of *Z. bailii* was obtained (Palou et al., 1997a). Ogawa et al. (1992) reported that for *Saccharomyces cerevisiae* inoculated in concentrated fruit juices, the number of surviving microorganisms depends on the juice soluble solids concentration, and that the inactivation effect at pressures $\leq$ 200 MPa decreased as juice concentration increased. Hashizume et al. (1995) also reported an increase in the number of *S. cerevisiae* surviving cells with increasing concentrations of sucrose (0–30% w/w) when pressurized at 260 MPa for 20 min at 25°C.

In addition, a baroprotective effect of reduced a_w for organisms that can grow under these conditions has been observed (Palou et al., 1997a, b; 1999). Oxen and Knorr (1993) observed that a HHP treatment at room temperature and 400 MPa for 15 min inactivated the yeast *Rhodotorula rubra* when the a_w of the suspension media was higher than 0.96, while the number of survivors was higher when the a_w was depressed. The resistance to inhibition at reduced a_w values may be attributed to cell shrinkage, which probably causes a thickening in the cell membrane that reduces membrane permeability and fluidity (Palou et al., 1997a). The increased baroresistance of microorganisms at low a_w may also be attributed to a partial cell dehydration due to osmotic pressure gradients between the internal and external fluids which may render smaller cells and thicker membranes and an increased pressure resistance (Knorr, 1993). The baroprotective effect of reduced a_w reveals that inhibition of microorganisms by high pressure depends not only on pressure and extent of treatment, but also the interactions with other intrinsic and extrinsic variables that influence microbial response (Palou et al., 1997a).

Inactivation Kinetics

Determination of microbial pressure inactivation kinetics depends on the pressure increase and decrease rates, and includes a decompression step for sampling. Therefore, sampling, treatment evaluation, and survivor enumeration are not continuous. However, pressure increase and decrease rates are not always reported and the come-up time to reach the pressure (or pressure build-up velocity) may not be included in the logarithmic representation of the survivors. In many reported HHP survival curves it is not clear if time zero experiments are growth controls without pressure treatment or pressure treatments which only take into account the come-up time to reach the working pressure. Table 18-3 presents the come-up times and *S. cerevisiae* survival fractions after HHP treatments at 175, 241, and 345 MPa by Pérez et al. (1997).

Table 18-2 Effect of soluble concentration and a_w on decimal reductions of *Z. bailii* after a 345 MPa pressure treatment for 5 min.

Soluble Solids Concentration (w/w)	Water Activity	Decimal Reductions
2.5	0.997	4.58
13.0	0.992	4.48
24.0	0.980	3.50
32.0	0.971	3.04
44.5	0.949	2.15
50.2	0.934	1.60
51.0	0.931	1.56
52.7	0.926	1.45
53.7	0.917	0.68
56.3	0.908	0.32
58.6	0.902	0.26

Table 18-3 Effect of pressure and pressure come-up time on *S. cerevisiae* initial population reduction (N_0/N_i) suspended in model systems with a_w 0.98 and pH 3.5.

Pressure (MPa)	Come-up Time (min)	Log (N_0/N_i)
175	1.9	-0.0702
241	2.4	-0.1135
345	3.1	-0.4083

N_0 = yeast count (CFU/ml) after the come-up time for the working pressure
N_i = yeast initial population (CFU/ml)

Pressure come-up times are shown to exert an important effect on yeast survival fractions, as counts decrease along with the pressure increases, and at 345 MPa a 0.5 log cycle reduction can be observed. Cheftel (1995) and Palou et al. (1997b) reported that the rate of pressure increase and decrease is often neglected as an experimental variable in HHP microbial inactivation studies and the initial population (N_0) can be notably reduced during the come-up time (Table 18-3). Palou et al. (1997b, c) also noted an important effect of the come-up time at pressures of 345, 517 and 689 MPa on *Z. bailii* log reductions in food model systems (a_w 0.98 or 0.95, and pH 3.5), where total inhibition was observed with only the processing time to reach 689 MPa.

The patterns of HHP inactivation kinetics observed with different microorganisms are quite variable (Palou et al., 1999). Some investigators indicate first-order kinetics for several bacteria and yeast (Carlez et al., 1993; Smelt and Rijke, 1992; Hashizume et al., 1995). Other authors observe a change in the slope and a two-phase inactivation phenomenon, as the first fraction of the population is quickly inactivated, whereas the second fraction appears much more resistant (Cheftel, 1995). The pattern of inactivation kinetics is also influenced by pressure, temperature and composition of the medium (Palou et al., 1999).

The results of this study indicate that the logarithm of a *S. cerevisiae* survival fraction (N/N_0) exhibits first-order kinetics due to its linear decrease with time (Fig. 18-4). Yeast cells were inactivated more rapidly with increasing pressure treatments. The experimental points with a pressure treatment duration of "0 min" expressed the effect of the come-up time to reach the working pressure, and corresponded to the initial population (N_0) for the kinetic analysis. The death velocity constants or inactivation rates (k) were calculated from the reciprocal of the slope of the survival curves following a traditional kinetic analysis. Table 18-4 presents the inactivation rate values. Similarly, Hashizume et al. (1995) observed first-order pressure inactivation kinetics for *S. cerevisiae* at 25°C and $a_w \approx 0.99$. Although the inactivation rates (k) obtained by Pérez et al. (1997) were greater than those calculated (0.0245, 0.0665, 0.0936, and 0.1872 min^{-1} for pressure treatments of 210, 240, 250, and 270 MPa, respectively) from the data reported by Hashizume et al. (1995), this can be attributed to the pH of the model system (3.5) which could favor inactivation.

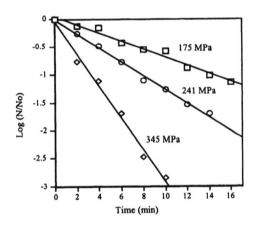

Figure 18-4 *Saccharomyces cerevisiae* first-order pressure inactivation kinetics in a laboratory model system with a_w 0.98 and pH 3.5.

Table 18-4 Effect of pressure on the inactivation rates (k) and decimal reduction times (D) of *S. cerevisiae* suspended in laboratory model systems with a_w 0.98 and pH 3.5.

Pressure (MPa)	k (min^{-1})	D (min)	r[a]
175	0.073 ± 0.008	31.8 ± 0.6	-0.96
241	0.126 ± 0.015	18.2 ± 0.3	-0.98
345	0.286 ± 0.086	8.0 ± 0.2	-0.95

[a]regression coefficient; figures ± are the standard deviations

To compare the effectiveness of pressure treatments and optimize process conditions, calculation of D values can be used to compare the resistance of microorganisms. The calculated D values for *S. cerevisiae* (2.303/k) presented in Table 18-4 represent the holding time needed at each pressure to reduce 90% of the initial population if the pressures are reached instantaneously.

The effects of pressure on inactivation kinetics can be analyzed using Equation 18-1 to obtain the activation volume. Palou et al. (1997c) reported activation volumes of -65.2 and -25.3 (cm^3mol^{-1}) for *Z. bailii* pressure inactivation at a_w 0.98 and 0.95, respectively. The values obtained differed depending on model system composition. The apparent ΔV^* which was calculated considering the activation process of microbial inhibition as one step indicated a variation in volume between the final and initial states of the yeast pressure inhibition "reaction" (Palou et al., 1997c). Since the negative activation volume represented a reaction favored by increased pressure, a reaction with a greater absolute ΔV^* value would indicate that increments in pressure can accelerate the response—which in this case is the yeast inactivation rate. By demonstrating the baroprotective effect of high sugar concentration or reduced a_w, these results imply there is an influence on microbial response under HHP treatments by media composition (Palou et al., 1997a, b, c; 1999).

In biological systems, volume changes associated with ionization can also be involved in the mechanism of microbial inactivation (Palou et al., 1997b; 1999) since enhanced ionization under high pressure treatments has been reported for water and acid molecules (Earnshaw et al., 1995). Palou et al. (1997c) mentioned that under certain conditions and the assumption that during pressurization a decrease in the pKa of the acids and pH is expected, a temporary reduction in pH and increase in the dissociated form of the acid may be present during pressurization. Accordingly, pH changes could enhance the effects of high pressure treatments on microorganisms and favor first-order kinetics in the pressure inactivation of *Z. bailii*.

Microbial Spores

Spores from yeast and molds are easily inactivated at pressures of 300 (*Aspergillus oryzae*) or 400 MPa (*Rhyzopus javanicus*) at ambient temperatures (Cheftel, 1995). However, Butz et al. (1996) and Palou et al. (1998b) demonstrated that ascospores of heat-resistant molds such as *Byssochlamys nivea* are extremely pressure-resistant. For the inactivation of *B. nivea* ascospores, pressures above 600 MPa and temperatures above 60°C were needed. No effects on spore viability after treatment at 70°C and 500 MPa for 60 min were observed by Butz et al. (1996); a slight inactivation (around 3 log cycle reduction) at 600 MPa after 60 min; approximately 5 log cycles reduced at 700 MPa; and total and rapid inactivation of *B. nivea* within a few minutes (<10) at 800 MPa.

Figure 18-5 shows *B. nivea* survivors during continuous pressurization at 21 and 60°C (Palou et al., 1998b). Although these ascospores survived pressure treatments at 21°C, the initial decrease in counts after a 5 min treatment can be explained by a certain portion of vegetative forms being present in the initial inocula. However, treatments for 15 or 25 min maintained or even increased the initial counts. Since high pressure treatments can break the mold asci and produce liberation or activation, the counts actually increase as a result of the increment in ascospore number. Nonetheless, a proportion of the activated ascospores can be inactivated during HHP treatment. Similar observations were made by Maggi et al. (1994) when *B. nivea* and *B. fulva* ascospores were pressurized at 900 MPa for 20 min at 20°C. The resulting survivor curves can be attributed to the summed rates of spore liberation from the asci and inactivation of the pressure-activated ascospores.

Temperature during pressurization can also have a significant effect on the inactivation of microbial cells. Several authors (Cheftel, 1995; Carlez et al., 1993; Palou et al., 1998b) observed that the resistance to pressure of endogenous or inoculated microbial strains is greatest at normal temperatures (15–30°C), but decreases significantly at higher or lower temperatures. When Palou et al. (1998b) elevated the temperature to 60°C during a treatment on *B. nivea*, the reduction of the initial inocula was more noticeable than at 21°C (Fig. 18-5). However, reductions after a 25 min treatment were only around 1 log-cycle. Maggi et al. (1994) reported no inactivation after a 7 min HHP treatment at 600 MPa and 60°C for *B. nivea* ascospores suspended in apricot nectar, but complete inactivation was obtained after 1 or 2 min once the pressure was raised to 700 MPa at the same temperature. Correspondingly, Butz et al. (1996) observed no effect on mold ascospore viability after a high pressure treatment at 500 MPa and 70°C for 60 min.

Figure 18-5 displays the results for pulsed HHP treatments on *B. nivea* ascospore viability. Inactivation of the initial spore inocula was achieved after three or five cycles of pulsed pressurization at 689 MPa and 60°C. In treatments at 21°C, no effect on spore viability was observed with pulsed HHP treatments. However, pulsed pressurization at 60°C was more effective on spore inactivation than continuous high pressure treatment (Palou et al., 1998b).

Enzymatic Reactions

Pressure inactivation of enzymes is influenced by the pH, substrate concentration, and subunit structure of the enzyme, as well as the temperature during pressurization (Hoover et al., 1989). The effects of pressure on enzyme activity are expected to occur upon substrate-enzyme interaction. However, if the substrate is a macromolecule the impact has more do to with conformation of the macromolecule, which can make the enzyme action easier or more difficult (Heremans, 1995). Enzyme inactivation via pressure can also be attributed to alteration of intermolecular structures or conformational changes at the active site. For example, inactivation of some enzymes pressurized to 100–300 MPa is reversible. Reactivation after decompression depends on the degree of molecule distortion, which becomes increasingly permanent beyond 300 MPa (Jaenicke, 1981).

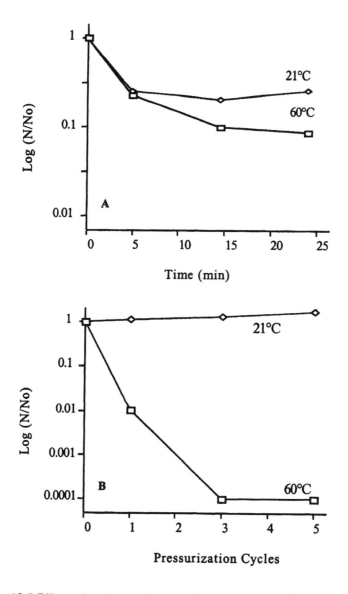

Figure 18-5 Effects of temperature (21 and 60°C) and HHP treatments: (A) continuous pressure (689 MPa) and (B) pulsed pressure (689 MPa, holding time 1 s) on survivors of *Byssochlamys nivea* ascospores suspended in apple juice with a_w 0.98 (adapted from Palou et al., 1998b).

Enzymes are generally inactivated in fruits and vegetables by hot water blanching, although this also produces thermal damage, leaching of nutrients, and possible environmental pollution due to the production of high biochemical oxygen demand effluent. High hydrostatic pressure treatment can fulfill the requirements of hot water blanching while avoiding its negative side-effects. Less effluent is produced because less water is required than in hot water blanching (Eshtiaghi and Knorr, 1993). In addition, Quaglia et al. (1996) reported that pressure treatments at 900 MPa for 10 min reduce 88% of the peroxidase activity in green peas. Furthermore, greater ascorbic acid and firmness retention also resulted. Lower pressure levels decreased less than 50% of the enzyme activity even when combined with moderate temperatures (39–60°C). For peroxidase from a carrot cell-free extract, Anese et al. (1995) observed that a complete loss of enzyme activity was achieved only when the pressure treatment was applied at 900 MPa for 1 min, but some activation was actually observed for treatments in the range of 300 to 500 MPa. For PPO from an apple cell-free extract at pH 7.0, 5.4, and 4.5, a significant reduction in enzyme activity occurred in pressure treatments at 900 MPa for 1 min. For both enzymes, a pH dependence on residual activity after the pressure treatment was found. Eshtiaghi and Knorr (1993) reported that the addition of citric acid could lead to increased PPO inactivation since pH reduction enhances the pressure effects on enzyme inactivation. Since denaturation and inactivation of enzymes occur only when very high pressure treatments are applied, the activation effects that could be presented at relatively low pressures could be attributed to reversible configuration and/or conformation changes on enzyme and/or substrate molecules (Anese et al., 1995). When Seyderhelm et al. (1996) evaluated the effects of HHP treatments on selected enzymes including catalase, phosphatase, lipase, pectinesterase, lipoxygenase, peroxidase, PPO, and lactoperoxidase, they reported that peroxidase was the most barostable enzyme with a 90% residual activity after 30 min treatment at 60°C and 600 MPa. Therefore, peroxidase could be selected as an enzyme indicator for HHP treatments.

Polyphenoloxidase and Color Changes

It is well established that PPOs from different sources may vary in molecular size and conformation. Thus, it is expected that the PPO may respond differently during and after high pressure treatments. It is also anticipated that important differences will occur when this type of enzymatic activity is analyzed in whole foods, extracts, or commercial enzymes. For example, in untreated onion cells, phenolic compounds are confined to vacuoles and spatially separated from the PPO by the tonoplast, but after pressurization (>100 MPa) the cell and tonoplast are disrupted and phenolic oxidation (browning) occurs because the PPO is no longer separated from the substrate (Butz et al., 1994). The activity of PPO was found to increase five times when slices of Bartlett pears were pressurized at 400 MPa and 25°C for 10 min, though further increase in pressure did not bring about any more enzyme activity. On the other hand, pressurization of homogenate apples, bananas, or sweet potatoes resulted in activation of PPO (Asaka and Hayashi, 1991). Gomes and Ledward (1996) reported a reduction in PPO activity from a crude potato extract with increasing pressure (400–800 MPa for 10 min), but when treated at 400 MPa for 10 min, an enhancement in the activity was observed. Cano et al. (1997) studied the combination of HHP and temperature on the peroxidase and PPO activities of fruit derived products and discovered that optimal inactivation of peroxidase in strawberry puree was achieved using 230 MPa and 43°C. Pressurization-depressurization treatments caused a significant loss of strawberry PPO up to 230 MPa, and combinations of high pressure and 35°C effectively reduced peroxidase in orange juice.

Although enzymes that cause color and flavor changes in avocado puree preserved as HMFP can be inactivated using thermal treatments, avocado is a highly sensible product to the development of undesirable flavors when thermally treated. López-Malo et al. (1999) have demonstrated that PPO in avocado puree can be partially inactivated to about 15 to 20% of its original activity without undesirable color changes when treated with HHP (689 MPa) at three pH levels (3.9, 4.1, and 4.3). Figure 18-6 presents the residual PPO activities for these evaluated HHP treatments. With treatments at pH 4.1 and 689 MPa, the PPO activity in avocado puree was reduced from 1,078 to 266, 235, or 168 units when the process time was 10, 20, or 30 min, respectively (representing a residual PPO activity decrease of 24.7, 21.8, and 15.6%, respectively). While the pressure and initial pH of the avocado puree significantly ($p < 0.05$) affected the residual PPO activity, the process time was not significant ($p < 0.05$) when 345 or 517 MPa was applied.

HHP treated avocado puree stored at 5, 15, and 25°C presented several color changes that contributed to the decline of its shelf-life. For a fixed shelf-life of 30 days, Figure 18-7 presents the possible combinations of pH and HHP treatments to attain the selected shelf-life at different storage temperatures for the avocado puree based on color changes during storage (López-Malo et al., 1999). Figure 18-7 also shows that there are few potentially successful

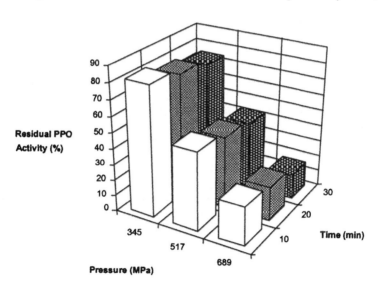

Figure 18-6 Effect of HHP treatments on the residual PPO activity in pH 4.1 avocado puree.

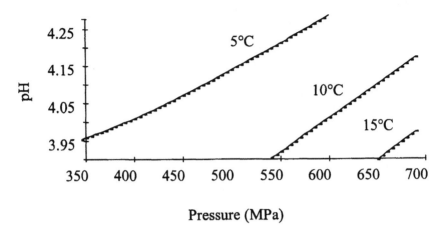

Pressure (MPa)

Figure 18-7 Predicted effect of high pressure (30 min treatment) and initial pH of avocado puree on the storage temperature required for a fixed shelf-life of 30 days.

combinations of initial pH and pressure to assure the 30 days with increasing storage temperature. In general, an acceptable shelf-life can be achieved with small initial pH and high pressures, as this produces the greatest PPO inactivation (Fig. 18-6). Residual PPO activity in avocado puree after HHP treatments suggests that inhibition of undesirable enzymatic reactions like browning requires the combination of pressurization with one or more additional factors such as low pH and refrigeration temperatures to inhibit enzyme activity (López-Malo et al., 1999). Accordingly, HHP seems to be a good option for the development of HMFP.

FINAL REMARKS

For a better and more efficient use of hurdle technology in HMFP development, further research is needed on the mechanisms of action for traditional and emerging preservation factors on microorganisms, enzymes, and deteriorative reactions. This continues to be an important issue for the development of minimal processing systems. In addition, a better understanding of these areas will help identify key factors and their combined effect on product safety, stability, and quality.

Since HHP in combination with other preservation factors has been proven to successfully inactivate microorganisms as well as deteriorative enzyme systems in HMFP, future work must be focused on the application of this alternative technology in the context of minimal processing. As such, studies on the interaction of HHP with other preservation factors such as pH, ionic strength, and temperature as they affect quality and stability should be encouraged.

REFERENCES

Ahvenainen R. (1996). New approaches in improving shelf life of minimally processed fruit and vegetables. *Trends Food Sci. Tech.* 7: 179–187.

Alzamora S.M., Gerschenson L.N., Cerruti P., and Rojas A.M. (1989). Shelf-stable pineapple for long-term non-refrigerated storage. *Lebensm. Wiss. Technol.* 22: 233–236.

Alzamora S.M., Tapia M.S., Argaiz A., and Welti J. (1993). Application of combined methods technology in minimally processed foods. *Food Res. Int.* 26: 125–130.

Alzamora S.M., Cerrutti P., Guerrero S., and López-Malo A. (1995). Minimally processed foods by combined methods. In: *Food Preservation by Moisture Control: Fundamentals and Applications.* Barbosa-Cánovas G.V. and Welti-Chanes J., eds. ISOPOW Practicum II. Lancaster, PA: Technomic Publishing Co. pp: 463–492.

Anese M., Nicoli M.C., Dall'Aglio G., and Lerici C.R. (1995). Effect of high pressure treatments on peroxidase and polyphenoloxidase activities. *J. Food Biochem.* 18: 285–293.

Argaiz A., López-Malo A., and Welti J. (1995). Considerations for the development and the stability of high moisture fruit products during storage. In: *Fundamentals and Applications of Food Preservation by Moisture Control.* Barbosa-Cánovas G.V. and Welti-Chanes J., Eds. ISOPOW Practicum II. Lancaster, PA: Technomic Publishing. pp: 729–760.

Asaka M. and Hayashi R. (1991). Activation of polyphenol oxidase in pear fruits by high pressure treatment. *Agric. Biol. Chem.* 55: 2439–2440.

Butz P., Funtenberger S., Haberdtzl T., and Tauscher B. (1996). High pressure inactivation of *Byssochlamys nivea* ascospores and other heat resistant moulds, *Lebensm.-Wiss. u.-Technol.* 29: 404–410.

Butz P., Koller W.D., Tauscher B., and Wolf S. (1994). Ultra-high pressure processing of onions: chemical and sensory changes. *Lebensm.-Wiss. u.-Technol.* 27: 463–467.

Cano M.P., Hernández A., and De Ancos B. (1997). High pressure and temperature effects on enzyme inactivation in strawberry and orange products. *J. Food Sci.* 62: 85–88.

Carlez A., Rosec J.P., Richard N., and Cheftel J.C. (1993). High pressure inactivation of *Citrobacter freundii, Pseudomonas fluorescens* and *Listeria innocua* in inoculated minced beef muscle. *Lebensm.-Wiss. u.-Technol.* 26: 357–363.

Cheftel J.C. (1995). High-pressure, microbial inactivation and food preservation. *Food Sci. Technol. Int.* 1: 75–90.

Deplace G. and Mertens B. (1992). The commercial application of high pressure technology in the food processing industry. In: *High Pressure and Biotechnology.* Balny C., Hayashi R., Heremans K., and Masson P., Eds. Montrouge, France: Colloque INSERM Vol. 224, John Libbey Eurotext.

Díaz de Tablante R.V., Tapia de Daza M.S., Montenegro G., and González I. (1993). Desarrollo de productos de mango y papaya de alta humedad estabilizados por métodos combinados. Boletín Internacional de Divulgación CYTED. Universidad de las Américas-Puebla, Mexico. 1: 5–21.

Earnshaw R.G. (1995). Kinetics of high pressure inactivation of microorganisms. In: *High Pressure Processing of Foods*. Ledward D.A., Johnston D.E., Earnshaw R.G., and Hasting A.P.M., Eds. Nottingham UK: Nottingham University Press.

Earnshaw R.G. (1996). High pressure food processing. *Nutr. Food Sci.* 2: 8–11.

Earnshaw R.G., Appleyard J, Hurst RM (1995). Understanding physical inactivation processes: combined preservation opportunities using heat, ultrasound and pressure. *Int. J. Food Microbiol.* 28: 197-219.

Eshtiaghi M.N., Knorr D. (1993). Potato cubes response to water blanching and high hydrostatic pressure. *J. Food Sci.* 58: 1371–1374.

Farr D. (1990). High pressure technology in the food industry. *Trends Food Sci. Technol.* 1: 14–16.

Fito P. and Chiralt A. (1997). Osmotic dehydration: An approach to the modeling of solid-liquid food operations. In: *Food Engineering 2000*. Fito P., Ortega-Rodríguez E., and Barbosa-Cánovas G.V., Eds. New York: International Thomson Publishing. pp: 231–252.

Gomes M.R.A. and Ledward D.A. (1996). Effect of high-pressure treatment on the activity of some polyphenoloxidases. *Food Chem.* 56: 1–5.

Gould G.W. (1992). Ecosystem approaches to food preservation. *J. Appl. Bacteriol.* 73: 58S–68S.

Gould G.W. (1995). The microbe as a high pressure target. In: *High Pressure Processing of Foods*. Ledward D.A., Johnston D.E., Earnshaw R.G., and Hasting A.P.M., Eds. Nottingham, UK: Nottingham University Press.

Gould G.W. and Jones M.V. (1989). Combination and synergistic effects. In: *Mechanisms of Action of Food Preservation Procedures*. Gould G.W., Ed. London: Elsevier.

Hashizume C., Kimura K., and Hayashi R. (1995). Kinetic analysis of yeast inactivation by high pressure treatment at low temperatures. *Biosci. Biotechnol. Biochem.* 59: 1455-1458.

Heremans K. (1995). High pressure effects on biomolecules. In: *High Pressure Processing of Foods*. Ledward DA, Johnston D.E., Earnshaw R.G., and Hasting A.P.M., Eds. Nottingham, UK: Nottingham University Press.

Hoover D.G., Metrick C., Papineau A.M., Farkas D.F., and Knorr D. (1989). Biological effects of high hydrostatic pressure on food microorganisms. *Food Technol.* 43(3): 99–107.

Huxsoll C.C. and Bollin H.R. (1989). Processing and distribution alternatives for minimally processed fruits and vegetables. *Food Technol.* 43(2): 132–138.

Jaenicke R. (1981). Enzymes under extreme conditions. *Ann. Rev. Biophys. Bioeng.* 10: 1–67.

Knorr D. (1993). Effects of high-hydrostatic pressure process on food safety and quality. *Food Technol.* 47(6): 156–161.

Ledward D.A. (1995). High pressure processing—the potential. In: *High Pressure Processing of Foods*. Ledward D.A., Johnston D.E., Earnshaw R.G., and Hasting A.P.M., Eds. Nottingham UK: Nottingham University Press.

Leistner L. (1995). Principles and applications of hurdle technology. In: *New Methods of Food Preservation*. Gould G.W., Ed. New York: Blackie Academic and Professional.

López-Malo A., Palou E., Welti J., Corte P., and Argaiz A. (1994). Shelf-stable high moisture papaya minimally processed by combined methods. *Food Res. Int.* 27: 545–553.

López-Malo A., Palou E., Barbosa-Cánovas G.V., Welti-Chanes J., and Swanson B.G. (1999). Polyphenoloxidase Activity and Color Changes During Storage in High Hydrostatic Pressure Treated Avocado Purée. *Food Res. Int.* 31(8): 549–556.

Maggi A., Gola S., Spotti E., Rovere P., Mutti P. (1994). Tratamenti ad alta pressione di ascospore di muffe termoresistenti e di patulina in nettare di albicocca e in acqua. *Indus. Cons.* 69: 26–29.

Mertens B. (1995). Hydrostatic pressure treatment of food: equipment and processing. In: *New Methods of Food Preservation*. Gould G.W., Ed. New York: Blackie Academic and Professional.

Monsalve-González A., Barbosa-Cánovas G.V., Cavalieri R.P., McEvily A., Iyengar R. (1993). Control of browning during storage of apple slices preserved by combined methods. *J. Food Sci.* 58: 797–800, 826.

Ogawa H., Fukuhisa K., and Fukumoto H. (1992). Effect of hydrostatic pressure on sterilization and preservation of citrus juice. In: *High Pressure and Biotechnology*. Balny C., Hayashi R., Heremans K., and Masson P., Eds. Montrouge, France: Colloque INSERM Vol. 224, John Libbey Eurotext.

Ohlsson T. (1994). Minimal processing-preservation methods of the future: An overview. *Trends Food Sci. Technol.* 5: 341–344.

Oxen P. and Knorr D. (1993). Baroprotective effects of high solute concentrations against inactivation of *Rhodotorula rubra*. *Lebensm.-Wiss. u.-Technol.* 26: 220–223.

Palou E, López-Malo A, Barbosa-Cánovas GV, Welti-Chanes J, Swanson BG (1997a). Effect of water activity on high hydrostatic pressure inhibition of *Zygosaccharomyces bailii*. *Lett. Appl. Microbiol.* 24: 417–420.

Palou E., López-Malo A., Barbosa-Cánovas G.V., Welti-Chanes J., and Swanson B.G. (1997b). High hydrostatic pressure as a hurdle for *Zygosaccharomyces bailii* inactivation. *J. Food Sci.* 62: 855–857.

Palou E., López-Malo A., Barbosa-Cánovas G.V., Welti-Chanes J., and Swanson B.G. (1997c). Kinetic analysis of *Zygosaccharomyces bailii* by high hydrostatic pressure. *Lebensm.-Wiss. u.-Technol.* 30: 703–708.

Palou E., López-Malo A., Barbosa-Cánovas G.V. and Swanson B.G. (1999). High Pressure treatment in food preservation. In: *Handbook of Food Preservation*. Rahman M.S., Ed. New York: Marcel Dekker Inc. pp: 533–576.

Palou E., López-Malo A., Barbosa-Cánovas G.V., Welti-Chanes J., Davidson P.M., and Swanson B.G. (1998a). High hydrostatic pressure come-up time and yeast viability. *J. Food Prot.* 61(12): 1657–1660.

Palou E., López-Malo A., Barbosa-Cánovas G.V., Welti-Chanes J., Davidson P.M., and Swanson B.G. (1998b). Effect of oscillatory high hydrostatic pressure treatments on *Byssochlamys nivea* ascospores suspended in fruit juice concentrates. *Lett. Appl. Microbiol.* 27(6): 375–378.

Patterson M.F., Quinn M., Simpson R., and Gilmour A. (1995). Effects of high pressure on vegetative pathogens. In: *High Pressure Processing of Foods.* Ledward D.A., Johnston D.E., Earnshaw R.G., and Hasting A.P.M., Eds. Nottingham, UK: Nottingham University Press.

Pérez J., López-Malo A., Palou E., Barbosa-Cánovas G.V., Swanson B.G., Vélez J., and Welti-Chanes J. (1997). Effect of high hydrostatic pressure treatments on *Saccharomyces cerevisiae* viability. Paper No. 59E-19. IFT Annual Meeting, June 14–18. Orlando, Florida.

Quaglia G.B., Gravina R., Paperi R., and Paoletti F. (1996). Effect of high pressure treatments on peroxidase activity, ascorbic acid content and texture in green peas. *Lebensm.-Wiss. u.-Technol.* 29: 552–555.

Rolle R.S. and Chism G.W. (1987). Physiological consequences of minimally processed fruits and vegetables. *J. Food Qual.* 10: 187–193.

Seyderhelm I., Boguslawski S., Michaelis G., and Knorr D. (1996). Pressure induced inactivation of selected enzymes. *J. Food Sci.* 61: 308–310.

Shewfelt R. (1987). Quality of minimally processed fruits and vegetables. *J. Food Qual.* 10: 143–156.

Smelt J. and Rijke G. (1992). High pressure treatment as a tool for pasteurization of foods. In: *High Pressure and Biotechnology.* Balny C., Hayashi R., Heremans K., and Masson P., Eds. Montrouge, France: Colloque INSERM Vol. 224, John Libbey Eurotext.

Tapia de Daza M.S., Argaiz A., López-Malo A., and Díaz R.V. (1995). Microbial stability assessment in high and intermediate moisture foods. Special emphasis on fruit products. In: *Food Preservation by Moisture Control: Fundamentals and Applications.* Barbosa-Cánovas G.V. and Welti-Chanes J., eds. ISOPOW Practicum II. Lancaster, PA: Technomic Publishing Co. pp: 575–601.

Tapia de Daza M.S., Alzamora S.M., and Welti-Chanes J. (1996). Combination of preservation factors applied to minimal processing of foods. *Crit. Rev. Food Sci. Nutr.* 36: 629–659.

Tauscher B. (1995). Pasteurization of food by hydrostatic pressure: Chemical aspects. *Z. Lebensm. Unters. Forsch.* 200: 3–13.

van Eldik R., Asano T., and Le Noble W.J. (1989). Activation and reaction volumes in solution. *Chem. Rev.* 89: 549–688.

Wiley R.C. (1994). Introduction to minimally processed refrigerated fruits and vegetables. In: *Minimally Processed Refrigerated Fruits and Vegetables.* Wiley R.C., Ed. New York: Chapman & Hall.

CHAPTER 19

Biocatalysts for the Food Industry

A. Illanes

INTRODUCTION

Catalysts are substances that reduce the energy barrier of chemical reactions by forming a low energy transition state with a reactant without undergoing any net change in structure. The kinetic consequence is a dramatic increase in reaction rate. Enzymes are biological catalysts that mediate all the chemical reactions of cell metabolism.

The concept of biocatalyst has expanded far beyond its physiological meaning. In the broadest of terms, any biological entity capable of catalyzing chemical reactions might be considered a biocatalyst. In the present context, the concept will be restricted to enzymes isolated from or conditioned within the cell system that produces them, so living cells will not be considered here as biocatalysts.

The potential advantages of biocatalysts are their high selectivity, activity at mild conditions, and turnover numbers that have profound effects on process economics. The label of "natural" is also relevant with respect to current environmental constraints. However, biocatalysts are also labile under harsh process conditions and expensive because they must be produced by fermentation with suitable microorganisms or extracted from tissues.

Process economics will ultimately determine whether a biocatalyst is a preferred option over a conventional chemical catalyst, except for the case where no chemical counterpart is possible. Odds will be in favor of biocatalysts when most (or all) of the following conditions are met (Tramper, 1996):

- higher selectivity so that the number of reaction steps can be drastically reduced
- higher product yields
- faster reaction rates
- milder process conditions
- fair stability and robustness
- fewer and tractable waste-streams
- environmental constraints over products

Biocatalysts are extremely evolved and versatile molecules so that almost any chemical reaction is likely to be influenced. As such, the potential for new technological breakthroughs is far greater than with current applications.

SOURCES OF BIOCATALYSTS

Biodiversity in nature represents a bountiful and certainly poorly exploited source of biocatalysts. Evolution over millions of years has perfectly tailored enzymes for extremely specific effects in a given environment. Moreover, genetic manipulation by conventional mutation and the more elaborate genetic engineering procedures amplify this natural potential. Modern protein engineering opens up the possibility of a rational design for biocatalysts by specific point modifications of the genes encoding them. Like catalytic RNA (ribozymes) and catalytic antibodies (abzymes), non-protein biocatalysts are promising novel types of biocatalysts (Hansen, 1991; Breaker and Joyce, 1994). Chemical synthesis of catalysts based on biological frameworks (enzyme mimics) might also be a future option (May, 1992).

Early biocatalysts were primarily obtained as crude preparations from plants and animal tissues, with the exception of Japan where microbial enzymes were produced traditionally. By 1960 about 70% of the enzymes in the market were from plant or animal origin. Twenty years later the situation reversed (Layman, 1986), so that today more than 80% of the biocatalyst market is microbial-based (Dordick, 1991).

Both strongly temperature-dependent, stability and activity are the most essential properties of the biocatalyst, their adequate balance being central to successful application (Illanes et al., 1996). These types of enzymes are stable but poorly active at low temperature, while the opposite holds at high temperatures. Major advancement has occurred in recent years in the extraction and characterization of extremely thermotolerant enzymes from thermophiles (Vieille and Zeikus, 1996) and psychrophiles which exhibit unusual activities at near-to-freezing temperatures (Marshall, 1997), making both types of biocatalysts of considerable technological potential.

Recombinant DNA technology has been a major breakthrough in biocatalyst production by allowing the synthesis of enzymes from different genetic sources in robust, safe, fast-growing microorganisms (White et al., 1984). It is estimated that 60% of industrial enzymes are now produced by recombinant microorganisms (Cowan, 1996), which is remarkable since the first recombinant enzymes were introduced in the market in the mid-1980s. Protein engineering came onto the scene soon after, and is now contributing to elucidate fundamental aspects of enzyme catalysis (Knowles, 1987). It also represents a major technological breakthrough in biocatalyst design by introducing specific modifications in the protein structure that may improve fundamental properties such as substrate specificity and stability (Nosoh and Sekiguchi, 1990).

TYPES OF BIOCATALYSTS

Biocatalysts can be cellular or non-cellular. Although sometimes considered as biocatalysts, viable, growing, or resting cells are beyond the scope of this chapter. In this context, cellular catalysts are fixed cells that have been subjected to conditions that disassemble their metabolic functioning even though particular enzymes remain active. Non-cellular biocatalysts are enzymes of varying purity that have been separated from the cell system that synthesized them.

Cellular biocatalysts will be the choice for the food industry when more than one reaction is involved, cofactor regeneration is needed, or unstable intracellular enzymes are prone to inactivation during cell rupture. Non-cellular biocatalysts are to be used in food processing and preservation when specificity of the reaction is mandatory, the enzyme is robust enough to readily recover after cell disruption, the substrate is a complex high molecular weight polymer, and the enzyme is extracellular.

Biocatalysts were traditionally used freely in solutions (or suspensions), but in the mid-1960s immobilization to solid carriers was developed. Immobilized biocatalysts have the potential advantage of higher stability, continuous processing with catalyst retention in the reactor, flexible reactor design, low residence time to avoid unwanted side reactions. Major drawbacks are the additional costs, activity losses during immobilization, and reduced reaction rates as a consequence of mass-transfer limitations. Immobilization will be a good option when low molecular weight substrates are involved, enzymes and reactants are unstable, and the cost of the catalyst is a significant fraction of the processing cost. Despite their potential advantages, few large-scale processes use immobilized biocatalysts, although the situation is changing.

CURRENT APPLICATIONS AND FUTURE TRENDS

Use of biocatalysts began early in the 20th century, mostly by the food industry. From selected patents issued from 1900 to 1940, 62% corresponded to food applications, the remaining 38% being mainly related to textile and leather (Neidelman, 1991). The situation did not change much until 1960, and major changes are still taking place. The main industrial applications of enzymes are still in the food area (amylases, glucose isomerase, pectinases, proteases, lipases), but now detergents (proteases and lipases), textiles (amylases and cellulases), pulp and paper (xylanases), pharmaceuticals (acylases, β-lactamases, steroid transforming), and chemicals (nitrile hydratase) are becoming more prevalent.

Initially, industrial biocatalysts were crude extracts or fermentation broth used mainly as additives in non-controlled environments. Most were and still are cheap, simple, extracellular hydrolase proteins. In the early 1960s, biocatalytic processes evolved. First came the Staley process which produced glucose syrups from starch (Michael-Sinclair, 1965), and then a method was developed to generate L-amino acids from racemates by using immobilized aminoacylase (Chibata, 1978). The most successful commercial application from then on was the massive production of high-fructose corn syrup using either immobilized glucose isomerase or glucose isomerase-containing whole cells (Carasik and O'Carroll, 1983). High expectations were based on immobilized biocatalyst technology, but only a handful of such processes proved commercially successful (Katchalsky-Katzir, 1993). Most processes were hydrolytic, and the

289

products did not have enough added value to compensate for the increased cost of the catalyst and R&D behind it.

Alternatively, organic synthesis presented a very different case where complex molecules are produced from simple precursors, allowing substantial profit margins. However, until very recently biocatalyst technology has been excluded from organic synthesis because synthetases are complex unstable intracellular cofactors requiring enzymes. While study of whole cells with internal cofactor regeneration cycles, enzyme derivatization, cofactor co-immobilization, and side enzymatic reactions for cofactor regeneration has brought about substantial improvements in membrane reactor design for such processes, the technology is still costly and far from straightforward (Turner, 1995). A more promising approach has been the use of conventional industrial enzymes in non-conventional reaction media even though this causes severely reduced enzyme activity. Non-aqueous biocatalysis represents the best potential for future development in the field based on such striking results as altered specificity, increased stability, and ability to reverse hydrolytic reactions (Klibanov, 1997). This latter aspect is of particular relevance since simple yet tough industrial hydrolases can perform synthesis reactions by pushing the equilibrium backward at low water activity. In such media, proteases will catalyze the synthesis of oligopeptides (Gill et al., 1996), lipases will catalyze all kinds of esterifications (John and Abraham, 1991), and glycosidases will catalyze the synthesis of oligosaccharides (Bucke, 1996). While biocatalysis in organic media is the most promising and studied of all non-aqueous media (Koskinen and Klibanov, 1996), non-conventional biocatalysis may have a greater range since it can be extended to gaseous reactions (Lamare and Legoy, 1993) and supercritical fluids (Randolph et al., 1991).

The trend seems to emphasize engineering the reaction medium rather than the catalyst. Cheap bulk industrial biocatalysts sold as commodities can now be tested in sophisticated organic synthesis to challenge conventional chemical synthesis or develop new products. Biocatalysis in non-conventional media is so flexible that many reaction schemes are possible, justifying the need for the recently proposed classification methodology (Davidson et al., 1997).

Although the industrial application of biocatalysts in organic synthesis is still in its infancy, advances in medium engineering and catalyst design will certainly help to pave the way for its increasing acceptance. The development of new forms of biocatalyst also deserves attention. For example, cross-linked enzyme crystals have excellent properties such as high specific activity, outstanding stability under harsh conditions, and compatibility with most organic compounds (Margolin, 1996). Should these catalysts be produced economically on a large scale, the industry will be eager to adopt them.

MARKET OVERVIEW

The annual world market for biocatalyts is somewhere between US $800 million and US $1 billion (Katchalsky-Katzir, 1993; Hodgson, 1994; Koskinen and Klibanov, 1996), although figures as high as US $1.4 billion have been claimed

(Cowan, 1996). The annual growth rate of the biocatalyst market has remained at a nearly constant 6% during the last decade (Polastro, 1989), although the increase in production volume is close to 10% due to the decrease in unit price resulting from increasing competition (Layman, 1986; Cowan, 1996). The market in the US alone was US $370 million in 1996 and close to US $400 million in 1997 (Wrotnowsky, 1997), with an average annual increase close to 7% in the last decade.

These figures represent bulk industrial biocatalysts sold as commodities as well as specialty biocatalysts for the pharmaceutical and fine-chemicals industries. Bulk industrial biocatalysts may now account for 80% of the market (Cowan, 1996). Over 60% of bulk industrial biocatalysts are claimed as recombinant products, some of which have been protein engineered. In addition, leading biocatalyst-producing companies are screening for novel enzymes from extremophiles and other unusual sources to achieve even better performance.

It is also important to note that market figures do not necessarily reflect the full impact of biocatalysis since many companies produce an unreported number of biocatalysts for their own use. Nevertheless, it is still helpful to get an idea of the bulk biocatalyst market distribution on a world basis, so this has been compiled in Table 19-1 (Koskinen and Klibanov, 1996; Wrotnowsky, 1997), along with the forecast for the US market. As seen, the food industry market is still the biggest, but with slow growth. Alternatively, substantial increase is expected in the detergent industry since enzyme liquid detergents with lipases and amylases have been introduced. The highest growth is expected in the paper and pulp industry where the use of xylanases and other depolymerases for biopulping and biobleaching is alleviating many environmental concerns by reducing toxic chemicals.

Some of the most recent developments in other industrial areas are the engineering of proteases and lipases for detergents and contact lens cleansers; improvements in the production of xylanases and ligninases for biobleaching, biopulping and waste treatment of effluents from paper mills; and the production and engineering of cellulases for enzymatic "stone-washing", softening, and biopolishing of fabrics, and laccase for the controlled decoloring of indigo. The application of biocatalysts in the fine-chemicals industry is also likely to develop extensively in the next few years, as chemicals like acrylamide and drugs like Ibuprofen and Ciprofloxacin are claimed to be produced by biocatalysis (Wrotnowsky, 1997).

Table 19-1 Bulk biocatalyst market distribution.

Area	World (%)	USA (%)	USA AAR* (%)
Food	42	46	2
Detergent	42	43	10
Leather & textiles	12	10	2
Pulp and paper	2	< 1	15
Others	2	1	10

*AAR: average annual growth rate for the period 1996–2006

CONVENTIONAL USE IN THE FOOD INDUSTRY

In this context conventional simply refers to applications in which biocatalysts are used as additives to produce a texture or nutritional improvement in foods or in a processing stage at rather uncontrolled conditions. Most biocatalysts of this sort are crude preparations of carbohydrases or proteases.

Among carbohydrases, α-amylases are common ingredients in baking powders for bread and crackers, β-glucanases are used traditionally in the production of beer, lactase is increasingly used in the production of delactosed milk for intolerants, and α-galactosidase is used in the removal of raffinose to aid in the recovery of sugar beets. Pectinases are a family of depolymerizing enzymes used in the maceration of fruits and vegetables and extraction and clarification of fruit juices and wine. Their production has become a very specialized technology in which companies produce different preparations according to specific customer needs. Some of these preparations contain variable amounts of cellulases and β-glucanases to aid in the extraction of edible oils from olive, rapeseed, and sunflower. Prospects also exist when these enzymes are mixed with xylanases and proteases to extract valuable cell components.

Another popular type of biocatalyst in the food industry is proteases. Their major application is in the production of cheese such as rennin and pepsin. Fungal and bacterial neutral proteases are often put in baking powders for bread and crackers, while neutral proteases aid in the production of protein hydrolyzates, meat recovery, and tenderization. Papain may be the most common enzyme of this group since it is extensively used in the chill-proofing of beer. Although developed for detergents, alkaline proteases have also been used successfully in the recovery of stickwater in fishmeal production (Schaffeld et al., 1989). Some minor applications also exist for non-hydrolytic enzymes in foods. Examples are the use of catalase in the dairy industry and glucose oxidase to preserve foods sensitive to oxidation or browning (Grampp, 1982). Since many of these proteases are plant or animal enzymes and their supply is limited, microbial counterparts have been developed. More recently, enzyme genes have been cloned for massive expression in suitable microbial hosts. Plant alkaline proteases are thermostable and active at very high pH (Illanes et al., 1985a; López et al., 1993), so their cloning into suitable hosts is very promising.

BIOCATALYTIC PROCESSES IN THE FOOD INDUSTRY

Since the evolution of biocatalytic processes in the 1960s the food industry has benefited by the production of inverted sucrose with soluble and later immobilized yeast invertase (Illanes et al., 1985b); these have since progressed to high-fructose syrups from starch. Enzymes are able to displace the chemical Staley process to produce glucose by liquefying starch in a continuous tubular reactor with soluble thermostable bacterial α-amylase so that gelatinization and hydrolysis occur simultaneously. The 10–12 dextrose equivalent (DE) syrup product is then treated batchwise with soluble glucoamylase to produce a 95–98 DE syrup (Michael-Sinclair, 1965). With the development of immobilized enzyme technology during the 1970s, a third enzymatic step was added in which immobilized glucose isomerase (enzymes or fixed-cells) converted the glucose syrup into an almost equimolar mixture of glucose and fructose known as high-fructose syrup (42% HFS) (Carasik and Carroll, 1983). Modern fractionation technology allows the enrichment to 90% fructose, expanding syrup use to 55% HFS as in cola beverages. This has been the most successful biocatalytic process until now, producing a noticeable impact in the world sugar market. Estimation of worldwide HFS production was 8 million tons in 1992 (Katchalsky-Katzir, 1993), which is now more than 10 million tons. A market of US $150 million has been estimated for all starch-processing enzymes (Crabb and Mitchinson, 1997), with glucose isomerase representing almost one third of this figure.

Additional opportunities exist in the animal and human nutrition markets when L-amino acids are produced from racemic mixtures by biocatalysis. The Japanese Tanabe Seiyaku Company has been conducting two of such biocatalytic processes for almost 30 years to produce L-aspartic acid and L-malic acid. The enzymes involved in these cases are intracellular, so immobilized cells are the preferred catalyst forms. L-aspartic acid is produced from fumaric acid with aspartase-containing *Escherichia coli* cells, while L-malic acid is produced from ammonium fumarate with fumarase-containing *Brevibacteriun ammoniagenes* cells (Hultin, 1983). The market for L-aspartic acid has increased dramatically in recent years since it was the precursor of aspartame, the leading non-caloric sweetener.

Lactase has been used for the hydrolysis of whey and whey permeate, as well as the production of delactosed milk and dairy products. Immobilized mold lactases have been used for the former, while free and immobilized yeast enzymes and cells are more suitable for the latter (Illanes et al., 1993). Low prices for sweeteners have precluded a higher impact of whey hydrolysis and permeate (Axelsson and Zacchi, 1990), although environmental constraints on effluents might make syrup an attractive option for enzymatic conversion into glucose or even HFS (Marwaha and Kennedy, 1988). The production of delactosed milk and even lactase tablets for intolerants is also promising.

NOVEL APPLICATIONS AND RECENT DEVELOPMENTS

The food industry is benefiting from all major technological advances in biocatalysis, including genetic and protein engineering, extremozymes, immobilized and crystal biocatalysts, and biocatalysis in non-conventional media. These advances are being applied to already established processes and new applications.

Recent advances in the starch processing industry have overcome the drawbacks from the commonly used thermostable α-amylases. Since these require Ca^{++} and are poorly active below pH 6, the α-amylase gene from *B. licheniformis* was cloned in *B. subtilis* strains and engineered to be active at lower pH and lower Ca^{++} concentrations, and even more stable at high

293

temperatures. α-Amylases from extremophiles are promising but not yet produced on a large scale. Since fungal glucoamylase is not as thermostable as α-amylase and is poorly active at pH over 5, efforts have been made to engineer such enzymes for thermostability and a wider pH range. Glucoamylase is poorly active over β1–6 linkages and produces reversion at high substrate concentrations, so microbial pullulanase has been introduced, which is active over such linkages at conditions compatible with those of glucoamylase (Crabb and Mitchinson, 1997). While genes for glucoamylase and α-amylase have been cloned in yeast and bacteria to produce ethanol directly from starch (Kennedy et al., 1988), most efforts are devoted to develop better glucose isomerase biocatalysts and improve isomerization yields. More robust isomerases are highly desirable since equilibrium moves towards fructose formation at higher temperatures and the presence of water miscible cosolvents. Thermostable isomerases have been engineered from natural sources and isolated from extremophiles (Crabb and Mitchinson, 1997). As a result, better immobilized enzymes and crosslinked glucose isomerase crystals are now produced (Margolin, 1996). Microbial β-amylases have been developed to displace plant enzymes in the production of maltose syrups for the food industry, and studies have been conducted to clone the plant enzyme into a suitable microbial host (Kennedy et al., 1988).

Chymosin illustrates the most recent advances in biocatalysis. Microbial chymosins were developed 20 years ago to replace calf rennet which can only be used to hydrolyze particular peptide bonds in κ-casein, but they were less specific and yield was lower. The same occurred when pepsin-rennet blends were used. Eventually however, the calf gene (cDNA) was cloned in molds (Cullen et al., 1987) and yeast (van den Berg and Koning, 1990), and the latter has been a complete commercial success. This recombinant chymosin (e.g., Maxiren®) is indistinguishable from calf rennet in its structure and functionality (Morris and Anderson, 1991) and has been in the market since the early 1990s.

New enzymes are also being included in baking powders. For example, lipase and xylanase-containing preparations have been recently introduced to improve dough quality by fractionating grains. In the animal-feed industry, the most interesting development is the introduction of microbial phytase that destroys antinutritional phytic acid in food while producing available phosphate for the animal. This not only reduces feeding cost but abates environmental pollution. While β-glucanases and xylanases are being used to remove antinutritional β-glucans and arabinoxylans in poultry and swine food, cellulases, related carbohydrases, amylases, and proteases are appearing in monogastric animal and fish food. Genetic and protein engineering continues to develop more stable enzymes produced by safe microorganisms (Walsh et al., 1993).

Biocatalysis in non-aqueous environments also has important applications in the food industry. Lipases are especially suited to perform in such media (Balcão et al., 1996) since under low water concentrations they will catalyze esterifications, transesterifications, and interesterifications. Perhaps the most relevant example is the interesterification of cheap oil fractions (e.g., palm oil mid-fraction, or POMF) into cocoa butter analogues with immobilized lipase. Lipase will catalyze interesterification at the 1 and 3 position in the triglyceride so that the balance of fatty acids in those positions can be modified to produce a cocoa butter analogue as seen in Table 19-2. To displace the equilibrium away from hydrolysis, water activity must be kept low by working with anhydrous solvents or solvent-free melted substrates (Mojovic et al., 1993). Other applications for lipases are the production of food emulsifiers in organic solvents and solvent-free substrates (Sarney and Vulfson, 1995). Interesterification with lipases in supercritical carbon dioxide seems very promising (Randolph et al., 1991), as does the production of "natural" fragrances by esterification of terpene alcohols with lipase in organic solvents (Koskinen and Klibanov, 1996).

Proteases acting in reverse within non-aqueous systems have been tested for the production of flavor peptides. The best example is the production of the leading non-caloric sweetener aspartame (L-aspartyl-L-phenylalanine methyl esther) with the commercial protease immobilized thermolysin in organic solvent (Katchalsky-Katzir, 1993). In this particular case, however, equilibrium can be driven into synthesis in aqueous media by promoting product precipitation, which is the current industrial practice (Koskinen and Klibanov, 1996).

FINAL REMARKS

Major advances are occurring in biocatalysis. Those robust biocatalysts produced by immobilization and crystallization are increasingly being used by the food industry. Genetic and protein engineering is producing more specific, stable, and safe biocatalysts. Screening of extremophiles as new sources of biocatalysts is also rendering interesting results. In addition, biocatalysis in non-conventional media are providing new opportunities for well-known industrial biocatalysts. Consequently, the future of biocatalysis will be supported both by catalyst and media engineering. All of these technological breakthroughs being applied to the food industry imply the prospects for this exciting innovation are better than ever before.

Table 19-2 Fatty acyl composition (%) of triglycerides in interesterified POMF.

Fatty Acyl Group	POMF	Cocoa Butter	Interesterified POMF
Palmityl	51.4	26.4	23.4
Stearyl	3.0	35.2	35.4
Oleyl	37.0	34.1	34.0
Linoleyl	7.2	3.0	6.5
Others	1.4	1.3	1.7

REFERENCES

Axelsson A. and Zacchi G. (1990). Economic evaluation of the hydrolysis of lactose using immobilized β-galactosidase. *Appl. Biochem. Biotechnol.* 24/25: 679–693.

Balcão V.M., Paiva A.L., and Malcata F.X. (1996). Bioreactors with immobilized lipase: state of the art. *Enzyme Microb. Technol.* 18: 392–416.

Breaker R.R. and Joyce G.F. (1994). Inventing and improving ribozyme function: rational design versus iterative selection methods. *Trends Biotechnol.* 12: 268–275.

Bucke C. (1996). Oligosaccharide synthesis using glycosidases. *J. Chem. Technol. Biotechnol.* 67: 217–220.

Carasik W. and O'Carroll J. (1983). Development of immobilized enzymes for the production of high-fructose corn syrup. *Food Technol.* 37(10): 85–90.

Chibata I. (1978). *Immobilized Enzymes: Research and Development.* Tokyo, Japan: Kodansha Ltd., pp. 168–176.

Cowan D. (1996). Industrial enzyme technology. *Trends Biotechnol.* 14:177–178.

Crabb W.D. and Mitchinson C. (1997). Enzymes involved in the processing of starch to sugars. *Trends Biotechnol.* 15: 349–352.

Cullen D., Gray G., Wilson L., Hayenga K., Lamsa M., Rey W., Norton S., and Berka R. (1987). Controlled expression and secretion of bovine chymosin in *Asper. nidulans. Biotechnol.* 5: 369–365.

Davidson B.H., Barton J.W., and Petersen G.R. (1997). Nomenclature and methodology for classification of nontraditional biocatalysis. *Biotechnol. Prog.* 13: 512–518.

Dordick J.S. (1991). An introduction to industrial biocatalysis. In: *Biocatalysts for Industry.* Dordick, J.S., Ed. New York: Plenum Press, pp. 3–19.

Gill I., López-Fandiño R., Jorba X., and Vulfson E.N. (1996). Biologically active peptides and enzymatic approaches to their production. *Enzyme Microb. Technol.* 18: 162–183.

Grampp E.G. (1982). Modification of certain foodstuffs by enzymes. *Proc. Biochem.* January/February, pp. 2–12.

Hansen D.E. (1991). Catalytic antibodies. In: *Biocatalysts for Industry.* Dordick, J.S., Ed. New York: Plenum Press, pp. 285–309.

Hodgson J. (1994). The changing bulk biocatalyst market. *Biotechnol.* 12: 789–790.

Hultin H.O. (1983). Current and potential uses of immobilized enzymes. *Food Technol.* 37(10): 66–82.

Illanes A., Schaffeld G., Schiappacasse M.C., Zuñiga M.E., Gonzáñez G., Curotto E., Tapia G., and O'Reilly S. (1985a). Some studies on the protease from a novel source: The plant *Cucurbita ficifolia. Biotechnol. Letters* 7: 669–672.

Illanes A., Chamy R., and Zuñiga M.E. (1985b). In: *Chitin in Nature and Technology.* Muzzarelli R., Jenniaux J., and Gooday G.W., Eds. New York: Plenum Press, pp. 411–415.

Illanes A., Ruiz A., and Zuñiga M.E. (1993). Análisis de dos lactasas microbianas. *Alimentos* 18(1): 26–34.

Illanes A., Altamirano C., and Zuñiga M.E. (1996). Thermal inactivation of penicillin acylase in the presence of substrate and products. *Biotechnol. Bioeng.* 50: 609–616.

John V.T. and Abraham G. (1991). Lipase catalysis and its applications. In: *Biocatalysts for Industry.* Dordick J.S., Ed. New York: Plenum Press, pp. 193–217.

Katchalsky-Katzir E. (1993). Immobilized enzymes: Learning from past successes and failures. *Trends Biotechnol.* 11: 471–478.

Kennedy J.F., Cablada V.M., and White C.A. (1988). Enzymatic utilization and genetic engineering. *Trends Biotechnol.* 6: 184–189.

Klibanov A.M. (1997). Why are enzymes less active in organic solvents than in water? *Trends Biotechnol.* 15: 97–101.

Knowles J.R. (1997). Tinkering with enzymes: What are we learning? *Sci.* 236: 1252–1258.

Koskinen A.M. and Klibanov A.M. (1996). *Enzymatic Reactions in Organic Media.* London, UK: Blackie Academic & Professional.

Lamare S. and Legoy M.D. (1993). Biocatalysis in the gas phase. *Trends Biotechnol.* 11: 413–418.

Layman P.L. (1986). Industrial enzymes battling to remain specialties. *Chem. Eng. News* 64(September 15): 11–14.

López L.M., Natallucci C.L., Caffini N.O., and Curotto E. (1993). Isolation and partial purification of serine proteinases in the latex of Maclura pomifera fruits. *Acta Aliment.* 22: 131–142.

Margolin AL (1996). Novel crystalline catalysts. *Trends Biotechnol.* 14: 223–230.

Marshall C.J. (1997). Cold-adapted enzymes. *Trends Biotechnol.* 15: 359–364.

Marwaha S. and Kennedy J. (1988). Whey-pollution problem and potential utilization. *Int. J. Food Sci. Technol.* 23: 323–336.

May S.W. (1992). Biocatalysis in the 1990s: A perspective. *Enzyme Microb. Technol.* 14:80–84

Michael-Sinclair P. (1965). Enzymes convert starch to dextrose. *Chem. Eng.* 72(18): 90–96.

Mojovic L., Siler-Marinkovic S., Kukic G., and Vunjak-Novakovic G. (1993). *Rhizopus arrhizus* lipase-catalyzed interesterification of the midfraction of palm oil to a cocoa butter equivalent fat. *Enzyme Microb. Technol.* 14: 80–84.

Morris H.A. and Andreson K. (1991). A compartive study of cheddar cheeses made with fermentation produced calf chymosin from *Kluyveromyces lactis* and with calf rennet. *Cult. Dairy Prod. J.* May, pp. 13–20.

Neidelman S.L. (1991). Historical perspective on the industrial uses of biocatalysts. In: *Biocatalysts for Industry.* Dordick J.S., Ed. New York: Plenum Press, pp. 21–33.

Nosoh Y. and Sekiguchi T. (1990). Protein engineering for thermostability. *Trends Biotechnol.* 8: 16–19.

Polastro E. (1989). Enzymes in the fine chemicals industry: Dreams and realities. *Biotechnol.* 7: 1238–1241.

Randolph T.W., Blanch H.W., and Clark D.S. (1991). Biocatalysis in supercritical fluids. In: *Biocatalysts for Industry*. Dordick J.S., Ed. New York: Plenum Press, pp. 219–237.

Sarney D.B. and Vulfson E.N. (1995). Application of enzymes to the synthesis of surfactants. *Trends Biotechnol.* 13: 164–177.

Schaffeld G., Bruzzone P., Illanes A., Curotto E., and Aguirre C. (1989). Enzymatic treatment of stickwater from fishmeal industry with the protease of *Cucurbita ficifolia*. *Biotechnol. Letters* 11(7): 521–525.

Tramper J. (1996). Chemical versus biochemical conversion: when and how to use biocatalysts. *Biotechnol. Bioeng.* 52: 290–295.

Turner M.K. (1995). Biocatalysis in organic chemistry. Part II: Present and future. *Trends Biotechnol.* 15: 285–386.

van den Berg G. and Koning P.J. (1990). Gouda cheesemaking with purified calf chymosin and microbially produced chymosin. *Neth. Milk Dairy J.* 44: 189–205.

Vieille C. and Zeikus J.C. (1996). Thermozymes: Identifying molecular determinants of protein structural and functional stability. *Trends Biotechnol.* 14: 183–190.

Walsh G.A., Power R.F., and Headon D.R. (1993). Enzymes in the animal-feed industry. *Trends Biotechnol.* 11: 424–430.

White T.J., Meade J.H., Shoemaker S.P., Koths K.E., and Innis M.A. (1984). Enzyme cloning for the food and fermentation industry. *Food Technol.* 38(2): 90–98.

Wrotnowsky C. (1997). Unexpected niche applications for industrial enzymes drives market growth. *Gen. Eng. News* 17(3): 14–30.

CHAPTER 20

Enzymatic Synthesis of Food Additives

A. van der Padt, F. Boon, N. Heinsman,
J. Sewalt, and K. van 't Riet

INTRODUCTION

Enzymes are the catalysts of many occurring chemical reactions in living organisms. *In vitro,* these heterogeneous catalysts can be used to selectively catalyze these chemical and similar reactions under mild conditions. Some enzyme-catalyzed processes are already being implemented in the food industry, but it is expected that there are more to come not only in the food industry but the chemical industry as well (Lalonde, 1997). This paper starts with an introduction on biocatalysis using enzymes, then briefly summarizes two commercial processes, and finally presents an overview of the research on enzymatic synthesis in food additives conducted in the authors' laboratory at Wageningen Agricultural University.

BIOCATALYSIS USING ENZYMES

Enzymes are categorized according to the Enzyme Commission (EC) system in six classes:

1. *Oxido-reductases* (Hydrogenases): Except for catalases and peroxidases, *in vivo* redox reactions use one of the co-factors as the other substrate of the redox couple. Co-factor regeneration is still the bottleneck in biotechnology (Turner, 1995).
2. *Transferases:* Enzymes like aspartate aminotransferase or cyclodextrin glycosyltransferase-transferring functional groups such as carbon, aldehydic, ketonic, acyl, glucosyl, phosphate, or S-containing groups.
3. *Hydrolases:* The majority of biotechnological processes apply to hydrolases such as glycosidase, proteases, amidases, lipases, and esterases acting on glycosides, peptides, ester bonds, as well as other C-N bonds and acid anhydrides.
4. *Lyases:* Additional reactions can be catalyzed by lyases, such as the stereospecific hydration of fumarate to L-malate catalyzed by fumerase.
5. *Isomerases:* This class of enzymes catalyzes structural rearrangements such as glucose isomerase for the conversion of glucose to fructose or

racemases used in the enantiomer recycle loop of optical resolution processes.

6. *Ligases:* This class gets attention in the pharmaceutical and fine chemical industries since these enzymes catalyze bond formation, in turn making carbon chain lengthening possible. Unfortunately, the coupled energy source is restricted to adenosine triphosphate (ATP) which is a costly fuel that becomes degraded upon forming adenosine diphosphate (ADP). Analogous to the co-factors used by oxido-reductases, regeneration of ADP is essential.

Since regeneration of cofactors and ATP is still too costly, most processes make use of enzymes from the transferase, hydrolase, isomerase, or lyase class. Out of these enzymes, at least 75% of the industrial enzymes are hydrolases.

Applying Enzymes

Since enzymes originate from living organisms, it was thought for a long period of time that enzymes could only be active in aqueous solutions. However, during the last two decades it has been proven that many enzymes do work in almost dry solvents (see Bell et al., 1995). Examples of solvents used in the literature are alkanes, alkenes, primary and secondary alcohols, ethers, mono-, di- and triglyme, acetonitrile, acetone, chloroform, methyl-pyrrolidone, tetrahydroftiran, and dioxane. Both enzyme stability and selectivity could differ from solvent to solvent (Cambou and Klibanov, 1984). For example, enzymes have been proven active at temperatures over 100°C in organic solvents (Zaks and Klibanov, 1984). Other non-conventional media for biocatalysis are gas phases (Lamare and Legoy, 1995), compressed gases (Carvalho et al., 1996), or supercritical gases (Kamat et al., 1992). Examples of supercritical gases used in the literature are CO_2, ethane, ethylhexanol, ethylene, and propane. Non-aqueous solvents offer the possibility of performing condensation reactions at reduced water activity.

Enzyme Stability

Because enzymes are proteins, they are fragile catalysts with a low heat and chemical resistance. This implies that the operational conditions for biocatalysis range from around 10°C up to 100°C at neutral pH. However, recent developments have enabled enzymes to become more and more stable.

Enzyme stability is conventionally improved by immobilization. From an engineering point of view, there are two additional benefits from immobilization. Since biocatalysts can be reused, enzyme costs are reduced, and once the enzyme is immobilized onto a carrier, implementation into a conventional reactor is guaranteed (Tramper, 1996). Examples of carriers are ion exchange resins, silica, and Accurel resins (Malcata et al., 1990; Oladepo et al., 1995). Unfortunately, support-enzyme interactions can force a change in the tertiary conformation of the active site. Hence, at different supports, an enzyme can have different catalytic activity and/or another selectivity (Bell et al., 1995; Oladepo et al., 1995). Also, since the addition of solvents can alter the conformation of the active site (Oladepo et al., 1995), enzyme screening should be done using immobilized enzymes in a desired solvent.

A second route to obtain more stable enzymes is to search for microorganisms growing under extreme conditions, such as pools around geysers or deep-sea geothermal vents. The enzyme system for these extremophiles is more heat-resistant and likely better solvent-resistant compared to the microorganisms living at ambient environments. Genetic engineering also enables a fast modification of microorganisms which leads to heat- and solvent-resistant enzymes (Herbert, 1992).

Once the right enzyme (system) to do the job is found, the enzyme (system) must be made ready to use in a reactor by downstream processing and immobilization onto a carrier. When a sequence of enzymes and co-factors (enzyme system) are needed, permeabilized cells could be used as the biocatalyst.

In the past, only naturally excreted enzymes were commercially available as biocatalysts. However, since genetic engineering has become routine, any enzyme can be expressed in a guest microorganism and used as a biocatalyst (Turner, 1995). The latest development is to cross-link the crystals of pure enzymes (CLEC) for direct use as a catalyst since this results in greater stability then with immobilized enzymes (Clair and Navia, 1992; Lee et al., 1995).

Pros and Cons of Biocatalysis

The advantages and disadvantages of enzymatic conversions over chemical processes are often discussed. For applications in the food industry the main advantages are:

1. *Green technique:* Products processed with this technique can be sold as natural ingredients.
2. *Mild process conditions:* These allow (poly) unsaturated substrates to be used and the formation of color bodies overcome.
3. *Selective catalysis:* This reduces the number of reaction steps since the enzyme catalyzes only the wanted reaction; therefore, protection of other reactive groups is no longer necessary.

Of course, one also has to cope with certain disadvantages:

1. *Low temperatures*: Often elevated temperatures are used to improve the yield at equilibrium, but this leads to a low yield or complex downstream processing.
2. *Low stability*: This leads to a narrow window of operation.
3. *New techniques:* This implies the need to train employees.

ENZYMATIC SYNTHESIS OF FOOD ADDITIVES

Out of the 3,000 known enzymes, only 300 are on the market now, with less than 30 produced and used on an industrial scale (Table 20-1). Well-known commercial processes involving synthesis of aspartame and high fructose syrups will be discussed below, followed by some examples of potential applications in the flavor, fragrances, functional food ingredient, and oils and fats industries.

Two Examples of Running Industrial Processes

Aspartame

Holland Sweetener, a joint venture between Tosoh and DSM, produces α-aspartame (α-L-aspartyl-L-phenylalanine methyl ester) on a commercial scale (Oyama, 1992). This product is a synthetic low-calorie sweetener 200 times sweeter than sucrose.

Chemical synthesis is based on the coupling of L-phenylalanine and an amino-protected aspartate in which the β-carboxylate group is blocked. Some disadvantages of this method are the necessity of expensive substrates and the production of 20 to 40% bitter-tasting β-aspartame. Using a protease as the biocatalyst, racemic phenylalanine methyl ester and the amino-protected z-aspartate are coupled stereo- and regioselectively to form pure α-aspartame (Fig. 20-1).

Proteases are naturally designed to hydrolyze peptides, but the condensation of amino acids can also be catalyzed by protease if the product is removed by precipitation. The natural selectivity of proteases guarantees stereo-specific peptide bond formation without the need for side chain protection of aspartate. An extensive screening program of commercially available proteases

Table 20-1 Enzymes used in the food industry.

Protease	Dairy, Bakery, Starch, Juice, Wine
α-Amylase	Starch, Dairy, Bakery, Juice, Wine
Lipase	Flavor, Fragrances, Oils and Fats, Dairy, Bakery
Hemicellulase	Bakery, Starch, Juice, Wine
Glucoamylase	Starch, Dairy, Bakery, Juice, Wine
Pectinase	Juice, Wine, Dairy
Lactase	Dairy P-Amylase Starch
Glucose isomerase	Starch, Dairy, Bakery, Juice, Wine
Pullinase	Starch

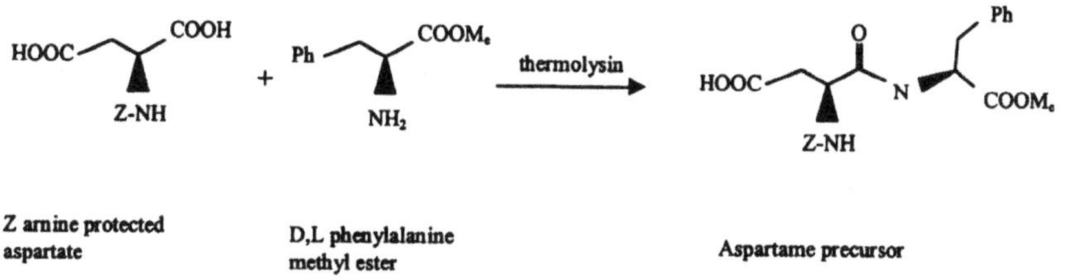

Figure 20-1 The biocatalyst thermolysin exhibits regio and stereoselectivity during the synthesis of a-aspartame.

demonstrated that the protease thermolysin efficiently catalyzed the coupling of the α-amino group of L-phenylalanine methyl ester with the α-carboxylate of Z-aspartate on an industrial scale. Because thermolysin originates from *Bacillus proteolyticus* Rokko, a bacterial strain found in the Rokko Hot Spring in Japan, the enzyme has an unusually high heat tolerance, showing optimum activity around 50–60°C (Oyama, 1992). Successful commercialization of the aspartame process demonstrates that enzyme-catalyzed reactions can be an effective method for the synthesis of short peptides.

High Fructose Syrups

Another example of a sweetener is high fructose syrup. A product containing 90% fructose syrup is some 20–30% sweeter than one with the same percentage content of sucrose. High fructose syrups are produced in large quantities (millions of tons a year). The basis of this process is the isomerization of glucose, derived from starch hydrolysis, to fructose catalyzed by glucose isomerase.

Starch is hydrolyzed in a sequence of reactors to high glucose syrup using thermotolerant α-amylase during the liquefaction and glucoamylase in the saccharification step (Fig. 20-2). The isomerization itself is a continuous process using immobilized glucose isomerase (Bentley and Williams, 1996) immobilized either in its host cell or on a polymeric resin. The process of starch hydrolysate isomerization was commercialized between 1965 and 1970 in the USA and Japan. The operational life of catalysts is now around six months, with productivity averaging 5–15 tons of product per kilo of catalyst.

Ongoing Research for the Enzymatic Synthesis of Food Additives

Several different enzyme-catalyzed synthesis reactions are the focus at our laboratory at Wageningen Agricultural University. In all cases, the reactions studied involve the synthesis of food additives such as flavors and fragrances, oligosaccharides, and tailor-made oils and fats. Two examples of reaction kinetics and one of reactor design are given below.

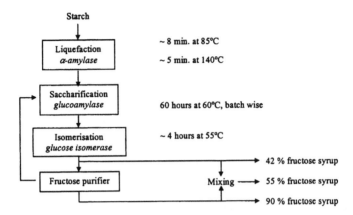

Figure 20-2 A typical route for the enzymatic production of fructose syrups.

Flavors and Fragrances

Homochiral branched chain fatty acids contribute to the characteristic smell of mutton and sheep-cheese, and the esters of these branched chain fatty acids are the main fragrances in dry cured ham and chardonnay wine. The enantiomers of branched chain fatty acid methyl esters can be separated using enzymatic kinetic resolution techniques in which one of the enantiomers of the racemate is converted faster to the product than the other. In addition, transesterification can be used to separate the enantiomers (Fig. 20-3).

Heinsman et al. (1998) tested 25 hydrolyses for both 4-methylhexanoic and 4-methyloctanoic acid methyl esters. Both *Rhizomucor miehei* lipase and *Candida antarctica* lipase B were found to catalyze this reaction, but only *Candida antarctica* lipase B was enantioselective for both substrates (Table 20-2). This modest stereoselectivity might be explained by the fact that the chiral center (C4-atom) was rather far away from the reaction center.

Table 20-2 shows that the enantiomeric ratio E, the ratio of the specificity constants, changes from enzyme to enzyme and substrate to substrate. Moreover, the enantiomeric ratio E is temperature-dependent, implying that both the reaction conditions as well as the substrates used during the screening program are very important. From a scientific point of view, it would be helpful to have a rationale for predicting the enantioselectivity of enzymes or at least the change in enantioselectivity upon a change in substrates.

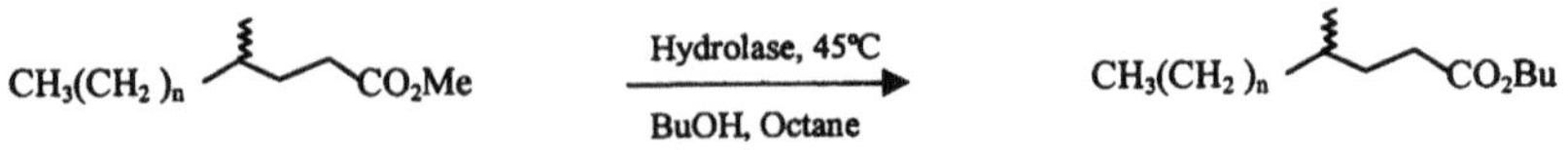

$$CH_3(CH_2)_n \quad CO_2Me \xrightarrow[\text{BuOH, Octane}]{\text{Hydrolase, 45°C}} CH_3(CH_2)_n \quad CO_2Bu$$

Figure 20-3 Kinetic resolution of 4-methyl hexanoic acid methyl ester (n = 1) and 4-methyl octanoic acid methyl ester (n = 3) using hydrolases. Only *Candida antarctica* lipase B showed enantioselectivity for both substrates.

Table 20-2 Enantiomeric ratio E in the transesterification of two seven-branched chain fatty acid methyl esters using two types of lipases.

	E(T= 45°C)	E(T= 27°C)
4-methylhexanoic acid methyl ester		
Rhizomucor miehei lipase B	2 (S/R)	
Candida antarctica lipase	5 (R/S)	
4-methyloctanoic acid methyl ester		
Rhizomucor miehei lipase	None	23 (R/S)
Candida antarctica lipase B	8 (R/S)	

Molecular modeling is one route to get more insight in the selectivity of enzymes. If the three-dimensional structure of the enzyme is known, the energy difference of various enzyme-substrate complexes can be calculated. For both R and S 4-methyloctanoic acid methyl esters, 10 substrate-enzyme complex structures were minimized for *Candida antarctica* lipase B. The R-structure was significantly lower in energy compared with the S-structure, indicating this complexity was favorable. This result was in agreement with the experimental data. In conclusion, molecular modeling studies support the measured stereochemical preference of *Candida antarctica* lipase B. The stereoselectivity varies from substrate to substrate and improves with decreasing temperature. Although the enantiomeric ratio E is rather low, the yield from a two-step kinetic resolution process can be calculated as an almost enantiomerically pure product.

Oligosaccharides

There is a growing interest in the synthesis of oligosaccharides for application in "functional foods" because these are non-digestible and effective for the growth of *Bifidobacterium* in the intestinal tract. However, during their enzymatic synthesis water is produced as a by-product and the equilibrium is unfavorable due to the hydrolysis of both substrate and product. In this study the thermostable β-glucosidase from *Pyrococcus furiosus* was used as a catalyst (Boon et al., 1998). Due to the extreme thermostability of this enzyme, high temperatures and subsequently high initial lactose concentrations were possible. The assumed reaction scheme is shown in Figure 20-4. For statistical reasons, the five parameters in the model were grouped into three parameters $1/k_1$, k_3/k_2, k_4/k_5. Apart from glucose inhibition, a first-order inactivation constant k_d was also included. The parameters were fit to data obtained from experiments with different initial lactose concentrations and temperatures in a batch reactor.

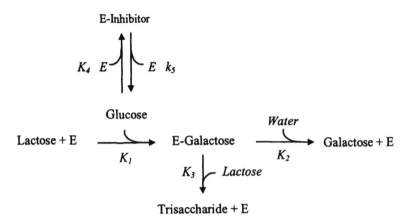

Figure 20-4 Reaction scheme for the enzymatic synthesis of trisaccharides.

Figure 20-5 provides an example of experimental data and the prediction of the model. Parameters k_d and k_3/k_2 increase with increasing temperatures (75°C to 95°C). The loss of enzyme activity (k_d) is probably caused by Maillard products in the reaction mixture which could be prevented by enzyme immobilization. Enlargement of the synthesis/hydrolysis ratio (k_3/k_2) from 4.5 to 73 min^{-1} points to a higher trisaccharide yield at higher temperature. Even when an excess of enzyme was added, this higher trisaccharide yield was not observed at higher temperatures. Although kinetics point out that at elevated temperatures the trisaccharide synthesis rate improves, the product yield belonging to it does not improve. This example clearly shows that both kinetics and product yield must be taken into account. It can be concluded that to produce a pure product, it must be removed from the reaction mixture.

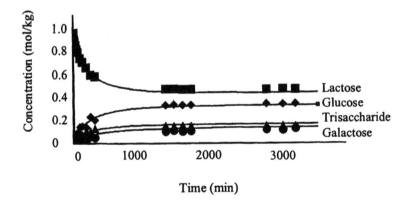

Figure 20-5 Conversion of lactose in a batch reactor at 95°C with measurements (symbols) and model predictions (lines) and an enzyme concentration of 75,000 units/kg.

Oils and Fats

The enzymatic esterification of glycerol with fatty acid for the synthesis of mono- and triacylglycerols is a sequence of three equilibrium reactions:

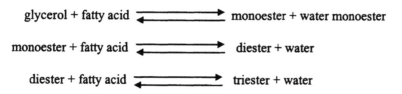

glycerol + fatty acid $\rightleftharpoons$ monoester + water monoester

monoester + fatty acid $\rightleftharpoons$ diester + water

diester + fatty acid $\rightleftharpoons$ triester + water

Reactionary intermediate monoesters (monoacylglycerol) are used as emulsifiers in food and cosmetics, while the final product triester (triacylglycerol) is used to adjust the melting range of foods and cosmetics.

A two-phase membrane reactor can be used for the lipase-catalyzed hydrolysed or synthesized glycerides. In this two-phase membrane bioreactor, lipase from *Candida rugosa* is immobilized at the inner fiber side of a cellulose hollow fiber module (van der Padt et al., 1990). The characteristic time for mass transfer in these types of membrane systems is about 180 s, while the reaction rate time constant exceeds 33,000 s (van der Padt et al., 1996). By placing a packed column with immobilized enzyme resins in a recirculation stream (Fig. 20-6), the characteristic time for reaction can be decreased without influencing the characteristic time for mass transfer. In this pertraction system, the water produced will accumulate in the non-polar phase so that the water activity of the non-polar phase increases. When the column outlet stream is led into the pertraction system, the surplus of water is extracted into the glycerol/water phase and the oil phase is saturated again with glycerol. Hence, the water activity is again reduced to the required level. The activity of the pertraction membrane bioreactor system is compared with other types of reactors used at our laboratory (Table 20-3).

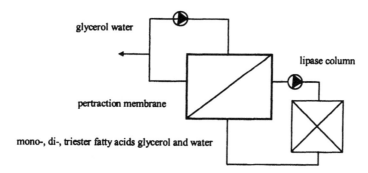

Figure 20-6 A pertraction system for the *Candida rugosa* lipase catalyzed esterification of glycerol and decanoic acid in hexadecane with an in-line enzyme column.

Table 20-3 Comparison of initial reaction rates $(a_w = 0.3)$ in different systems.

	Reaction rate	
	$(mol\ s^{-1}m^{-3})$	$(\mu mol\ s^{-1}g^{-1}enzyme)$
Emulsion	0.1	28
Membrane bioreactor	0.05	301
Pervaporation system	0.01	60
Pertraction membrane reactor *Candida rugosa* lipase	0.03	41
Novozym SP435 and an Accurel® spacer	1.15	94

Preliminary experiments showed that the packed bed filled with a combination of a spacer (Accurel® EP100) and Novozym SP435 as the catalyst gave a high esterification activity per system volume. However, the activity of the enzyme in the packed bed was not as good as in the other membrane reactors. This could be due to diffusion limitation, bad flow regime in the packed column, or less than optimal column volume.

CONDLUDING REMARKS

There are already some examples of successful industrial biocatalytic processes, but the increasing number of commercially available enzymes enables implementation of biocatalysis in more and more industrial processes. Screening, genetic engineering, crystallization, and immobilization techniques guarantee the development of robust catalysts. However, for successful process development, more insight is needed in enzyme kinetics and the thermodynamics of reactions at temperatures below 100°C.

REFERENCES

Bell G., Halling P.J., Moore B.D., Partridge J., and Rees D.G. (1995). Biocatalyst behavior in low-water systems. *Trends Biotechnol.* 13: 468–473.

Bentley I.S. and Williams E.C. (1996). Starch Conversion. In: *Industrial Enzymology*. Godfrey T., West S., Eds. London: MacMillan Press Ltd., pp. 339–357.

Boon M.A., van der Oost J., de Vos W.M., Janssen A.E.M., and van 't Riet K. (1998). Synthesis of oligosaccharides catalyzed by the thermostable β-glucosidase from *Pyrococcusfuriosus*. Appl. Biochem. Biotech. 75(2):269–281.

Cambou B. and Klibanov A.M. (1984). Unusual catalytic properties of usual enzymes. *Ann. N. Y. Acad. Sci.* 434: 219–223.

de Carvalho I.B., de Sampaio T.C., and Barreiros S. (1996). Solvent effects on the catalytic activity of subtilisin suspended in compressed gases. *Biotechnol. Bioeng.* 49(4): 399–404.

Clair N.L. and Navia M.A. (1992). Cross-linked enzyme crystals as robust biocatalysts. *J. Am. Chem. Soc.* 114:7314-7314.

Heinsman N.W.J.T., Orrenius S.C., Marcelis C.L.M., de Sousa Teixeira A., Franssen M.C.R., van der Padt A., Jongejan J.A., and de Groot A. (1998). Lipase-mediated resolution of y-branched chain fatty acid methyl esters. Accepted for publication by *Biocatal. Biotransform.* 16(2):145–155.

Herbert R.A. (1992). A perspective on the biotechnological potential of extremophiles. *Tibtech.* 10: 395–402.

Kamat S., Burrera J., Beckman E.J., and Russel A.J. (1992). Biocatalysis synthesis of acrylates in organic solvents and supercritical fluids I: optimization of enzyme environment. *Biotechnol. Bioeng.* 40: 158–166.

Lalonde J. (1997). Enzyme catalysis: Cleaner, safer, energy efficient. *Chem. Eng. News* September: 108–112.

Lamare S. and Legoy M.D. (1995). Working at controlled water activity in a continuous process: The gas/solid system as a solution. *Biotechnol. Bioeng.* 45: 387–397.

Lee C.-K., Chem C.-S., and Fan C.-H. (1995). Enzyme crystal embedded in latex filin as immobilized enzyme. *Biotechnol. Technol.* 9: 827–832.

Malcata F.X., Reyes H.R., Garcia H.S., Hill C.G., and Amundson C.H. (1990). Immobilized lipase reactors for modification of fats and oils: A review. *J. Am. Oil Chem. Soc.* 67: 890–910.

Oladepo D.K., Hailing P.J., and Larsen V.F. (1995). Effect of different supports on the reaction rate of *Rhizomucor miehei* lipase in organic media. *Biocatal. Biotransform.* 12: 47–54.

Oyama K. (1992). The industrial production of aspartame. In: *Chirality in Industry.* Collins A.N., Sheldrake G.N., and Crosby J., Eds. New York: John Wiley & Sons, pp. 237–247.

van der Padt A., Edema M.J., Sewalt J.J.W., and van I Riet K. (1990). Enzymatic acylglycerol synthesis in a membrane bioreactor. *J. Am. Oil Chem. Soc.* 67: 347–352.

van der Padt A., Sewalt J.J.W., van I Riet K. (1993). On-line water removal during enzymatic triacylglycerol synthesis by means of pervaporation. *J. Memb. Sci.* 80: 199–208.

van der Padt A., Sewalt J.J.W., and van I Riet K. (1996). Membrane bioreactor design to force equilibrium towards a favorable product yield. In: *Engineering of 1 with Lipase.* Malcata F.X., Ed. Kluwer Academic Publishers, pp. 473–481.

Tramper J. (1996). Chemical versus biochemical conversion; when and how to use biocatalysts. *Biotechnol. Bioeng.* 52: 290–295.

Turner M.K. (1995). Biocatalysis in organic chemistry. Part II: Present and future. Trends Biotechnol. 13: 253–258.

Zaks A. and Klibanov A.M. (1984). Enzymatic catalysis in organic media at 100°C. Sci. 224: 1249–1251.

CHAPTER 21

Plastic Materials for Modified Atmosphere Packaging

R. Catalá and R. Gavara

INTRODUCTION

Throughout the last few decades there has been an increasing demand for fresh or minimally processed foods free of additives and preservatives. This trend has promoted the development of technologies to provide a reasonable shelf-life attending to the above requirements. Modified atmosphere packaging (MAP) is one of these technologies with an increasing presence in the market.

The modification and control of the environment surrounding food products has been used to limit their biological activity since the beginning of the 20th century under the assumption that a change in this gaseous composition may help to reduce microbial growth, the rate of internal chemical reactions, and the exchange of substances with the environment. Each of these processes affects product quality, but is slowed by decreasing temperature. Consequently, products with MAP are usually stored under refrigeration.

According to the most common definition of MAP, a container is sealed after all the air present has been exchanged by a mixture of different gases with the appropriate composition to extend the shelf-life of the food product (Hintlian and Hotchkiss, 1986). No control or gas exchange is carried out during distribution or storage. In commercial practice, the composition of the modified atmosphere presents a higher concentration of carbon dioxide (CO_2), a reduced concentration of oxygen (O_2), and the required nitrogen to obtain atmospheric pressure. However, it is often convenient to control the atmosphere or maintain a continuous flow of the modified gas during storage. This technology is called controlled atmosphere packaging (CAP), which although very useful for the preservation of foods in large stores, is neither practical nor economical for consumption packaging. Moreover, maintaining a fixed gas composition during distribution and storage is not generally necessary. Thus, MAP is more convenient for consumer-ready food packaging.

Both MAP and CAP are alternative solutions to the well-established vacuum packaging in which a food's internal atmosphere composition is not altered. By the synergistic effect of the appropriate atmosphere and refrigeration, the two offer technological solutions to double or triple the shelf-lives of many fresh and processed foods without compromising their organoleptic qualities. In this chapter the most relevant characteristics of MAP, its use for diverse food products, the plastic materials available, and their possibilities in accordance with product requirements are presented and discussed.

APPLICATIONS

The use of modified atmosphere in foods is not new, as the preservative effect of CO_2 on foods has been known for a century. The first basic research experiences on modified atmosphere were focused on fruits and fresh meats. In 1922, Brown studied the effect of temperature and O_2 and CO_2 concentrations on the fungi responsible for fruit rot. Kid and West (1927) analyzed the effect of modified atmosphere on apple shelf-life and carried out the first commercial controlled atmosphere storage in 1929. Killefer (1930) showed that pork and lamb meats double their shelf-lives when air is replaced by 100% CO_2. Haines (1933) observed that some bacteria present on meats needed more time to grow when stored in an atmosphere containing 10% CO_2, which was applied in the commercialization of Australian meat in the 1930s to preserve products for 40–50 days. Coyne (1932, 1933) verified that the shelf-lives of raw fish and fish fillets doubled when stored under an atmosphere with a minimum of 25% CO_2. However, fish products showed undesirable changes both in appearance and texture with high concentrations of CO_2.

The industrial development of modified atmosphere technologies for fresh product preservation began in the late 1950s, although they weren't commercially introduced until the 1970s when chicken in large containers was stored under refrigeration for three weeks by the use of a continuous stream of air enriched with CO_2. The application of CAP has been growing since then, and is now used extensively for the distribution and preservation of raw foods in large stores. Red meats, poultry, cured meats, seafood, fruits, vegetables, nuts, snacks, pasta, bakery products, and cheese are all currently being successfully preserved by MAP.

The attractive advantages of this technology for both consumers and the food processing industry have continued to facilitate its development and application. The most relevant advantages of MAP are:

- Significant increases in the commercial shelf-lives of fresh and minimally processed foods
- Reduction of excessive production, defective products, and waste
- Excellent commercial presentation by maintaining product freshness
- Cost reduction by the substitution of expensive treatments or addition of chemicals
- Better commercial distribution due to larger distribution areas and turnover periods

Despite the benefits of this technology, there are also some disadvantages that should be mentioned:

- Investments in packaging equipment

312

- High cost of gases and packaging materials
- Cost of quality control and quality assurance (high in comparison to other preservation technologies)
- Costs of refrigeration
- High costs for transportation and storage of packages and finished products
- Development of pathogens due to loss of atmosphere, breakage of temperature chains, and poor quality of raw products

Overall, however, MAP represents a revolution in food processing and commercialization. Some estimates indicate that in 10 years 50% of all food products will be treated with this technology.

MODIFIED ATMOSPHERE PACKAGING FACTORS

All biological materials including food products tend to deteriorate and alter as time goes by due to microbiological activity or chemical or biochemical reactions (Table 21-1). Fortunately these may be prevented without aggressive treatments and can be significantly altered by environmental conditions (temperature, gas composition, and relative humidity).

In industrial practice, atmosphere modification usually implies the reduction of O_2 and the increase of CO_2 concentrations. The respiration rate of fresh produce as well as all aerobic activity decreases by low O_2 and high CO_2 partial pressures. Total O_2 removal is beneficial in some cases since it restricts all oxidative processes. Nevertheless, it may result in the development of off-flavors and other undesirable effects as occurs in fruits and vegetables. Anaerobic conditions also allow the growth of microorganisms which may induce toxicological effects. Hence, the appropriate atmosphere has to be carefully selected to limit all undesirable activities and expand product shelf-life. It is obvious then that the best gas composition will not only depend on product characteristics, but production and storage conditions as well.

Table 21-1 Principal mechanisms of food deterioration.

Microbiological activity	Spoilage, poisoning
Senescence	Loss of fresh like attributes
Chemical alteration	Enzymatic degradation
	Non-enzymatic browning
	Rancidity
	Lipid and polysaccharide hydrolysis
	Protein denaturalization
	Vitamin oxidation
	Pigment degradation
Physical alteration	Physical damage
	Texture changes
	Water loss/uptake

The main factors to be considered in the application of MAP are:

- Food characteristics and composition (nature, ripeness, water activity, and basic components such as carbohydrates and fats)
- Sanitary conditions and initial microbiological load
- Storage temperature (product temperature susceptibility, specific microorganisms)
- Atmospheric composition (O_2, CO_2, ethylene, humidity, pressure)
- Packaging materials and technology

FOOD REQUIREMENTS

The gases applied in MAP/CAP technologies are basically O_2, N_2, and CO_2. Although other gases such as carbon monoxide, sulfide dioxide, ethylene, and ozone have been studied, they are hardly ever applied commercially (Church, 1994).

Carbon dioxide acts as a bacteriostatic and fungistatic agent which delays and reduces the multiplication of aerobic microorganisms. It also delays the senescence of fresh produce. However, since CO_2 is very soluble in water and fats, it induces acidity and retracts or collapses flexible containers. The presence of O_2 is generally considered undesirable since it is involved in biochemical reactions and the development of aerobic microorganisms. Nevertheless, O_2 preserves fresh meat pigmentation, maintains vegetable metabolism, and prevents the growth of anaerobic microorganisms in fish and meat products. Nitrogen is basically an inert gas with no significant direct effect on food products. It is used to complete package atmosphere and as a substitute for O_2 and CO_2.

Although they can be adapted to product requirements and preservation technologies, there are three basic technologies for the application of gases:

- Gas injection after vacuum
- Stripping with a gas stream
- Generation of atmosphere within a closed package (passive modification by product respiration, O_2, CO_2, ethylene scavengers, CO_2, or ethanol emitters)

The effect of gases on foodstuffs is dependent of factors such as food nature, product origin, and maturity. In spite of this variability, there are some general guidelines that can be used as a first step. Table 21-2 includes the recommended range of atmospheric composition for some products presently commercialized in MAP, although for the best application of this technique the gas environment must be specifically studied.

Table 21-2 Recommended atmospheric composition for some food products.

Product	% O$_2$	% CO$_2$	% N$_2$	Source
Bakery products	—	50	50	Black et al., 1993; Pedrelli et al., 1994; Piergiovanni and Fava, 1997; Rodriguez-Castilla and Jordano, 1997
Fresh red meats	40–60	10–20	Rest	Paleari et al., 1987; Sorheim et al., 1997
Fresh poultry	—	20–30	70–80	Grandison and Jennings, 1993; Marshall et al., 1992, Sawaya et al., 1995; Soffer et al., 1994
Sausages	—	20–30	70–80	Feng-Sheng et al., 1995; Viallon et al., 1996
Fresh blue fish	—	40–60	40–60	Moral-Rama, 1993; Nychas and Tassou, 1996; Oka et al., 1993; Randell et al., 1995
Fresh white fish	10–20	30–40	Rest	Moral-Rama, 1993; Nychas and Tassou, 1996; Oka et al., 1993; Randell et al., 1995
Fresh pasta	—	—	100	Castelvetri, 1991; Giavedoni et al., 1994,
Cheese	—	30–40	60–70	Piergiovanni et al., 1993; Mannheim and Soffer, 1996; Sarantopoulos et al., 1995; Vercelino-Alves et al., 1996
Apples	1–3	1–5	Rest	Evelo and Boerrigter, 1994; Geeson et al., 1994; Gran and Beaudry, 1993
Avocado	2–5	3–10	75–80	Gerdes and Parrino, 1995; Meir et al., 1997
Strawberry	5–10	15–20	50–70	Brecht et al., 1992; Chambroy et al., 1993; Renault et al., 1994
Vegetable mix	2–4	2–4	Rest	Buick and Damoglou, 1989; Kwang et al., 1996; Lopez-Galvez et al., 1997
Broccoli	1–2	5–10	90	Barth et al., 1993; Talasila et al., 1994

MATERIALS AND CONTAINERS

As outlined above, MAP technology implies a gas exchange prior to container closure and without any further modification or control during distribution and storage. The system (external environment/container/internal environment/food) will progress towards a stationary state or equilibrium by itself since the container is the only element controlling the composition of the internal environment. Traditional packaging materials such as tinplate and glass are not adequate for the MAP of many products due to their total barrier to gases. The versatility of plastic materials, however, is not only more suitable for MAP, but has actually made its development possible. Current technology offers a wide variety of packages to cover the requirements of specific products.

Modified atmosphere packaging requires an appropriate container designed or selected according to food characteristics, distribution and storage conditions. Flexible or semirigid containers can be made of mono- or multilayer structures with the required gas permeation rate and mechanical characteristics. According to container manufacture technology, MAP packages can be classified either as 1) flexible pouches with three (pillow type) or four (envelope type) seals or 2) rigid or semirigid trays closed with a rigid lid or thermosealed flexible film wrapped with a retractile film inside a pouch.

Also available are a wide variety of equipment for modifying the atmosphere in different containers ranging from manual vacuum or bell systems to automatic machinery (form/fill/seal). Depending on the container manufacturer and product dispensing automatic equipment, these can be classified as:

- Vertical packaging machinery for three or four-seal pouches
- Horizontal packaging machinery for wrapping or pre-formed trays in pouches (flow-pack)
- Horizontal tray thermoforming machinery with thermoformed rigid lids or flexible film seals

The selection of materials and container types should be based on different properties as defined by the packaged product or marketing strategy. These properties include mechanical, optical, thermal, and barrier characteristics. The mechanical properties of the finished container should assure packaging integrity during packaging distribution and storage. Abrasion, perforation, tear, impact, and flexion resistance are commonly measured. Transparency is often a marketing requirement since it is attractive to consumers. However, food requirements may demand opacity since light causes lipid oxidation and degradation of colorants and vitamins.

The above characteristics are very important in packaging design but not as critical as the material barrier properties. Since permeation to gases will determine the initial atmospheric composition and exchange rate throughout a product's shelf-life, the exchange of gases through container walls as defined by the permeability of materials must be selected according to food consumption or generation of gases. For instance, fresh vegetables and fruits breathe during the distribution period, so they consume O_2 and generate CO_2. The package should allow the entrance of O_2 and the exit of CO_2 to avoid anaerobic respiration and CO_2 damage. Other food products do not exchange significant amounts of gases and therefore a low permeable material is adequate for a long shelf-life.

Packaging Foods with Low/Null Exchange Gases

Those food products that do not develop any important biological activity (dehydrated foods, snacks, nuts, meats) should be packaged in modified atmospheres to limit chemical deterioration such as the oxidation of lipids, pigments, vitamins, or textural changes by humidity. The atmospheric composition for such products should be that which minimizes all chemical activities. This requires a package designed to limit gas exchange with the external atmosphere. The best package would be impermeable, which means it cannot be made of plastic materials. However, a very low permeable container is adequate to maintain product quality during most required shelf-lives.

An adequate container for MAP of non-exchanging gas products must be constituted by materials offering a high barrier to O_2 and CO_2 based on food characteristics. It is thus essential to know the minimum and maximum partial pressure values of each atmospheric component necessary to maintain product quality. In the case of a product packaged in an atmosphere poor in O_2, the initial composition will define the minimum partial pressure (p_i) and the shelf-life of the product (t) will expire when the atmosphere reaches the maximum permissible partial pressure (final pressure p_f). Assuming that the size and shape of the container have been given, the dimensions (volume V and surface area A) can also be determined. Therefore, an adequate structure must have a maximum permeance ($\wp$) of:

$$\wp = \frac{P}{\ell} = \frac{(p_f - p_i)V}{0.21At} \qquad (21\text{-}1)$$

where P is the permeability of the material to the gas and ℓ is the film thickness. Equation 21-1 also shows that the selected structure must either have a thin high barrier or a thick medium barrier.

There are a few plastic materials that can provide a high barrier to gases (Catalá and Gavara, 1996). Vinylidene chloride copolymers (PVdC) and ethylene-vinyl alcohol copolymers (EVOH) are the most commonly used. Vinylidene chloride copolymers make an excellent material from a packaging point of view since they are thermosealable and offer a high barrier to all gases and vapors irrespective of atmospheric conditions. In addition, a high barrier pouch can be manufactured with a single PVdC layer. However, since PVdC is used in industrial practice in complex structures obtained by lamination or coating, this presents environmental problems due to difficulties with recycling and emission of toxic products during incineration, which in turn has limited the presence of PVdC in the packaging market.

Polyvinyl alcohol (PVOH) provides the highest polymeric barrier to gases when dry, although it is difficult to process and extremely sensitive to humidity. As an alternative, EVOH improves processability and reduces water sensitivity, but is not such a good barrier to gases and cannot be thermosealed. Consequently, EVOH is always used as part of a multilayer structure obtained by coextrusion in which EVOH is sandwiched for water protection. Ethylene-vinyl alcohol copolymer structures can also be thermoformed, which makes them the most used option for high barrier trays. Other materials that provide medium to high barriers are some polyamides (PA) and polyacrylonitrile (PAN) that can be used in MAP when the product is not very susceptible to gas variation.

The use of non-polymeric materials in conjunction with plastics can be a means of obtaining high levels of impermeability. Traditionally, high barrier flexible containers were constructed with aluminum foil, but these were eventually substituted with metallized films which provide a high barrier with a large reduction of raw materials so that the finished product is recyclable without previous layer separation. However, opacity limits the effectiveness of both foil and metallized films, so aluminum oxide and silicon oxide have been developed (Catala and Gavara, 1996). Finished films with these new materials are not only transparent but also able to maintain an excellent barrier.

In practice, high barrier containers are made of multilayer structures in which each layer contributes to specific requirements of the packaging. These structures can be very complex, although the basic scheme is (Catalá and Gavara, 1996):

- An external layer responsible for structural and printing properties (polyamide, polyesters, polystyrene, polypropylene or polyethylene)
- A central layer providing the barrier (polymeric materials such as EVOH, PVOH, PVdC, or non-polymeric such as coating with aluminum or metal oxides)
- An internal layer allowing thermosealing (polyolefins)

Packaging Foods that Exchange Gases

A number of food products maintain significant biological activity during distribution and storage. Examples are fresh fruits and vegetables which breathe after harvesting by consuming O_2 and generating CO_2. By breathing, fresh products continue ripening until they naturally decay. A prolonged shelf-life can be obtained by reducing the food respiration rate or achieving low O_2 and high CO_2 partial pressures. An adequate container for these products thus needs to be permeable to gases to avoid total O_2 consumption and anaerobic respiration.

When a fresh food product is packaged in a permeable plastic container a gas exchange between internal and external atmospheres is established. The O_2 concentration at any time ($[O_2]_t$, Eq. 21-2) is a function of the O_2 volumes permeated and breathed plus the total volume of the package:

$$[O_2]_t = [O_2]_{initial} + \frac{V(O_2)_{permeated} - V(O_2)_{breathed}}{V_{package}} \qquad (21\text{-}2)$$

In a similar way, the concentration of CO_2 can be expressed by:

$$[CO_2]_t = [CO_2]_{initial} + \frac{V(CO_2)_{breathed} - V(CO_2)_{permeated}}{V_{package}} \qquad (21\text{-}3)$$

During the breathing process, each O_2 molecule consumed produces a CO_2 molecule. Therefore, an ideal container should allow the entrance/exit of O_2/CO_2 at the same rate they are consumed/produced although this is not possible because containers are either not permeable enough or too permeable. In the first case the composition of the internal atmosphere will change towards low O_2 content, whereas in the second case the O_2 partial pressure builds up with time.

Among those packaging materials used for the MAP of fresh produce, flexible polyvinyl chloride (PVC) and oriented polypropylene (OPP) have the largest shares. The thickness of these plastic films is selected depending on the respiration rate of the product so that the higher the respiration rates, the thinner the film. Polyvinyl chloride is used to wrap the product itself or as wrapping for the rigid trays containing the product. Oriented polypropylene films are used in pouches with different manufacturing technologies such as vertical or horizontal form-fill-seal machines or flow-pack equipment.

Besides the common PVC and OPP, polyethylene is also used for pouches. Rigid trays that can be wrapped or contained in a bag are made of polystyrene (PS) and polyethylene terephthalate (PET) (as well as rigid PVC), and sometimes these trays are closed with a rigid lid made of the same polymer. However, since these types of trays often provide too much of a barrier, pinholes are made on their walls to increase gas exchange.

Low Permeable Non-porous Containers

Most of the packaging materials used in MAP present a lower gas exchange than the respiration rate of the contained product. The O_2 level in a package's headspace decreases with time while the level of CO_2 increases, and the respiration rate of the product decreases until a stationary state is achieved. Product shelf-life is prolonged when this pseudo-equilibrium is sufficient to maintain product freshness. Once the O_2 concentration becomes too low (or an excess of CO_2 accumulates), the product will quickly deteriorate if the container is inappropriate for that particular product. Figure 21-1 shows the internal atmosphere evolution of a commercial polypropylene pouch containing cut romaine lettuce. Initially packaged in air, the low permeation rate of the film and the high respiration rate of the product result in a depression of O_2 partial pressure and an increment of CO_2.

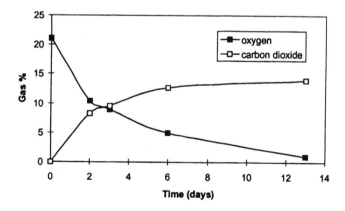

Figure 21-1 Internal atmosphere composition of a polypropylene pouch containing 250 g of cut romaine lettuce.

High Permeable Non-porous Containers

When a product is packaged in a highly permeable container, the gases consumed and produced by product respiration may be inferior to the permeation rate through the package walls in that the internal atmosphere will approach air composition with time. In these cases the product is initially packaged with an atmosphere rich in CO_2 and poor in O_2. Product deterioration will thus consist only of ripening since anaerobic respiration and CO_2 damage are not possible. These types of high permeable non-porous containers are very useful in products with high respiration rates for which a low permeable container will result in a short shelf-life.

It is also important to note that permeabilities to O_2 and CO_2 through common plastics are not equal. Normally, the permeability to CO_2 is four to five times higher than that to O_2 (Anonymous, 1995). According to the definition of permeability (P), the transmission rate of O_2 and CO_2 through a pouch type container with a simple film of thickness ℓ and surface area A would be:

$$P(CO_2) = K.P(O_2)$$
$$\frac{v(O_2)}{t} = \frac{P(O_2).A.\Delta p(O_2)}{\ell} \tag{21-4}$$
$$\frac{v(CO_2)}{t} = \frac{K.P(O_2).A.\Delta p(CO_2)}{\ell}$$

Equation 21-4 shows that the entrance of O_2 and exit of CO_2 would be equal only when the ratio of partial pressure gradients is inverse to the ratio of permeabilities. Therefore, a MAP container accomplishing this is basically theoretical because containers cannot yet be designed so precisely. However, newly developed materials have shown reduced permeability differences with respect to O_2 and CO_2. Some biodegradable/edible films actually present $P(CO_2)/P(O_2) \sim 2.5$ (Gontard et al., 1996; McHugh and Krochta, 1994). Extensive research is being carried out on these materials with the intention of obtaining films with adequate mechanical characteristics to provide adequate food protection. So far edible materials are exclusively used as coatings.

New synthetic materials have also been developed for MAP applications. The most interesting are based on rubber and metallocene technologies. Butadiene-styrene (BS) copolymers have been used for many years as substitutes for natural rubber, but recently these copolymers have been modified to allow film forming. Varying the composition of the co-monomers, the resulting polymer can be either flexible (rich in butadiene) or rigid (rich in styrene). But probably the most interesting characteristic of these rubbers is that they are more permeable to O_2 than CO_2 (Anonymous, 1995). By lamination or co-extrusion of BS with common plastics, the ratio $P(CO_2)/P(O_2)$ can be designed for specific applications.

Metallocene polyethylene and polypropylene with improved characteristics have already entered the packaging market (Dixon, 1997) because they allow the design of polymers with specific properties. Metallocene catalyzed polymers are being developed at an especially fast rate since they present low densities, improved thermal resistance, and high hot-tack. It is difficult to say how far this technology can go, but the next attempts will most likely be focused on the substitution of flexible and rigid PVC which are being questioned due to environmental reasons. Although not much information has been published, these interpolymers would theoretically have varying flexibility, good thermal resistance, and adequate barrier properties.

High Permeable Porous Containers

Some food products breathe at such a high rate that there is no plastic material that could supply enough O_2 by molecular permeation to avoid anaerobic respiration or CO_2 damage. A popular solution in such cases is to perforate the film to increase the gas exchange between the external and internal environments. Permeation through pores differs in that it occurs through non-porous plastics by molecular transport via eventual itineraries with the collaboration of the plastic. Therefore, polymer characteristics as well as permeants affect the transport process itself. Transport through pores occurs through permanent itineraries (holes), although this process is affected by the permeant and dimensions of the pores irrespective of the packaging material, because the film is a passive substrate. According to Knudsen (Hernandez, 1997), flux (J) through a pore depends on pore dimensions (d diameter and ℓ length), pressure gradient, and the thermal energy of the permeant (speed v), which are all magnitudes in the CGS unit system:

$$J = 7e - 5\frac{d^3.v.\Delta p}{\ell.T} \tag{21-5}$$

where

$$v = \sqrt{\frac{3RT}{M}} \tag{21-6}$$

As Equations 21-5 and 21-6 show, the only parameter related to packaging material is thickness since the permeant affects the process through its molecular weight. However, because the molecular weight of O_2 is similar to that of CO_2, the permeabilities of porous films to these gases are very close.

It is also important to note that permeation through a porous film is much higher than through a plastic film. For instance, the O_2 flow through an OPP pouch (25 μm thick, 300 cm^2 surface area) with an initially anaerobic internal atmosphere is about 40 cm^3/day, while the flow through a 25 μm pore is about 1,250 cm^3/day, which demonstrates that when porous permeation is present, molecular diffusion can be neglected.

With the development of laser technology, microporous plastic materials can be manufactured with the required permeability and fixed film thickness to suit a specific product. Higher permeabilities are obtained by increasing pore diameters or the number of pores. By microperforation (1 pore/cm^2, 10 μm diameter) of the above described OPP pouch, the entrance of O$_2$ can be changed from 40 to 24,000 cm^3/day. Applications of this technology are becoming very popular for the packaging of products with high respiration rates such as corn on the cob, strawberries, cauliflower, and mushrooms (Morales-Castro et al., 1994; Burton, 1988).

Active Packaging

Although great advances in packaging design have been made with regard to plastic materials, the newest concept in packaging science is not related to mechanical materials or barrier properties but the retention (scavengers) or addition of (emitters) substances related to product deterioration that are produced or consumed during shelf-life. Because this type of packaging does not suffer passive product interaction, it is called active packaging. Such systems are obtained either by 1) adding the active product in a pouch inside the container or 2) adding the active product to the polymer during the transformation procedure. The other types of active packaging (Church, 1994) as applied to MAP are related to O$_2$ elimination (scavengers), CO$_2$ or ethylene control (scavengers or emitters), ethanol emitters, and humidity control (desiccants).

REFERENCES

Anonymous (1995). *Permeability and Other Film Properties of Plastics*. PDL Handbook Series. New York: Plastics Design Library.

Barth M.M., Kerbel E.L., Broussard S., and Schmidt S.J. (1993). Modified atmosphere packaging protects market quality in broccoli spears under ambient temperature storage. *J. Food Sci.* 58: 1070–1072.

Black R.G., Quail K.J., Reyes V., Kuzyk M., and Ruddick L. (1993). Shelf-life extension of pita bread by modified atmosphere packaging. *Food Aust.* 45: 387–391.

Brecht J.K., Sargent S.A., Bartz J.A., Chau K.V., and Emond J.P. (1992). Irradiation plus modified atmosphere for storage of strawberries. *Proc. Fla. State Hortic. Soc.* 105: 97–100.

Brown W. (1922). On the germination and growth of fungi at various temperatures and in various concentrations of oxygen and carbon dioxide. *Ann. Bot.* 36:257–283.

Buick R.K. and Damoglou A.P. (1989) Effect of modified atmosphere packaging on the microbial development and visible shelf life of a mayonnaise-based vegetable salad. *J. Sci. Food Agric.* 46: 339–347.

Burton K.S. (1988). Modified atmosphere packaging: A new technology for extending mushroom storage-life. *Mushroom J.* 183: 510–507.

Castelvetri F. (1991). Pasteurization, vacuum, water activity, pH and temperature for improvement of the quality of pasta. *Tec. Molitoria* 42: 952–956.

Catala R. and Gavara R. (1996). Review: Alternative high barrier polymers for food packaging. *Food Sci. Technol. Int.* 2: 281–291.

Chambroy Y., Guinebretiere M.H., Jacquemin G., Reich M., Breuils L., and Souty M. (1993). Effects of carbon dioxide on shelf-life and post harvest decay of strawberries fruit. *Sci. Aliments.* 13: 409–423.

Church N. (1994). Developments in modified-atmosphere packaging and related technologies. *Trends Food Sci. Technol.* 5: 345–352.

Coyne F.P. (1932). The effect of carbon dioxide on bacterial growth with special reference to the preservation of fish. *J. Soc. Chem. Ind.* 51: 119T–121T.

Coyne F.P. (1933). The effect of carbon dioxide on bacterial growth. *Proc. Roy. Soc. Ser. B* 113: 196–217.

Dixon J. (1997). Assessing future trends in Europe. *Can. Packag.* 50: 53–54.

Evelo R.G. and Boerrigter H.A.M. (1994). Integral approach for optimizing modified atmosphere packages for Elstar apples: A combination of product and packaging constraints. *Packag. Technol. Sci.* 7: 195–204.

Feng-Sheng W., Yang-Nian J., and Chin-Wen L. (1995). Lipid and cholesterol oxidation in Chinese-style sausage using vacuum and modified atmosphere packaging. *Meat Sci.* 40: 93–101.

Geeson J.D., Genge P.M., and Sharples R.O. (1994). The application of polymeric film lining systems for modified atmosphere box packaging of English apples. *Postharv. Biol. Technol.* 4: 35–48.

Gerdes D.L. and Parrino L.V. (1995). Modified atmosphere packaging (MAP) of Fuerte avocado halves. *Lebensm-Wiss Technol.* 28: 12–16.

Giavedoni P., Roedel W., and Dresel J. (1994). Study on the stability of Italian-filled pasta packaged in modified atmosphere and in an ethanol vapour pressure environment. *Fleischwirtschaft* 74: 639–640, 643–646.

Gontard N., Thibault R., Cuq B., and Guilbert S. (1996). Influence of relative humidity and film composition on oxygen and carbon dioxide permeabilities of edible films. *J. Agric. Food Chem.* 44: 1064–1069.

Gran C.D. and Beaudry R.M. (1993). Determination of the low oxygen limit for several commercial apple cultivars by respiratory quotient breakpoint. *Postharv. Biol. Technol.* 3: 259–267.

Grandison A.S. and Jennings A. (1993). Extension of the shelf life of fresh minced chicken meat by electron beam irradiation combined with modified atmosphere packaging. *Food Control* 4: 83–88.

Haines R.B. (1933). The influence of carbon dioxide preservation on the rate of multiplication of certain bacteria as judged by viable counts. *J. Soc. Chem. Ind.* 52: 13T–17T.

Hernandez R.J. (1997). Food packaging materials, barrier properties, and selection. In: *Handbook of Food Engineering Practice.* Valentas K.J., Rotstein E., and Singh R.P., Eds. Boca Raton: CRC Press, pp. 291–360.

Hintlian C. and Hotchkiss J. (1986). The safety of modified atmosphere packaging. A review. *Food Technol.* 40(1): 70–76.

Kidd F. and West C. (1927). *Gas Storage of Fruit.* Food Investigation Special Report No. 30. Department of Science and Industrial Research.

Killefer D.H. (1930). Carbon dioxide preservation of meat and fish. *Ind. Eng. Chem.* 22: 140–143.

Kwang S.L., In S.P., and Dong S.L. (1996) Modified atmosphere packaging of a mixed prepared vegetable salad dish. *Int. J. Food Sci. Technol.* 31: 7–13.

Lopez-Galvez G., Peiser G., Xunli N., and Cantwell M. (1997). Quality changes in packaged salad products during storage. *Z. Lebensm-Unters Forsch* 205: 64–72

Mannheim C.H. and Soffer T. (1996). Shelf-life extension of cottage cheese by modified atmosphere packaging. *Lebensm-Wiss Technol.* 29: 767–771.

Marshall D.L., Andrews L.S., Wells J.H., and Farr A.J. (1992). Influence of modified atmosphere packaging on the competitive growth of *Listeria monocytogenes* and *Pseudomonas fluorescens* on precooked chicken. *Food Microbiol.* 9: 303–309.

McHugh T.H. and Krochta J.M. (1994). Permeability properties of edible films. In: *Edible Coatings and Films to Improve Food Quality*. Krochta J.M., Baldwin E.A., and Nisperos-Carriedo M., Eds. Lancaster, PA: Technomic Publishing, pp. 139–188.

Meir S., Naiman D., Akerman M., Hyman J.Y., Zauberman G., and Fuchs Y. (1997). Prolonged storage of Hass avocado fruit using modified atmosphere packaging. *Postharv. Biol. Technol.* 12:51–60.

Moral-Rama A. (1993). Refrigeration and preservation of fish packaged under modified atmosphere for retail sale. *Aliment. Equip. Tecnol.* 12: 101–104.

Morales-Castro J., Rao M.A., Hotchkiss J.H., and Downing D.L. (1994) Modified atmosphere packaging of sweet corn on cob. *J. Food Proc. Pres.* 18: 279–293.

Nychas G.J.E. and Tassou C.C. (1996). Growth/survival of *Salmonella enteritidis* on fresh poultry and fish stored under vacuum or modified atmosphere. *Lett. Appl. Microbiol.* 23: 115–119.

Oka S., Fukunaga K., Ito H., and Takama K. (1993). Growth of histamine producing bacteria in fish fillets under modified atmospheres. *Hokkaido Daigaku Suisangakubu Kenkyu Iho* 44: 46–54.

Paleari M.A., Soncini G., and Beretta G. (1987). Observations on modified atmosphere packaging of meat products. *Ind. Aliment.* 26: 1003–1008.

Pedrelli T., Vicini L., Spotti E., Mutti P., Bianco S., and Dall'Aglio G. (1994). Extension of shelf-life of bakery products packed under modified atmosphere. *Ind. Aliment.* 33: 988–995.

Piergiovanni L., Fava P., and Moro M. (1993). Shelf-life extension of Taleggio cheese by modified atmosphere packaging. *Ital. J. Food Sci.* 5: 115–127.

Piergiovanni L. and Fava P. (1997). Minimizing the residual oxygen in modified atmosphere packaging of bakery products. *Food Add. Contam.* 14: 765–773.

Randell K., Ahvenainen R., and Hattula T. (1995). Effect of the gas/product ratio and CO_2 concentration on the shelf-life of MA packed fish. *Packag. Technol. Sci.* 8: 205–218.

Renault P., Houal L., Jacquemin G., and Chambroy Y. (1994). Gas exchange in modified atmosphere packaging. II. Experimental results with strawberries. *Int. J. Food Sci. Technol.* 29: 379–394.

Rodriguez-Castilla M.V. and Jordano R. (1997). Envasado de pan y productos de panadería. I. Composición de la atmosfera de envasado, films, equipos y efectos en el pan y productos de panadería. *Aliment.* 285: 79–89.

Sarantopoulos C.I.G.L., Alves R.M.V., and Mori E.E.M. (1995). Effects of modified atmosphere packaging on the preservation of grated parmesan cheese. *Colet. Inst. Tecnol. Aliment.* 25: 67–79.

Sawaya W.N., Elnawawy A.S., Abu-Ruwaida A.S., Khalafawi S., and Dashti B. (1995). Influence of modified atmosphere packaging on shelf-life of chicken carcasses under refrigerated storage conditions. *J. Food Safety* 15: 35–51.

Soffer T., Margalith P., and Mannheim C.H. (1994). Shelf-life of chicken liver/egg pate in modified atmosphere packages. *Int. J. Food Sci. Technol.* 29: 161–166.

Sorheim O., Aune T., and Nesbakken T. (1997). Technological, hygienic and toxicological aspects of carbon monoxide used in modified-atmosphere packaging of meat. *Trends Food Sci. Technol.* 8: 307–312.

Talasila P.C., Cameron A.C., and Joles D.W. (1994). Frequency distribution of steady-state oxygen partial pressures in modified-atmosphere packages of cut broccoli. *J. Amer. Soc. Hortic. Sci.* 119: 556–562.

Vercelino-Alves R.M., de-Luca G., Sarantopoulos C.I., Gimenes-Fernandes-van D.A., and Assis-Fonseca F.J. (1996) Stability of sliced mozzarella cheese in modified-atmosphere packaging. *J. Food Prot.* 59: 838–844.

Viallon C., Berdague J.L., Montel M.C., Talon R., Martin J.F., Kondjoyan N., and Denoyer C. (1996). The effect of stage of ripening and packaging on volatile content and flavor of dry sausage. *Food Res. Int.* 29: 667–674.

CHAPTER 22

Gelation of Soybean Proteins at Acidic pH

M.C. Puppo and M.C. Añón

INTRODUCTION

Gelation is a very important property of several proteins, and plays a major role in the preparation of many foods. Gelation is utilized not only for the formation of solid viscoelastic gels, but improved thickening, adhesion, water capacity, emulsion, and foam-stabilizing effects. The gelation process, which is generally achieved by heating, requires a protein that unfolds to expose its functional groups. However, gelation also requires the formation of a three-dimensional matrix of partially associated polypeptides. It is therefore a complex process involving denaturation, dissociation-association, and aggregation reactions (Hermansson, 1986; Damodaran, 1989).

The formation of a protein network is regarded as the result of a balance established by protein-protein and protein-water interactions and the repulsion-attraction forces acting between adjacent polypeptidic chains (Cheftel et al., 1993). The characteristics of the gel formed depend on the molecular properties of the proteins in an unfolded state and process conditions such as protein concentration, cooling regime, pH, and ionic strength of the medium and type of ions present in the solution (Damodaran, 1988).

Various studies have shown that gel properties such as viscoelasticity, texture, and water holding capacity are closely related to gel microstructure (Furukawa and Ohta, 1982; Foegeding et al., 1995; Chronakis, 1996). Soybean proteins contain β-conglycinin (7 S fraction) and glycinin (11 S fraction), two major globulins that have different structures (Wolf et al., 1961; Brooks and Morr, 1985), thermal behavior (German et al., 1982; Damodaran 1988), and gel forming properties (Hermansson, 1986; Morr, 1990). The effects of salts, reducing agents, and denaturants have been studied to explain the nature of the molecular forces involved in the heat-induced gelation of soybean proteins (Shimada and Matsushita, 1981; Utsumi and Kinsella, 1985a).

Soy protein isolates are usually employed as functional food ingredients. Their gelation requirements are different from that of their respective principal protein fractions, as are the characteristics of the gels obtained (Utsumi and Kinsella, 1985a; German et al., 1982; Damodaran and Kinsella, 1982; Utsumi et al., 1984). Both protein isolates and mixtures having equal proportions of glycinin and β-conglycinin show better gelation characteristics than their constituent parts (Babajimopoulos et al., 1983).

A thorough understanding of the effect of conformational changes on functional properties is required to optimize the use of soy protein isolates as functional ingredients in foods. Given that proteins experience structural changes both during extraction and thermal gelation, the degree of denaturation and nature of the exposed functional groups can affect the extent to which their matrices are formed, as well as the physical characteristics of the gels.

The functional properties of soybean proteins have been widely studied at pHs higher than the isoelectric point (pI) (Shimada and Cheftel, 1988; Utsumi and Kinsella, 1985a, b), but the information available in the acidic pH range below the pI is scarce. This study therefore concentrated on the effect of acidic pH on the conformational state of soybean proteins and the physical, rheological, and structural properties of gels.

PROTEIN CONFORMATION AT ACIDIC PH

Treatment of soy protein isolates at acidic pHs below the pI (2.5 to 3.5) results in a decrease of thermal stability and gradual increase of acidity (Figs. 22-1a–c). For pHs less than 3.0 the isolates present a high degree of denaturation (approximately 80% with respect to the native pH 8.0 isolate), whereas closer to the pI protein denaturation is less intense (about 35% with respect to the native isolate). Decrease of the transition temperature (Td) and denaturation enthalpy (ΔH) indicate a partial conformational change in the 7 S and 11 S globulins. Hermansson (1978) observed similar results in pH 2.0 and 3.0 soy protein isolates, as well as and other proteins such as oats (Harwalkar and Ma, 1987) and broad beans (Arntfield and Murray, 1981).

In this study the changes detected in the isolates reflected the transformations underwent by the β-conglycinin and glycinin. DSC thermograms (Fig. 22-1) corresponding to the 7 S fraction showed a shift of the denaturation endotherm toward lower temperatures and a decrease of the area (ΔH) as the globulin preparation pH decreased. On the pH 3.5 glycinin, the narrow and symmetric peak detected at pH 8.0 (Fig. 22-1a, b) split and shifted toward the lower temperatures as a consequence of the dodecameric 11 S form dissociating into two hexameric forms (endotherm of lower Td) (Peng et al., 1984). At pH 3.0 a sole endotherm was detected which belonged to the undissociated hexameric form. In turn, these hexameric forms dissociated to 3 S forms which did not exhibit cooperative transitions as indicated by the absence of endotherms in the pH 2.75 sample endotherm (Fig. 22-1c).

The preceding results allowed the determination of the contribution of each fraction (7 S and 11 S) to the acidic soy protein isolate thermogram. In the pH 3.5 sample, the first part of the thermogram (temperatures below 80°C) was found to correspond to both the denaturation of the β-conglycinin and hexameric forms of glycinin, whereas the second part (temperatures above 80°C) was ascribed only to the 11 S fraction. At pH 2.75 the glycinin was found totally dissociated and so did not represent a thermal transition, but the endotherm observed at that pH in the protein isolate would therefore correspond to the denaturation of the β-conglycinin subunits. These results also indicate that the 7 S fraction was more resistant to denaturation than the glycinin.

Structural Properties of Gels

Figure 22-2 shows gels (10% w/w) prepared at acidic pH. The pH 2.5 gel is the most transparent and at the same time the least firm. The gels become more opaque and rigid as the protein and pH approach the gel isoelectric point.

The pH 3.5 gel exhibited a more particulate, less homogeneous, and more open structure than the more acidic gel (Fig. 22-3a, b). The protein molecules were positively charged at pH 2.75 due to the predominating repulsion forces which prevented aggregation and the formation of fine polypeptide chains (Stading and Hermansson, 1991; Heertje and Van Kleef, 1986). The alkaline gel matrix was similar to that of the more acidic gel (Fig. 22-3c).

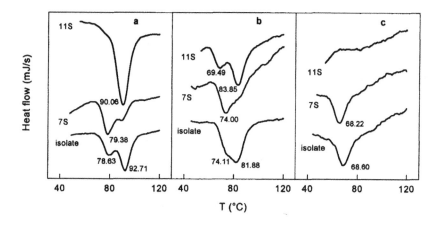

Figure 22-1 DSC thermograms of soybean protein isolates 7 S and 11 S fractions of: (a) pH 8.0, (b) pH 3.5, and (c) pH 2.75.

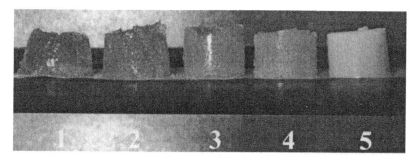

Figure 22-2 Soybean protein gels (10% w/w) at: (a) pH 2.5, (b) pH 2.75, (c) 3.0, 3.25, and (d) 3.5.

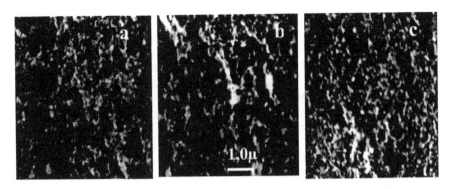

Figure 22-3 SEM of soybean protein gels (10% w/w) of at: (a) pH 2.75, (b) pH 3.50, and (c) pH 8.0 (c). Magnification is 12,000×.

Solubilization tests of acidic pH gels in distilled water (DW), Tris-glycine pH 8.0 μ 0.1 buffer (B), and Tris-glycine-SDS-urea pH 8.0 buffer (BSU) have provided useful data for identifying the nature of the bonds involved in the formation and stabilization of protein matrices. In this study the solubility in DW was maximum (about 60–70%) for the more acidic gel, but decreased strongly as the pH was increased from 2.5 to 3.5 (Fig. 22-4). The observed trend was as expected since the net charge decreased as the pH got closer to the pI, which favored protein-protein interaction. The presence of 30–40% soluble protein at pH 2.5 suggested that the existence of a high amount of molecules would be weakly linked to the gel matrix. The solubility of all the gels in buffer B was about 20% regardless of their pH, suggesting a low contribution of electrostatic forces during their formation and stabilization. With the BSU, solubility was almost total at acidic pH and about 80% at pH 8.0. As this buffer possessed denaturing agents which disturbed the hydrogen bond-type and hydrophobic linkages, the results revealed a strong contribution of non-covalent type bonds and absence of disulfide bonds (unlike at pH 8.0) (Shimada and Cheftel, 1988) during the formation and stabilization of the protein matrices in the acidic gels. The increase in protein concentration from 10 to 14% w/w decreased the solubility of the acidic pH gels in the DW and buffer B, particularly at pH 3.50.

The SDS-PAGE electrophoretic analysis carried out in the presence and absence of the β-mercaptoethanol for the DW, B, and BSU soluble fractions provided evidence to identify the proteins involved in protein network formation and those not linked or weakly linked to the matrix. In the acidic gels, the DW soluble fraction consisted mainly of an AB glycinin subunit and the 7 S fraction-β subunit, as well as minor amounts of A and B-11 S polypeptides. Dimeric forms of the B polypeptide and low molecular mass (less than 20 kDa) peptides were also observed (Fig. 22-5d, f). The DW extract of the pH 8.0 gel presented high amounts of the A-11 S polypeptides in the monomeric form and polymers stabilized by disulfide bonds. Similar results were obtained by

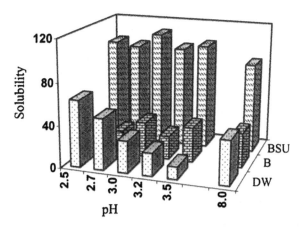

Figure 22-4 Solubility of soybean protein gels at different pHs. Distilled water, DW, buffer pH 8.0, B, and buffer pH 8.0–SDS–urea, BSU .

Yamauchi et al. (1991). Also detected were very small quantities of the subunits that constituted the 7 S fraction (Fig. 22-5a, b). The BSU-soluble proteins of all the acidic pH gels were equivalent. The presence of soluble aggregates composed of α and α'-7 S subunits were detected, as was that of β-conglycinin subunits, the AB subunit and the A and B-11 S polypeptides (Fig. 22-5c, e). In the pH 8.0 gel, the BSU-soluble fraction possessed a greater proportion of high molecular mass aggregates, which in turn were composed of α and α'-7 S subunits and minor amounts of A and B-11 S polypeptides, 7 S fraction subunits, and monomeric and polymerized B polypeptides. These results agreed with those reported by Utsumi et al. (1984).

Based on these findings, it was concluded that the acidic pH gel matrix basically consisted of the proteins making up the 7 S fraction, whereas the 11 S fraction (AB subunits) and the β-7 S subunits were the principal components of the pH 8.0 gel structure.

Hydration and Rheological Properties of Gels

The structural differences discussed previously for acidic pH gels were expected to be reflected in their hydration, viscoelastic, and textural properties. Acidic pH gels possess a high water holding capacity (WHC) that decreases as the pH increases from 2.5 to 3.5. Accordingly, the WHC of the pH 8.0 gels in this study was similar to that of the more acidic gels (Table 22-1). Values of pH around the protein pI value favor protein-protein and decrease protein-water interactions as a consequence of diminished electrostatic repulsions and a lower degree of functional group exposure. These groups have the capacity to retain water in multilayers by hydrogen bond interactions, which become stronger during the gel cooling stage. An increase in gel protein concentration from 10 to 14% w/w decreases the WHC of gels whose pH is close to the pI (Table 22-1).

The gelation process consists of two stages: the first is heating that promotes denaturation and interactions among protein molecules, and the second is cooling. This study concentrated on the viscoelastic behavior of acidic gels by oscillatory measurements carried out during the first gelation stage, as well as scans for deformation, frequency, time, and temperature. The deformation analysis provided data to establish the linear viscoelasticity range used in the remaining test (1 Hz and 10% deformation).

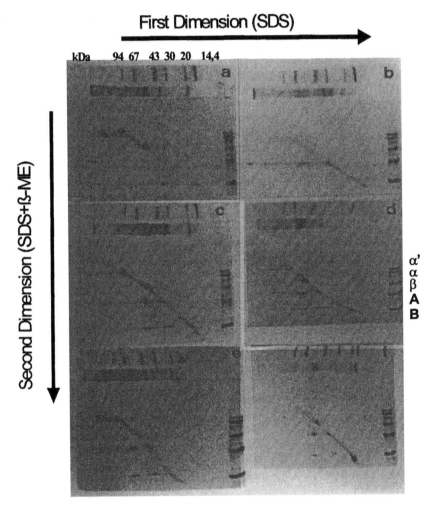

Figure 22-5 SDS-PAGE of DW (**a, c, e**) and BSU (**b, d, f**) extracts of soybean protein gels at pH 2.75 (**a, b**), pH 3.5 (**c, d**), and pH 8.0 (**e, f**).

Table 22-1 Water holding capacity (WHC) of soybean protein gels.

pH	2.5	2.75	3.0	3.25	3.5	8.0
10% w/w (protein/water)	98.59	98.05	95.59	92.2	85.95	98.29
14% w/w (protein/water)	—	99.39	—	—	85.18	—

A series of experiments were done to determine the gelation temperature of 10% w/w protein dispersions at different pHs. The elastic (G') and viscous (G") moduli values were close to zero at 60°C in both the acidic pH and pH 8.0 dispersions (Fig. 22-6). The dispersions become more elastic as the temperature increased. For acidic pH, the pronounced increase of G' occurring from 60°C on coincided with the denaturation temperature of the 7 S fraction (71.3 ± 1.4) and 11 S hexameric form (66.5 ± 0.5). The increase of G' that took place from 70°C on was less pronounced at pH 8.0, which was concurrent with the denaturation value of the 7 S fraction (78.6 ± 0.3).

The dynamic spectra (G', G", and η* as a function of the oscillation frequency) of pH 2.75, pH 3.5, and pH 8.0 dispersions indicated a viscoelastic liquid behavior that followed the Maxwell model (Fig. 22-7a–c). According to Giboreau et al. (1994), the three samples behaved as semi-diluted macromolecular dispersions whose elasticity increased with increasing oscillation frequencies. Clark and Ross-Murphy (1987) and Ross-Murphy (1987, 1995) have classified these sample types as structured network systems. Such systems can be described as "highly elastic solutions" that show the predominance of the association phenomenon between ordered protein chain segments (Cuvelier and Launay, 1986) where non-covalent interactions (particularly of the hydrophobic type) may be present (Baird, 1981).

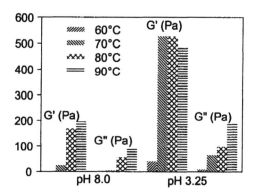

Figure 22-6 G' and G" (1 Hz) values of pH 8.0 and 3.25 dispersions (10% w/w) heated at different temperatures.

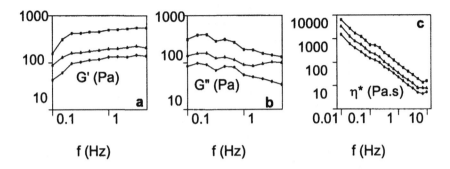

Figure 22-7 Variation of: (a) G', (b) G", and (c) η* with oscillation frequency of soybean protein dispersions (10% w/w) at pH 2.75 (●), pH 3.5 (■), and pH 8.0 (▲).

According to Cooney et al. (1993), viscoelasticity is represented by the tangent of the deformation angle (tan δ = G"/ G'). The dispersions studied gave tan δ values less than 0.5, which revealed that these gels were more liquid-like than typical gels prepared from other biopolymers such as agar (Nishinari, 1991), gelatin (Nishinari et al., 1991), or ovoalbumin (Van Kleef, 1986). It must be recalled that tan δ values may decrease after the gel cooling period because already existing bonds linking polypeptidic chains become stronger and new bonds form, producing behavior that becomes more similar to that of a solid. The tan δ increases with the proportion of the disperse phase in filled gelatin gels (Kohyama and Nishinari, 1992). The high tan δ values found in this case would also be a consequence of the presence of proteins that are unlinked or weakly linked to the matrix, as discussed previously.

The increase in protein content from 10 to 14% w/w in the acidic and alkaline dispersions resulted in the formation of a viscoelastic solid obeying the Kelvin-Voigt model. The elastic modulus was higher than the viscous, and the values of the former were constant over the entire frequency range tested (Fig. 22-8). Chronakis (1996) obtained similar results in soy dispersions at 20 and 24% w/w by Baird (1981), and at 14 and 20% w/w. This behavior was also reflected in the calculated tan δ values that varied from 0.2 to 0.3 (i.e., below those obtained for the 10% w/w protein dispersions). Despite the fact that these values belong to the first gelation stage (heating), they clearly indicate that increased protein concentrations favor interactions between polypeptidic chains, and therefore the formation of a three-dimensional network resembling a more solid structure.

Figure 22-9 shows the gelation curves at 90°C of 10% w/w acidic soy protein isolates at 2.75, 3.5, and 8.0 pH. It can also be observed that the elastic modulus increases gradually and continuously with the heating time, following a first-order kinetics. A similar behavior was observed for 7 S and 11 S fractions heated at 80°C (Nishinari et al., 1991; Yoshida et al., 1992; Nagano et al., 1994). The G' values reached by the dispersions prepared at extreme pH (2.75 and 8.0) were below those of the dispersion prepared at a pH close to the pI, which revealed important differences in the degree of protein chain interaction. The rate constant values ($k \times 10^4$) calculated by fitting the equation

$$G'(t) = G'sat \, [1 - exp \, (-k \, t)] \tag{22-1}$$

indicated fastest gelation in the pH 3.5 dispersions (k values were 4.08, 7.55, and 7.17 s^{-1} for pH 2.75, 3.5, and 8.0, respectively).

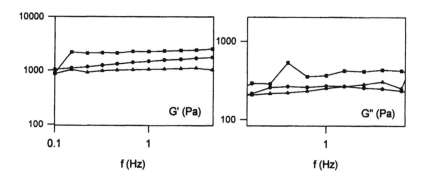

Figure 22-8 Variation of G' and G" with oscillation frequency soybean protein dispersions (14% w/w) at pH 2.75 (●), pH 3.5 (■), and pH 8.0 (▲).

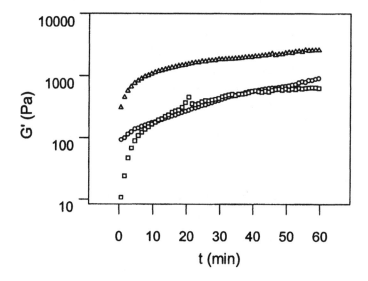

Figure 22-9 Variation of storage moduli (G') with the heating (90°C) time of soybean protein dispersions (10% w/w) at pH 2.75 (O), pH 3.5 (Δ), and pH 8.0 (□).

High temperature treatment favors two opposite phenomena. On the one hand it causes rupture of hydrogen bonds and decrease in electrostatic interactions which weaken molecular association. On the other hand, since it stabilizes the hydrophobic interactions it also favors protein interactions and formation of a three-dimensional structure. The value of G' is the result of this balance (Chronakis, 1996). Accordingly, the results obtained indicate that the thermal treatment applied to both the acidic and alkaline soy dispersions favored molecular interactions, and especially in pH close to the pI. Under these conditions, proteins from the 7 S and 11 S fractions were found partially denatured and dissociated with a net charge close to zero before starting the first gelation stage. Proteins completed their unfolding during treatment at 90°C, and were able to interact with each other from the beginning. Even at pH 2.75 when they were partially (7 S fraction) or totally (11 S fraction) denatured, their high positive charge delayed interactions and reduced their number so that longer times were required to form the three-dimensional network.

Texture tests revealed that the rupture force (F) and hardness (H) in gels obtained at acidic pH were significantly lower than in the pH 8.0 gel. In acidic pH, differences detected are significant only when comparing gels prepared at extreme pHs (Table 22-2). Such variations can be due to the diverse forces responsible for the stabilization of acidic and alkaline gels. The greater hardness of alkaline gels has been assigned to the existence of disulfide bonds in their matrix. The deformability modulus (Ev) exhibited a pronounced decrease between pH 2.5 and 2.75, with a minimum at pH 3.5 (Table 22-2). Concerning the pH 8.0 gel, its resistance to deformation was comparable to that of a more acidic gel. The rupture relative deformability (Dr) displayed a moderate decrease as the pH approached the pI.

The texture profile of the acidic gels changed as the gel protein concentration increased from 10 to 14% w/w. At pH 2.75, a structure was detected that showed independent hardness and fracturability, parameters that increased with protein concentration. Alkaline pH values favored the SS/SH exchange and made the increase of F and H more noticeable (Table 22-2). Similar results were obtained by Kang et al. (1991) for soy protein isolates and Nio et al. (1985) and Utsumi et al. (1982) for 7 S fraction and 11 S fraction gels. The modulus of deformability also got higher with increasing protein concentration, and this raise was more pronounced at pH 3.5. Deformability at rupture decreased in pH 2.75 and pH 8.0 gels as the protein concentration increased. For pH values that were far from the pI, gel behavior was like that observed for increasing protein concentration. It can thus be assumed that non-covalent interactions of equally-charged molecules predominate at such pHs, and that they bear a significant influence on gel texture. Close to the pI, where the protein net charge is near zero, the increase of protein-protein interactions gives rise to gels whose structure is more resistant to deformation and whose relative deformability is higher than that in gels of pH distant from the pI.

Table 22-2 Textural properties of soybean protein gels. (PC: protein concentration, F: fracturability, H: hardness, E_D: deformability modulus, Dr: relative deformability at rupture.)

	pH 2.75		pH 3.5		pH 8.0	
PC (w/w)	10%	14%	10%	14%	10%	14%
F (N)	2.1 ± 0.04	7.2 ± 1.50	3.0 ± 1.00	10.3 ± 1.20	4.7 ± 1.10	23.8 ± 2.40
H (N)	2.1 ± 0.04	4.8 ± 1.52	2.5 ± 1.01	6.5 ± 2.86	4.7 ± 1.10	23.8 ± 2.40
E_D (kPa)	3.1 ± 0.90	4.8 ± 1.49	1.2 ± 0.70	14.9 ± 3.50	2.5 ± 1.30	5.2 ± 0.00
Dr	0.6 ± 0.08	0.3 ± 0.05	0.5 ± 0.07	0.5 ± 0.04	0.4 ± 0.02	0.2 ± 0.03

CONCLUSIONS

Even with equivalent WHC, gels prepared at the most acidic pH (2.75) are more transparent, softer, less sensitive to fracture, more resistant to deformation, and more elastic than alkaline gels (pH 8.0). The former gels present a fine structure consisting of pH-dissociated and totally denatured glycinin and β-conglycinin which are partially denatured during acidic treatment (The denaturation of this last constituent is completed during the heating required for gelation).

Gels prepared at near-pI values of pH were observed to possess a particulate matrix of low WHC which contained the pH-dissociated 11 S fraction in a partial denaturation state, just as the 7 S fraction. The complete loss of native conformation occurred during the heating stage. The α and α'-7 S subunits were the main constituents involved in the formation of the three-dimensional structure of the acidic gels, where the β-7 S and AB-11 S subunits were found in a solubilized state in the gel interstices or weakly linked to the gel matrix. Only non-covalent bonds were responsible for stabilizing the gel matrix. At pH 8.0, fine-structure gels were formed from native proteins which denatured during the thermal treatment. All protein subunits from the 7 S and 11 S fractions participated in the three-dimensional structure, which was established by non-covalent interactions and SS bonds.

REFERENCES

Arntfield S.D. and Murray E.D. (1981). The influence of processing parameters on food protein functionality. I. Differential scanning calorimetry as an indicator of protein denaturation. *Can. Inst. Food Sci. Technol. J.* 14: 289–294.

Babajimopoulos M., Damodaran S., Rizvi S.S.H., and Kinsella J.E. (1983). Effects of various anions on the rheological and gelling behavior of soy proteins: Thermodynamic observations. *J. Agric. Food Chem.* 31: 1270–1275.

Baird D.G. (1981). Dynamic viscoelastic properties of soy isolate doughs. *J. Tex. Stud.* 12: 1–16.

Brooks J.R. and Morr C.V. (1985). Current aspect of soy protein fractionation and nomenclature. *J. Am. Oil Chem. Soc.* 62: 1347–1354.

Cheftel J.C., Cuq, J.L. and Lorient D. (1993). Aminoácidos, péptidos y proteínas. In: *Química de los Alimentos*. Fennema, O.R., Ed. Acribia, España, pp. 275–414.

Chronakis I.S. (1996). Network formation and viscoelastic properties of commercial soy protein dispersions: Effect of heat treatment, pH and calcium ions. *Food Res. Intern.* 29: 123–134.

Clark A.H and Ross-Murphy S.B. (1987). Structural and mechanical properties of biopolymer gels. *Adv. Polym. Sci.* 83: 60–192.

Cooney M.J., Rosenberg M., and Shoemaker C.F. (1993). Rheological properties of whey protein concentrate gels. *J. Tex. Stud.* 24: 325–334.

Cuvelier G. and Launay B. (1986). Concentration regimes in xanthan gum solutions deduced from flow and viscoelastic properties. *Carbohydr. Polym.* 6: 321–333.

Damodaran S. (1988). Refolding of thermally unfolded soy proteins during the cooling regime of the gelation process: Effect on gelation. *J. Agric. Food Chem.* 36: 262–269.

Damodaran S. (1989). Interrelationship of molecular and functional properties of food proteins. In: *Food Proteins*. Kinsella J.E. and Soucie W.G., Eds. The American Oil Chemists Society, pp. 21–51.

Damodaran S. and Kinsella J.E. (1982). Effect of conglycinin on the thermal aggregation of glycinin. *J. Agric. Food Chem.* 30: 812–817.

Foegeding E.A., Bowland E.L., and Hardin C.C. (1995). Factors that determine the fracture properties and microstructure of globular protein gels. *Food Hydrocol.* 9: 237–249.

Furukawa T. and Ohta S. (1982). Mechanical and water-holding properties of heat-induced soy protein gels as related to their structural aspects. *J. Tex. Stud.* 13: 59–69.

German B., Damodaran S., and Kinsella J.E. (1982). Thermal dissociation and association behavior of soy proteins. *J. Agric. Food Chem.* 30: 807–811.

Giboreau A., Cuvelier G., and Launay B. (1994). Rheological behaviour of three biopolymer/water systems, with emphasis on yield stress and viscoelastic properties. *J. Tex. Stud.* 25: 119–137.

Heertje I. and Van Kleef F.S.M. (1986). Observation on the microstructure and rheology of ovalbumin gels. *Food Microstruc.* 5: 91–98.

Hermansson A.M. (1978). Physico-chemical aspects of soy proteins structure formation. *J. Tex. Stud.* 9: 33–58.

Hermansson A.M. (1986). Soy protein gelation. *J. Am. Oil Chem. Soc.* 63,: 658–666.

Kang I.J., Matsumura Y., and Mori T. (1991). Characterization of texture and mechanical properties of heat-induced soy protein gels. *J. Am. Oil Chem. Soc.* 68: 339–345.

Kohyama K. and Nishinari K. (1992). The effect of glucono-δ-lactone on the gelation time of soybean 11S protein: concentration dependence. *Food Hydrocol.* 6: 263–274.

Morr C.V. (1990). Current status of soy protein functionality in food systems. *J. Am. Oil Chem. Soc.* 67: 265–271.

Nagano T., Mori H., and Nishinari K. (1994). Rheological properties and conformational states of ß-conglicinin gels at acidic pH. *Biopolym.* 34: 293–298.

Nio N., Motoki M., and Takinami K. (1985). Gelation of casein and soybean globulins by transglutaminase. *Agric. Biol. Chem.* 49: 2283–2286.

Nishinari K., Kohyama K., Zhang Y., Kitamura K., Sugimoto T., Saio K., and Kawamura Y. (1991). Rheological study on the effect of the A_5 subunit on the gelation characteristics of soybean proteins. *Agric. Biol. Chem.* 55: 351–355.

Peng I.C., Quass D.W., Dayton W.R., and Allen C.F. (1984). The physicochemical and functional properties of soybean 11 S globulin-A review. *Cereal Chem.* 61: 480–490.

Ross-Murphy S.B. (1987). Physical gelation of biopolymers. *Food Hydrocol.* 1: 485–495.

Ross-Murphy S.B. (1995). Rheology of biopolymer solutions and gels. In: *New Physico-Chemical Techniques for the Characterization of Complex Food Systems.* Dickinson E., Ed. Glasgow, UK: Blackie Academic & Professional, pp. 130–156.

Shimada K. and Matsushita S. (1981). Effects of salts and denaturants on thermocoagulation of proteins. *J. Agric. Food Chem.* 29: 15–20.

Shimada K. and Cheftel J.C. (1988). Determination of sulfhydryl groups and disulfide bonds in heat-induced gels of soy protein isolate. *J. Agric. Food Chem.*, 36:147–153.

Stading M. and Hermansson A.M. (1991). Large deformation properties of ß-lactoglobulin gel structures. *Food Hydrocol.* 5: 339–352.

Utsumi S., Nakamura T., and Mori T. (1982). A micro-method for the measurement of gel properties of soybean 11 S globulin. *Agric. Biol. Chem.* 46: 1923–1924.

Utsumi S., Damodaran S., and Kinsella J.E. (1984). Heat-induced interactions between soybean proteins: Preferential association of 11 S basic subunits and ß-subunits of 7 S. *J. Agric. Food Chem.* 32: 1406–1412.

Utsumi S. and Kinsella J.E. (1985a). Forces involved in soy protein gelation: effects of various reagents on the formation, hardness and solubility of heat-induced gels made from 7 S, 11 S, and soy isolate. *J. Food Sci.* 50: 1278–1282.

Utsumi S. and Kinsella J.E. (1985b). Structure-function relationships in food proteins: Subunit interactions in heat-induced gelation of 7 S, 11 S, and soy isolate proteins. *J. Agric. Food Chem.* 33: 297–303.

Van Kleef F.S.M. (1986). Thermally induced protein gelation: gelation and rheological characterization of highly concentrated ovalbumin and soybean protein gels. *Biopolym.* 25: 31–59.

Wolf W.J., Babcock G.E., and Smith A.K. (1961). Ultracentrifugal differences in soybean protein composition. *Nature* 191: 1395–1396.

Yoshida M., Kohyama K., and Nishinari K. (1992). Gelation properties of soymilk and soybean 11 S globulin from Japanese-grown soybeans. *Biosci. Biotech. Biochem.* 56: 725–728.

Yamauchi F., Yamagishi T., and Iwabuchi S. (1991). Molecular understanding of heat-induced phenomena of soybean protein. *Food Rev. Intern.* 7: 283–322.

INDEX